TEXTUAL AGENCY

Writing Culture and Social Networks in Fifteenth-Century Spain

Textual Agency examines the massive proliferation of poetic texts in fifteenth-century Spain, focusing on the important yet little-known *cancionero* poetry – the largest poetic corpus of the European Middle Ages. Ana M. Gómez-Bravo situates this cultural production within its social, political, and material contexts. She places the different forms of document production fostered by a shifting political and urban model alongside the rise in literacy and access to reading materials and spaces.

At the core of this book lies an examination of both the materials of writing and how human agents used and transformed them, giving way to a textual agency that pertains not only to writers but to the inscribed paper. Gómez-Bravo also explores how authorial and textual agency were competing forces in the midst of an era marked by the institution of the Inquisition, the advent of the absolutist state, the growth of cities, and the constitution of the Spanish nation.

(Toronto Iberic)

ANA M. GÓMEZ-BRAVO is an associate professor in the Department of Spanish and Portuguese Studies at the University of Washington.

Textual Agency

Writing Culture and Social Networks in Fifteenth-Century Spain

ANA M. GÓMEZ-BRAVO

UNIVERSITY OF TORONTO PRESS
Toronto Buffalo London

© University of Toronto Press 2013
Toronto Buffalo London
utorontopress.com

Reprinted in paperback 2022

ISBN 978-1-4426-4720-6 (cloth)
ISBN 978-1-4875-5892-5 (paper)

Toronto Iberic

Publication cataloguing information is available from Library and Archives Canada.

We wish to acknowledge the land on which the University of Toronto Press operates. This land is the traditional territory of the Wendat, the Anishnaabeg, the Haudenosaunee, the Métis, and the Mississaugas of the Credit First Nation.

University of Toronto Press acknowledges the financial support of the Government of Canada, the Canada Council for the Arts, and the Ontario Arts Council, an agency of the Government of Ontario, for its publishing activities.

Contents

Introduction 3

1 Poetry, Bureaucracy, and the Social Order 15

2 Escribano Culture and Socio-professional Contiguity 33

3 Pervasive Papers 59

4 The Hands Have It 77

5 Papers Unite 92

6 Paper Politics 123

7 Books as Memory 148

8 Arranging the Compilation 164

9 The Book of Fragments 187

Conclusion 214

Notes 219
References 269
Index 321

TEXTUAL AGENCY

Writing Culture and Social Networks
in Fifteenth-Century Spain

Introduction

In a 1928 letter to Jorge Guillén, Pedro Salinas (1992), a fellow member of the self-conscious Generación del 27, congratulated Guillén on his forthcoming book of poetry: "Me has dado una alegría sin igual. Ver tus poemas así en conjunto, casi al borde del libro es lo que yo quería. Porque así se te ve en toda tu estatura de poeta, completo, variado, fidelísimo y rico a la vez" (90).¹

The single poems circulated in unpublished form among the members of the group and were published in newspapers and other periodicals, largely due to economic necessity. The important goal of publishing in book form, anxiously pursued by Salinas and the other members of this poetic circle, is emphasized throughout their writings and is particularly evident in their mutual correspondence. The book constitutes a publication venue that helps consolidate the figure of the author and is therefore an end in itself, bringing closure to the creative and transmission processes.² The book also functions as a repository for individual texts, now gathered together in the production of meaning under a larger format and marking a new beginning in the circulation of the compiled texts.

The print medium has successfully appropriated the term "book" and created an implied value judgment, helping perpetuate the medieval/Renaissance divide. In Febvre and Martin's (1976) influential *L'apparition du livre*, rendered as *La aparición del libro* in the Spanish translation and in slightly messianic undertones in the English *The Coming of the Book*, the book appeared at the same time as the printing press. Bouza Álvarez's (2001) important work on sixteenth- and seventeenth-century Spain, a book meaningfully entitled *Corre manuscrito: una historia cultural del Siglo de Oro*, counters this perception by bringing to light the extensive use of the manuscript medium just at a time when the printing press was supposed to have effected its revolution. While print may supersede manuscript for commercial purposes, it does not override it at many of its other moments or points of production and dissemination; the book in its codex form in fact spans a continuum that intertwines print and manuscript and anchors one to the other. The emphasis on the book as an important medium that is a result of, among other steps, progressive selection and aggregation brings out the role of its components and the question of their own production.

As the chapters that follow attempt to show, the fifteenth century showcases the loaded interplay of the book with other material supports and exemplifies its uses as a serviceable, though often unstable, codicological unit as well as a dynamic archive of sundry texts and supports. Telling instances such as fifteenth-century poet Juan Álvarez Gato's *libro* have much to show about how a book – any book – comes to fruition. Though the focus here is mainly on manuscript practices in the Spanish fifteenth century, this analysis is pertinent to other periods, since the production of the personal text displays a continuous tension with the book in its more institutional or public uses throughout a wide chronological span. The term "material" as used in these pages is aligned with studies in material culture and modes of textual reproduction and thus refers to physical and technological aspects of the texts, contexts, and human networks as they appear in daily life. In its extended semantic field, material also opposes "ideal." In this sense, the emphasis on materiality pulls away from Lachmannian models that guide philological efforts towards the recovery of a non-extant ideal text, a pure and sanctioned authorial exemplar. This ideal, presumed, or (to use Benedict Anderson's term) imagined text continues to weigh heavily in *cancionero* criticism and produces textual genealogies that are more like constellations than tidy stemmata, attesting to the need for an analytical model that reflects the intricate production and dissemination patterns.

The material aspects of textual production and reproduction in fifteenth-century Spain must be placed in their proper sociocultural context. As McGann (1991) has rightfully noted, to a large extent we imagine texts as read, leading to an emphasis on hermeneutics (4). According to this view, a text therefore is something we read and interpret, not something we do. This understanding further conceals the role of the agents involved in textual production and rests on an understanding of the book that is to a degree a self-generated fallacy that relies on the invisibility of other material supports. The emphasis on the book as an object that is the end repository of other objects and processes brings to the fore the importance of such processes and the hands, objects, and circumstances that contribute to its making. Inquiries into the constitution of a text point theoretically to a more inclusive approach, leading to an epistemological turn to the extratextual, embodied by interdisciplinary practice and solidly grounded in the material context.

Materialist approaches help bridge the gap between exclusive textuality and that which exists outside language (Hennessy 1993; Landry and MacLean 1993). At the same time, the destabilizing effect that material situation has on cultural production, noted by Walkowitz, Jehlen, and Chevigny (1989: 31), makes the study of the interplay between the text and the extratextual a crucial endeavour.[3] This interplay grants the text an agency that stems both from human agents and situations, and from the motility of the text itself as a discrete artefact in a constant process of becoming, making relevant the issue of textual situation to which Andrew Taylor (2002) has persuasively drawn attention. The text appears within interfacing networks of other texts – documentary, literary, or otherwise – as well as of human and external agents that are complexly situated. The situated text, as articulated by LaCapra (1982) (also Spiegel 1997: 25), emphasizes the relation of the word to a material world that it helps communicate and name but that exists alongside language, thus grounding it in a coordinate system of social, political, economic, and material axes.

As the focus on a contextualized study of the material text helps point out, material textuality emphasizes the need to contextualize the text within its bibliographic and cultural situation (Bornstein 2001), including the relation of the book to other material supports (Chartier 2007). The careful consideration of the paratext (Genette 1997) needs to be expanded in order to incorporate the wider contextualization of the book, the "bibliographic coding" (Bornstein 2001: 1–7, and passim; McGann 1991), and a broader sociology of the text (McKenzie 1999).[4] The critical look at the different methodological approaches to book culture suggested by Howsam (2006: esp. 46–64) must lead to an inclusion of the manuscript component of book culture and consider, along with bibliography and the sociology of texts, the contribution of palaeography, codicology, and diplomatics, in an integrative approach akin to the "textual scholarship" proposed by Greetham (1994, 1999). Of further importance for the study of manuscripts are the sociocultural approaches to manuscript culture by Oliveira (1994), and the combined study of the manuscript in its social and codicological context (J. Taylor 2007), alongside the "material philology and codicology" showcased in the studies published by Ferrari (1998) or the materialist philology proposed by Nichols and Wenzel (1996). In line with the work of these scholars, my study attempts to highlight some of the tangible consequences of the material nature of the text and its interplay with the agents and networks that make its production and dissemination possible.

The concept of the book as object can obliterate the traces of its fragmentary and more volatile constituents. The written text lives a pervasive and flexible existence in various supports, which far from being a subsidiary medium, interact with the print or manuscript book, itself lying at the end of a variegated media spectrum. One of the problems that arises from the acceptance of the primacy of the book through the centuries is that other material supports are either destroyed or left unstudied and therefore rendered invisible. This is a situation common not only in the medieval and early modern period, but also through later centuries. However, sometimes an author will "let the seams show," allowing an important insight into the life of texts and human agents before the book is shaped into its final form. One such author is Carmen Martín Gaite (2000), whose continuous references to her medium show how she regularly wrote on pieces of paper or in notebooks, the ubiquitous "Centauro" brand that she often mentions (in, for example, *Nubosidad variable*) and that those who grew up in Spain in the 1970s and 1980s used on a regular basis for school work or drafts, and that continues to be in use today:

> La cómoda la cambié de sitio el lunes, que es cuando vino a visitarme Soledad; después de marcharse ella, porque la conversación que tuvimos me removió muchas cosas. Podría convertirse en un capítulo del cuaderno [...] Cogí luego unos apuntes de la conversación y les puse fecha. Creo que es la última vez que he escrito algo. En el cuaderno no, en papeles, sueltos. ¿Dónde los pondría? Es fatal lo de los papeles sueltos. "Los debería pasar a limpio —recuerdo que pensaba, mientras cambiaba la cómoda de sitio—. Todo consiste en seguir escribiendo despacio, puntada a puntada." (120–1)[5]

Book writing here is revealed as a complex process tightly dependent on its material support and closely intertwined with its material context and the physicality of writing. Drafts, loose pieces of paper, booklets, walls, desk surfaces, and palms of hands provide a common medium for the text, which is then recopied and edited through its integration into the greater entity of the book. The "book" is then revealed as a plural entity in dynamic relationship with various material supports. These may be seen as constitutive elements of the book, but they also follow their own diverse paths, running courses that are parallel to or far removed from those of the book. Loose papers are easily lost but favoured for their flexibility, availability, and portability. They allow the slow stitching of a text and do not impose continuous writing. Notebooks have some of the flexibility and portability of loose papers while providing a more permanent material support, though they too can be easily destroyed by loss or by fire (Martín Gaite 2000: 167). The margins are the site for reader's notes that are then intended to be read by the author, who is in turn transformed into a reader of a now jointly composed text (e.g., 388). For Martín Gaite, writing is often measured in booklets or parts of booklets, pages, and "chapters" in notebooks which were posthumously compiled and published under the title *Cuadernos de todo* (Martín Gaite 2002). The pages purposefully left blank in many manuscript books, both pre-modern and later, encourage the prolongation of the writing and reading processes, thus forming, in practice, open texts in continuous production. Albeit concealed under the mirage of the edition and the printed text, the book was and still is the process by which it comes into being. It also relies on the agents that bring it to fruition as well as the concrete situation that enables its production.

If the book is where loose papers and other independent textual units survive, it is also where they are buried. The monumentalizing effect that the book may have is effective in hiding the erasure of the material life of its independent textual components before being absorbed or phagocytized into a larger macrotext whose formative layers must be recovered through an archaeological search. This search, often involving archives, libraries, and other repositories, can also take place within the texts themselves, since materials that go into book composition leave traces. There are self-conscious texts and authors that bring the book into three-dimensional relief by way of a transparent account of the materials of writing. In this way the book may be seen as resisting the erasure of its material constitutive elements and processes by narrating itself as a self-conscious material support and object. The book itself is a story that becomes narrated and narratable. So are many other supports.

The study of the relation of the textual and the extratextual entails an emphasis on production, producers/receivers, material support, circulation, choice of signifiers, material culture and artefacts, and the sociocultural developments that facilitate as well as necessitate their use in their complex interrelations. Writing, particularly poetic writing, takes place in social interaction, within networks of exchange. McGann (1991) has emphasized the social condition of textuality, which is constantly changing according to the variables of time, places, and agents involved, and which poetry openly displays (9–10, 16, and passim). As an eminently self-conscious discourse, poetry serves to "give an account of oneself" in a dialogic situation, taking place, as Judith Butler (2005) has established, in a crucible of social relations (132), and establishing a relationship that, for Levinas (1994, 1998), has by definition an ethical basis and is therefore imbued with a "semantics of proximity" (1998: esp. ch. 4; 1994: 93). These are key notions to consider when we undertake the study of the human networks that produce texts and cooperate in their exchange. The close relation of texts and agents encourages the use of the body as conveyor of texts. This bears different implications for the hands that write and the sleeves, pockets, and chests that keep them close and for the consciousness that relies on a material support as an extension of memory. The materials of writing are in the hands of the human agents who use and transform them. These agents work within socio-political and cultural situations that both shape and are shaped by textuality and the materials that convey it.

Chapters 1 and 2 examine the implications of studying literary production within the context of socio-professional networks and power in the fifteenth century. The *escribanos* (scriveners, copyists, or scribes) embody the power of writing in an increasingly complex and ambitious state. The nobility, the *medianos* or middle-class artisans and merchants, as well as some upwardly mobile members of the lower classes negotiated their status within a shifting urban and political model using administrative and literary production as a way to accrue economic and personal worth. The production of poetry and bureaucratic documentation by the same human agents facilitated an interface between literature and law, a significant phenomenon in Europe during the period (Cátedra 2009, Knapp 2001, Steiner 2003). But this was not solely a conceptual contact, as sharing textual strategies and material supports led to literary innovation and to different paths of interaction between textuality and authorship. The emphasis on textual production had an impact on both documentary as well as literary production, a phenomenon with a strong bearing on poetry, a privileged discourse and status mark. Contacts within and among the different groups fostered both solidarity and conflict, both of which were heightened by the presence of converted Jews or *conversos* among the medianos, the nobility, and bureaucratic personnel.

The flexibility of textuality owes much to the plasticity of the material supports that convey it. Chapters 3 and 4 focus on the use of the single piece of paper for the purposes of providing both flexibility and portability, thus ensuring textual motility and pervasiveness. The text thus lives as a discrete entity as enduring as the materials that support it and the agents that write, fold, tear, or pass it on. The emphasis on the hands as a main instrument leads to a focus on chirography, the authenticating imprint of the writing hand.

Motility and flexibility are favoured traits of textual supports, but they can at the same time bring a precariousness that the single text needs to resist. Single texts may join other texts in the creation of a larger discourse or for the purpose of ensuring longevity. Chapter 5 looks at rolls, wrappers or *envoltorios*, and *cuadernos* (booklets, or quires), which provide a still flexible and mobile support while taking advantage of the greater security of bundling or grouping single papers for the purpose of transmission. Individuals, households, municipalities, and the state all share a need to keep important texts safe. Chapter 6 examines the register as it provides a useful support for single or grouped or bundled papers. The larger book format attempts to preserve the flexibility of individual papers by means of the threaded book model, while the blank book promises a safe haven for short texts. The book is in this way devised as a textual archive, both textual repository and an opportunity to form a larger textual unit out of individual contributions. Organizing and compiling the individual texts calls for methodologies that will help imbue the larger repository with a meaning of its own. The personal and political uses that the register enables are fully taken advantage of by individuals looking to create a personal or family memory and by the state wanting to organize and document power.

The book as compilation exposes its nature as memory in that it is both remembered and documented. Chapter 7 delves into the monastic memorial book, political and economic reports known as *memoriales*, and memory books, all of which show the use of writing as a form of memory. Personal memory books ensure the portability of such memory aids as well as their close attachment or relation to the body. The potential offered by flexible means of textual support encouraged a continuous inscription of consciousness onto a portable writing medium.

The macrotext facilitated the arrangement of assorted individual texts into a coherent whole. This arrangement involved gathering and organizing the materials to often create a meaningful narrative thread that reinterpreted earlier texts in the light of the time of compilation. Chapter 8 examines the ways in which explanatory titles or rubrics helped create a unifying narrative, as well as identify and situate the text. Rubrication provided an explanation of an otherwise decontextualized piece that was removed from an elusive extratext. By providing hermeneutic access, rubrics proved a useful tool in creating a meaningful whole out of independent texts by an author or compiler. Poets of different social groups such as working noble

Gómez Manrique, bureaucrat Juan Álvarez Gato, or mediano Antón de Montoro show ways in which writing, reading, compiling, and circulating poetry played up the poets' self-generated worth in their socio-political context. The examination of these authors' compilational strategies and their meaning is the subject of chapter 9. The analysis of these cancioneros invites in turn the consideration of the book as a fluid process constituted by textual fragments. The material memory of this process, preserved within the texts well through the establishment of the printing press and the institution of copyright, never lets the "book" forget that it is always a manuscript.

Background

When, on a cold 13 December 1474, Isabel proclaimed herself Queen of Castile, it was obvious that the long fight for the throne was not over. Her half-brother Enrique IV had died two days earlier, ending a reign that had started in 1454 and which Isabeline historiography would characterize as unstable and conflictive, blaming a weak and problematic monarch who yielded to the interests of a powerful nobility.[6] The historiography of the period provides vivid accounts of the important political role played in the fifteenth century not only by nobles but also by the king's favourites. These famously include Álvaro de Luna in the case of Isabel's father, King Juan II (who ruled 1406 to 1454), and don Beltrán de la Cueva for King Enrique IV (who ruled 1454–74). Both favourites gained important titles, but Luna was the less fortunate of the two since in 1453 he was executed on orders from Juan II, who yielded to pressures of a nobility alarmed at Luna's seemingly unlimited (and competing) power. Civic wars plagued the fifteenth century, including the one waged in the early years of Isabel's reign. Already by the fourteenth century, the nobility had undergone what has been represented as a process of renewal of its main families, which entailed a rise in relevance of some of their side branches. The tendencies towards the accumulation and preservation of wealth within the families had given great power to the model of *mayorazgo* (primogeniture) since 1369. This aristocratic structure had also placed great emphasis on the idea of *linaje* (lineage), at times widened into *bandos* (alliances) that included others members of urban society. The Trastamaran dynasty had secured the Castilian throne through bloodshed, as Enrique II (1369–79), the illegitimate son of King Alfonso XI and Leonor Núñez de Guzmán, fought and killed his half-brother, King Pedro I. Isabel's dynastic battle was based on the claim to legitimacy after her brother Alfonso died, and a fraction of the nobles sided with her in an attempt to overthrow King Enrique and to claim the illegitimacy of his daughter Juana. It was alleged that Enrique's marriage was not valid and, further, building on the presumption of the King's impotence, that Juana was actually the daughter of don Beltrán de la Cueva and the Queen. Isabel successfully navigated the many efforts to make her a political pawn, choosing a partner in marriage who better suited her own political interests. In 1469, after slipping away from her guardian Juan Pacheco, Isabel married Fernando, who had been able to make a safe trip to Valladolid in disguise. When Fernando ascended to the Aragonese throne in 1479, the two kingdoms were united and together joined forces, defending Aragon's interests in the Mediterranean, which included Naples, Sicily, and Sardinia, and continuing Castile's move to the west across the Atlantic on to the well-known colonizing enterprises in the New World. The wars created by Isabel's proclamation as queen were fed by part of the nobility, as well as by the interests of neighbouring Portugal and France. In 1480, the Cortes in Toledo helped position the politics of what would prove to be a powerful monarchic rule.

Periodic waves of plague starting in 1348 decimated the Iberian population, wiping out one-quarter of Catalonia's population. The demographic crisis affected the cities and had a clear impact on agrarian growth. The ill effects that the plague had on agricultural production in the fourteenth century were eventually surmounted and periods of growth followed. Due to increased demand, the fifteenth century also saw economic growth in some areas, such as the great expansion in sheep raising, which was regulated by the Mesta, and of the wool industry. In spite of periods of decimation, migratory, political, and commercial forces had an effect on the growth of cities such as Barcelona, Córdoba, Jaén, Lisbon, Medina del Campo, Murcia, Toledo, Seville, Valladolid, and Valencia, which increased significantly in size. The comparatively high number of populous towns in Andalucía made up approximately 20 per cent of the total population in Castile by the end of the fifteenth century. In 1492 close to 70 per cent of the population in the Iberian Peninsula lived in Castile, which represented over 64 per cent of the Iberian soil.

In constant power struggles with the nobility, the emergent absolutist monarchy moved towards increasing royal control, which necessitated the existence of a solid bureaucratic system that would ensure that laws and royal orders would reach all peoples in the realm. There was also a need for a solid system of both financial and legislative control. Two main figures in urban government were the *regidores* (aldermen), often members of the local aristocracy, and the *corregidores* (chief magistrates), who were sent by the monarchy. Both figures were pivotal in the struggle between cities and monarchy for municipal control. Many members of the upper nobility had their permanent residence in cities and tried to dominate urban government and attract the loyalty of local aristocracies that formed what has been called "urban oligarchies" or "urban patriciates." Urban oligarchies were heterogeneous, consisting of *hidalgos* (nobles), *caballeros* (knights), royal administrators, lesser nobles, and wealthier merchants and artisans, as well as rural owners with urban residence. A considerable number of these were *judeoconversos*. Power equilibrium in the cities depended not only on noble rank and royal intervention but also on economic capital. The upper level of the *común* (commons) had the means to be on a par with the elites and played an important part in city affairs. *Caballeros* and *escuderos* (knights and squires) were key members of the local government (*concejo*), which the *regimiento* increasingly dominated as the fifteenth century progressed.

An important urban group was that of the medianos, middle-class professionals, artisans, and merchants, some of whom could be relatively wealthy. Wealthier medianos were by design also caballeros, as they had the obligation of maintaining a horse and arms and of being ready to participate in warfare. They were known as *caballeros de cuantía* or *de premia*, but did not have *hidalgo* (noble) status. They had to demonstrate readiness for war by participating in periodic parades or *alardes*, where they were obligated to appear with horse and arms. The "war in Granada" is a common reference in many literary texts of the period. Exemption from most taxes was a caballero privilege, but it did not quite make fighting worthwhile, for which reason many caballeros contracted with others who would replace them in battle by means of a *contrato de reemplazo*. As the following chapters will show, nobles, as well as caballeros and wealthier medianos were active in textual production by virtue of their administrative and professional duties, and because of the opportunities for sociability that urban contact supplied.

At the state level, the monarchy relied on government institutions for administrative, judicial, and legislative purposes. These included the wide-reaching Consejo Real (Royal Council) and the Cortes (held with the participation of cities), while the Cancillería Real (Royal Chancery) was entrusted with the administering of documents. Newly centralized under Isabel and Fernando, the Santa Hermandad provided police control of the cities. The Casa de las Cuentas, headed by the *contadores mayores,* or chief royal accountants, was in charge of the royal taxes and finances. State and municipal government necessitated the employment of a small army of bureaucrats and administrators that helped generate and organize the over 300,000 documents that have survived from the reign of Isabel and Fernando alone. The bulk of record production lay in the hands of secretaries and escribanos. University-trained *letrados* were particularly useful in the administration of legal matters, but were not as ubiquitous as the escribanos. Overall, the Catholic monarchs tightened royal power over state and municipal governments and institutions. They negotiated their relationship with the nobles by exercising control but also guaranteeing some of the nobility's privileges, and they handsomely rewarded some of their closer collaborators.

The establishment of the Inquisition in 1478 represented yet another form of control. The pogroms against Jewish communities in 1391 had marked the end of a century filled with conflict and opened an equally violent one. The preachings of Ferrant Martínez and Vicente Ferrer, and the dispute of Tortosa, show the earlier pressures for conversion and assimilation of non-Christians that derived in the statutes of purity of blood, the first of which was issued in Toledo in 1449. These statutes were fully in effect by the sixteenth century and were intended to banish anyone with Jewish or Muslim ancestry from positions of political, social, or economic power. The multitudinous conversions of the fourteenth and fifteenth centuries brought about the growth of a substantial number of converts, or conversos, which proved to have a destabilizing effect on both religion and society. Often characterized as hybrids (as in the anonymous Tratado del Alborayque), neither Jews nor Christians, judeoconversos were part of the general population and were employed in a variety of jobs. For the purposes of the present study, it must be noted that a number of conversos participated in municipal government and held jobs in royal administration, working for a monarchy that they hoped would offer protection. The obvious benefits drawn by close proximity to the monarchy were the cause of much competition among groups and individuals and at times were the source of animosity against conversos, who were portrayed as illegitimate arrivistes. In discussing the contributions of individual conversos in the chapters that follow, I make no original claims in relation to their converso status. For this, I refer to the studies by historians and literary scholars cited in the text and notes. At the same time, my purpose here is not to raise any claims as to the pertinence of notions such as a "converso code," a "converso voice," or a converso-specific writing. Although these Américo Castro-inflected propositions have merited discussion since at least Gilman, my aim here is to situate poetry in the socio-professional, cultural, and political milieu where poets and bureaucrats produced their texts.

According to PhiloBiblon, approximately three-quarters of all extant medieval Spanish texts date from the fifteenth century.[7] Among other types of texts, the fifteenth century provided fertile ground for the many nobiliaries and individual noble and royal chronicles, which helped in the efforts to legitimize the noble or monarch, and conversely in their discredit, as in the case of the chronicles describing the rule of Enrique IV written during the Isabeline reign. Prose and poetry often referred to Isabel as a divine figure and to both Isabel and Fernando in messianic terms. Other prose works that constitute new literary models or genres in the period such as *Celestina* or *Cárcel de amor* portrayed the interests and anxieties of the nobility in their interface with other social groups in an urban and courtly environment. The itinerant nature of the court, which would occupy a number of local houses for the duration of its stay, facilitated active contact between state and municipal life. The same century witnessed an unparalleled growth in poetry. Collective or personal anthologies, miscellanies, and different kinds of texts have preserved well over 7,000 poems (more if fragments and different versions are taken into account) involving the work of around 700 poets (Dutton 1990–1, 1: vi). Some of the manuscripts are luxury presentation copies, but most are modest in production. In addition, many poems have been preserved in non-literary manuscripts, proof of the pervasive nature of the poetry of the period. As it will appear throughout the pages that follow, poetic papers, like their administrative counterparts, appeared more frequently in areas and among groups that had a stronger investment in textual production. In the court and the cities, these included nobles, administrators, bureaucrats, and members of other groups such as the medianos with social membership in textual circles. The study of the forces behind the massive textual production of the period are the subject of the chapters that follow.

1 Poetry, Bureaucracy, and the Social Order

In one of his letters (Letra X), royal chronicler, secretary, and escribano Fernando de Pulgar (1982) identified ink-stained hands as a mark of the pen professional and a bloody foot as that of the warring nobility, both being part of their respective jobs ("oficio"): "usando vuestra merced de su oficio e yo del mío, no es maravilla que mi mano esté de tinta e vuestro pie sangriento" (61).[1] Don Enrique Enríquez, King Fernando's uncle and his chief *mayordomo* (major-domo), had been fighting in the war of Granada when he received the injury, caused by a firearm. Pulgar's image presents his own hand stained in a liquid similar to that which covered Don Enrique's foot, making ink his own escribano's blood. Pulgar shed such blood while writing about the Granada war in his *Crónica de los Reyes Católicos*, which concerned those who, like Enrique Enríquez, were taking active military part in it. Neither of the attributes bore a negative social connotation in the letter, where Pulgar portrayed them as "occupational hazards." However, these hazards could be, depending on the socio-political value applied to either occupation by a particular author, either a sign of prestige or a hindrance. For those who derived much of their livelihood from the exercise of the pen, such as Pulgar or Juan Alfonso de Baena, poet and likewise escribano, writing was a skill that gave social salience over those lacking writing ability. The hand and the instruments it wielded were a sign of social class and as such a synecdoche for intellectual and political power. Juan Alfonso de Baena placed clear emphasis on the implications of social class in his detailed description of the escribano's toil and materials: "El muy sotil escrivano / que trabaja noche e día / con su linda escrivanía, / en papel liso toscano, / … / en la su rica escritura / de letra tajante e pura, / bien escripta con su mano, / non por çierto de villano" (PN1, ID1320, vv. 10–13, 15–18).[2] The "well written" "pure handwriting," in Baena's view, takes the "subtle scribe" who works night and day far from the lower-class "villano" and closer to the "gracious courtier" (don Álvaro de Luna in this case), who is in near proximity to the king, and whose praise will earn him wealth and advantage: "Por ende, pues onra gano / con riqueza e mejoría / por loar tan gran valía / del graçioso e palaçiano / que del Rey está çercano" (vv. 19–23).[3] Juan de Mena similarly acknowledged the political power of the writing hand when he offered his to the king: "porque yo, señor, escriva, / el menor de vuestra gente, / cosas con que vos contente / mi mano, vuestra cativa" (ID0328, vv. 74–7).[4]

Like Pulgar, Juan de Lucena (1892a), probably a converso, was interested in comparing the active life of the man of arms with that of the pen professional in his *Epístola exhortatoria a las letras*, although he presented them as competing occupations when pondering which of the two wielded more power and better served the state. Without completely discounting military force, he proclaimed the pen and the three fingers needed to sustain it to be more powerful than the fist holding the sword, reasoning that the pen cuts the knife more often than the opposite: "Más veces corta la péñola al cuchillo, que el cuchillo a la péñola" [More often the pen cuts the knife than the knife the pen] (211). In a letter dated in 1447, Diego de Valera, converso chronicler, poet and prolific author, as well as noble caballero, used a more inclusive approach. He placed value on both his physical and intellectual abilities when situating his role in the body politic, which he presented as working fluidly as a mystical body: "e yo, aunque el menor destos mienbros, sé esforçarme servir mi Príncipe, no solamente con las fuerças corporales, mas aun con las mentales e intelletuales" (text in Penna 1959: 7).[5] These authors exemplify the discourse of professional pride among pen professionals that coexisted with the common association of *caballería* with nobility and which in turn claimed for the noble caballería superiority over other socio-professional levels.

In contrast, the noble caballería had to keep dispelling negative associations in relation to the debasing physical labour that it commonly involved. Lucena's (1892b) description of Santillana as the ideal combination of two worthy occupations (warfare and intellectual work) is prefaced in his *Libro de vida beata* by a long description of the hardships of the caballero's life – which is said to be bitter and bothersome ("amara y permolesta") – together with a notice of Santillana's own pained reaction to the mention of warfare (esp. 131–3). While here Lucena praises caballería as an occupation of great profit and ornament ("La cauallería es de gran prouecho, y mayor ornamento" 131), the description of the warring caballero's hard life is filled with discomforting graphic detail: "[el guerrero] ni come, ni duerme, ni jamás un hora fuelga en reposo, expiando a otros, o con miedo de ser insidiado: siruen los armígeros y nunca enriquecen: nunca medran, y siempre trabajan: en trabacas, alfaneques, pauellones o tiendas portátiles, y a las vezes so enramadas, o debaxo el escudo, fazen su morar de continuo: quando carambalados de frío, y quando del sol tostados de fuego: llagados de ligeras feridas, por falta de remedio, se mueren por los Reales" (131–2).[6] The fact that this topic was mentioned by classical authors such as Salustius does not deter from the harshness of an experience that some caballeros sought to avoid. Resistance to military service was a serious enough problem since earlier centuries and had prompted King Alfonso XI to punish it with death (document in Romero Tallafigo et al. 1995: 148–9, plate 27). Some urban caballeros (caballeros de premia, on whom see below) contracted individuals who would serve in war in their stead via the contratos de reemplazo (replacement contracts) (Córdoba de la Llave and del Pino García 1988). Access to caballero status became limited after Isabel and Fernando significantly raised the amount of wealth necessary to claim tax-exempt status, causing a decrease in the number of people who thought it worthwhile to maintain a horse. It resulted in an increased number of mules and a decreased number of horses, a trend that alarmed the monarchs because it translated into a diminished army. The order they issued in 1499 was an attempt to reverse the trend (document in Domingo Palacio 1907: 489–95).

Many of the hardships and the physical nature of the occupation in fact brought the caballero away from gentility and closer to the toil of peasant labourers and members of the lower classes. The description of Juan de Mena's appearance in Lucena's (1892b) work is typical of that of a man of letters. Mena's long night vigils, which were spent studying, marked his thin body and pale, worn-out face. The model he is being compared to is ostensibly that of a battle-worn caballero, but this is worded in such a way that some aspects of the reference could also point to a peasant worn from working in the fields: "De grand ánimo te muestras, mi Ioan de Mena, que las armas tanto exaltas. Trahes magrescidas las carnes por las grandes vigilias tras el libro, mas no durescidas ni callosas de dormir en el campo: el uulto pálido, gastado del studio, mas no roto ni recosido por encuentros de lança" (131).[7] The face, broken and "stitched up" as a result of encounters with the spear, can only belong to a caballero, but the flesh, hard and callous from sleeping in the fields, belong to both a caballero and a peasant. The primary use of the fist, in contrast with the escribano's trinity of fingers, was also common to both caballero and peasant, being useful to hold both the sword and farming equipment.

In fact, in spite of the honour that caballería allegedly commanded, legislation subordinated the military caballería to escribanos and other city officials that were ostensibly less expendable. State laws de facto conferred protection and therefore importance to the life and work of escribanos over the caballeros by exempting the former from participation in warfare. King Juan II at the 1432 *cortes* in Zamora had granted city escribanos in charge of teaching reading and writing, as well with public escribanos and other public officials, exemption from military duty.[8] The ruling was incorporated into the 1484 *Ordenanzas reales de Castilla* (Díaz de Montalvo 1990: fol. 71v), with the whole text from the Zamora cortes copied verbatim in fol. 118v. The ruling was included in later legislative works such as Hugo de Celso's (2000, 1553) *Repertorio Universal* (fol. CXXVR). This exemption grouped the escribanos together with officials involved in city government such as *jurados* and regidores as well as with other professionals, such as doctors, who had no direct role in government but worked in positions of public interest.[9] The message was clear that government and bureaucratic positions, and those involved in literacy services deserved privileges that would not be extended to the warring caballeros.

The competition between the caballeros and other emerging urban groups is evident in the fifteenth-century chronicle *El Victorial* by Gutierre Díaz de Games (1997), which goes into further detail when praising the noble caballería as the most honourable work and highlighting the degrading physical labour involved. The caballeros, argues the author, work very hard under extreme physical duress, earn their bread with pain, amid sweat and hard labour; they eat mouldy bread and unpalatable food, and drink stagnant water, with little or no wine; they stay in poor quarters, their beds made up of rags or twigs and leaves. In contrast, the middle class, those holding the "commoner jobs," have easy and comfortable lives; they have their bread at hand, dress in delicate clothing, eat delicious food, sleep in soft, scented beds, and enjoy good houses in the company of their wives and children, having everything easy at hand, suffering no fear, getting fat, living a good life (278–9). The extreme dichotomies put forth by these texts not only opposed the man of letters to the military caballero, but also the caballero as a noble ideal to the middle-class artisan or merchant, the medianos, whose "ofiçios baxos & viles" were forbidden to the noble caballeros.[10] In addition, these socio-economic divisions became imbued with a new addition to the medianos and the nobility, that of the conversos, which had been steadily growing since the pogroms and ensuing mass conversions of 1391.

The nature of the caballero was further complicated by the existence of an urban caballería. Belying the sharply drawn contrast between noble caballería and middle class, the caballería de cuantía or de premia was composed of members of the middle classes, artisans and merchants with a large enough income to be able to maintain a horse and arms. They were therefore required to fulfil military duties and go to war when summoned. These caballeros were part of a continuum that placed the noble caballero of old noble family lineage at one end and that was constituted by nobles, caballeros, and even wealthy peasants (*labradores ricos*) of various economic levels and origins. In addition, cities were incorporating eligible Jews to the urban caballería in the fifteenth century (Torres Fontes 1966). Jara Fuente (2001) has argued that all these developments caused the caballería to lose prestige towards the end of the fifteenth century and therefore its function as a clear status marker, giving way instead to an emphasis on lineage and nobility.[11] In addition, other developments, together with the common association of caballería and nobility, prevented the nobility from being untouched.

Learning Nobility

Two important developments in the fourteenth and fifteenth centuries fed the instability of the concept of nobility. One was the growth of an urban middle class with warfaring duties, resulting in a different kind of caballero, the caballero de cuantía or de premia discussed above. Heusch (2000) has persuasively shown that this development fed the ongoing debate on caballería and nobility, which argued over whether they were traits inherited through lineage or performative exercises of virtue (149–200).[12] The copious literature debating the nature of caballería and nobility (Gerli 1996, Heusch 2000, Rodríguez Velasco 1996) reflects the crucial importance of these matters to the writers of the period. Looming over any discussion over nobility as a birth trait was the fact that the king had the right to bestow the noble rank at his discretion. In his *cédula* bestowing Miguel Lucas de Iranzo the title of *condestable*, the king states his power to "make noble" anyone who behaves as one (text in Ceballos-Escalera y Gila 1993: 285–98, particularly 295). The second momentous development was that of a shift in the role played by the nobility in relation to the administration of the state, which encouraged learning and the exercise of letters, putting nobles in proximity to other collectives working for urban or state government. New roles for the nobility included administrative duties in monarchic service as well as a key presence in urban government.

The association of caballería with the ideals of nobility, namely with military duties as the *bellatores* of the feudal system highlighted by Ladero Quesada (1996–7: 36–9), had to be negotiated in light of this changing role of the nobility. Historians such as Valdeón Baruque have pointed that the pre-eminence of urban oligarchies rested on wealth as a foundation and caballería as an ideal (1990: 516).[13] However, it is important to realize that state bureaucracy and urban life also increased the number and importance of men of letters (Saenger 1975: 409). Further pressure came from the growing professional and merchant readership that in the thirteenth and fourteenth centuries had helped spread literacy, as well as private reading and writing practices dealing with non-public legal transactions and documents (Petrucci 1992). These practices were flourishing in the fifteenth century.

A telling illustration of some of the notions associated with the impact of learning on the unstable notion of nobility may be found in a letter that converso secretary Fernando de Pulgar (1982) wrote to a friend in Toledo (Letra XIV). In the letter, Pulgar notes that men of lowly blood ("de baxa sangre") leave the lowly jobs of their parents ("los oficios baxos de los padres") in order to follow their natural inclination ("su natural inclinación") and become great letrados ("ser grandes letrados"). On the contrary, Pulgar goes on to remark as a well-known fact that there are sons and descendants of kings and nobles ("fijos y decentientes de muchos reyes e notables") who fall into oblivion because they are unapt and of lowly condition ("inábiles e de baxa condición"). Pulgar repeats the dictum that "God made men but did not make lineages" ("Dios fizo ommes e no fizo linajes"), stating that by his power God made everyone noble upon birth ("todos fizo nobles en su nacimiento"), while lineage and blood become stained when the individual strays from that early path of virtue. This leads him to believe that virtue bestows true nobility ("la virtud, que da la verdadera nobleza") and encourages the good deeds that make nobility ("buenas obras, que facen la verdadera fidalguía") (67–71). In his *Crónica de los Reyes Católicos*, Pulgar (1943) put similar or identical statements in Gómez Manrique's words – though they may be his own – in a speech to the people of Toledo (1: esp. 348–50). In his *Coplas de virtudes y vicios* or *Setecientas* (ID0072), noble Fernán Pérez de Guzmán similarly states that good education procures more virtue than nature, having seen the sons of rustic servants show themselves educated and intelligent through their stay at court and, conversely, children of noble birth become slow and "viles" through negligence.[14] The pressure was all the greater in light of views of authors such as Diego de Valera, for whom the move from commoner to noble effectively purged rusticity and other *villano* traits, while the noble could become "denatured" or "de-generated" by unbefitting behaviour (Di Camillo 1996: 237). According to this argument, a formal noble title would follow or be preceded not so much by a self-fashioning in the manner noted by Greenblatt (1980) for the Renaissance, but rather by a self-generative or self-transformative process via suitable deeds and behaviour, a process that emphasized the radically transformative nature of actions. In the language of the period, an individual was "made" (*hacer*) by actions alone. As seen in the details of caballero, escribano, and mediano life quoted above, these actions were tightly dependent on the physical impact of the individual's occupation. Learning played a key role in carrying out the generative process, which went beyond controlling the perception of the self by others, and instead was grounded in a retooling of the self from the inside. It is not coincidental that converso bureaucrats such as Fernando de Pulgar and converso nobles such as Diego de Valera showed a strong interest in these matters, given the significant growth of converso population through the fifteenth century and the incorporation of conversos to noble ranks and to the "power elite" as studied by Rábade Obradó (1993).

The potential effect of these developments on the notion of nobility was not lost on members of the more established noble families, such as Pérez de Guzmán, who had themselves been undergoing important shifts since the previous centuries and for whom this new addition proved even more destabilizing. Stressing the importance of learning for political purposes, while offering himself implicitly as example, Gómez Manrique (2003) had argued in the introductory letter to his cancionero that knowledge did not dull the blade of the sword, nor did it weaken the arms and hearts of the caballeros (99–100).[15] The difficulty of the process made the transformation all the more valuable and deserving of lasting fame. As Fernán Pérez de Guzmán clearly stated in the stanzas subtitled "De sciencia e cavallería" in his *Coplas de virtudes y vicios* or *Setecientas* (ID0072), joining knowledge ("sciencia") and caballería in sweet brotherhood ("dulçe hermandat") was most valuable because it was most difficult ("es muy grave de juntar") (Severin 1990: 25). Although knowledge was prized by the monarchy since earlier centuries, as is evident from Alfonso X's example, Beceiro Pita (2000) has pointed out that it was mainly towards the fourteenth and fully in the fifteenth century, starting with the reigns of Enrique III and Juan II, that particular courtly knowledge and behaviour became associated with opportunity for social mobility and proximity to the monarch; this knowledge involved not only basic learning, but also manners, games, dance, music, and poetry (esp. 194–5). Further, the link between culture and government intensified towards the end of the fifteenth century (Beceiro Pita 2000: 199).[16] Knowledgeable individuals were most useful to an increasingly complex state that leaned its heavy weight on a well-organized bureaucracy. In addition, a strong mastery over language, particularly specialized or stylized discourse, was considered key both for administrative purposes and for those associated with interaction within social networks. For this reason, the debate on the role of poetry in constitutive notions of the self went hand in hand with the debate on nobility, and both debates with the processes of becoming (*hacerse*). In this too escribanos could claim an advantage over the noble, warring caballeros.

Working Titles: Nobles and Bureaucrats

Tightly intertwined with the shifts in the model of government was the curialization of the nobility, a phenomenon closely related to the physical migration of nobility to the court and urban residences that was particularly noteworthy towards the fifteenth century.[17] The migration from the country to the cities led to the establishment of large family residences in key towns, spawning beautiful architecture and ornamentation, often imprinted with the noble's device (Yarza Luaces 1993: 233, and passim). Aristocrats gravitated towards the stronger monarchy and positions of urban power in what was becoming an absolutist state.[18] The move towards the bureaucratization of administrative occupations encouraged the emergence of specialized professionals, as was already evident in the court of Alfonso XI (Moxó y Ortiz de Villajos 1975). This created a system that would work towards the progressive strengthening of the monarchy and the construction of the monumental apparatus of the fifteenth-century state (García Marín 1974; González Alonso 1981; Torres Sanz 1982). Administrative personnel played a seminal role in generating and maintaining the paperwork needed to build the bureaucratic system that sustained the state.[19] The ensuing increase in the number of court and city officials caused public outcries that they were detrimental to the economy of court and cities.[20] Members of the middle and lower ranks of the "working nobility" often had limited means and depended on the Crown and urban governments for their livelihood, filling posts at court remunerated by a yearly pay ("quitaciones"), or working as secretaries or judges or in other bureaucratic positions. Higher-ranked nobles received the more important positions, largely bestowed as honour titles. The honorary nature of these jobs led to the growth of a large body of subalterns who would be in charge of fulfilling the material aspects of the jobs for a lower pay. These *lugartenencias* or substitute system did not free the higher-ranking officials from involvement in the preparation of documents since they still had supervisory duties concerning the issuing and archiving of official records. The lugartenencias were also the source of conflict because of the practical nature of the job. Those substitutes lacked the official title, though not necessarily the skills, to draw the documents and tended to overcharge for their services as a way to increase their earning power (Rábade Obradó 1992). The potential for career advancement and the acquisition of wealth through the exercise of administrative posts offered in turn the possibility of gaining access to noble titles (Maravall 1986: 497–8).

The Catholic monarchs reputedly promoted individuals on the basis of merit, including those of the lower ranks of the nobility, to service and administrative jobs. These individuals were given not only monetary rewards, but also enhanced possibilities of upward social mobility.[21] The practice of choosing medianos or members of the middle classes and members of the low nobility for royal service had already been clearly espoused in the *Siete Partidas* of King Alfonso X (2004). The king reasoned that, although the high nobles should be given positions of great importance where they could bring honour to the court, they would have conflict of interest in other posts and be too powerful to be useful to the monarch, while the poor would be "viles" and greedy (II, 12v). Since their work on sensitive issues regarding the state brought them close to the monarchs, bureaucrats effectively became advisers and close collaborators (Torres Sanz 1982: 116–17), constituting an administrative middle class that had textual production at its core. The role of pen-professionals as close royal collaborators had a long tradition. The mid-thirteenth-century *Poridat de Poridades* emphasized the importance of the careful production of documents, of which the escribanos were the body, their content the soul, and the handwriting the embellishment. The need for the escribanos to be trustworthy and able secret-keepers should compel the monarch to reward them handsomely and to put them in positions of power (text in Kasten 1957: 50). Employed in posts that required intimate involvement in textual production, secretaries, escribanos, the heterogeneous group of the letrados, and bureaucrats – whom Rucquoi (1997: 63–85, 187–95) has termed "the other nobility" – functioned as the true right hand of the monarchs.[22] All these formed power structures reminiscent of those that incorporated what Bourdieu (1996) termed "State Nobility." Even though the university-trained letrados became increasingly useful players in the administration of the absolutist monarchy, as did other faithful collaborators, they did not make up the bulk of bureaucratic personnel either in courts or cities (Phillips 1986). Overall, university graduates made up a little over 10 per cent of the officials working under Isabel and Fernando

(Phillips 1986: 482). Salaries did not correspond to degree of education (Phillips 1986: 480–1). Nobles outnumbered university graduates in all offices in royal service, except for those dealing with medicine and law. There was a visible growth for both groups in the reign of Isabel and Fernando that was greater for nobles, noticeably in household (11.5 per cent), executive and military (12.4 per cent), and administrative offices (21.5 per cent) (Phillips 1986: esp. table 1). However, with the exception of law and medicine, non-noble and non-academically trained personnel constituted the majority of those holding a royal office.

The needs of the administration contributed to the ever-increasing number of public officials, including the ubiquitous escribanos,[23] and the development of a group identity. The post of escribano could be the point of entry into the urban nobility for members of the lower classes. The son of an artisan could apprentice with an escribano or notary and, after establishing himself as an escribano in his own right, could try to access upper social levels, most notably the nobility. In this path he was competing with the sons of already-established escribanos, who tended to be given preferential treatment (Cruselles Gómez 1998: 169–86). A related development, the patrimonialization (*patrimonialización*) of public offices, developed strong roots during the fifteenth century (García Marín 1974: 341–4). This necessarily contributed to the strengthening of family power and group integrity through the creation of interfamilial ties by marriage or shared business ventures (Val Valdivieso 1995). All these developments fed the growth of group solidarities as well as competition among their various constituents.

Competing Newcomers

Beyond theoretical musings over what constituted a noble, the instability of the concept of nobility was the result of the disappearance of old lineages and the (re)constitution of new ones. In addition, many individuals "without lineage" gained access to the higher echelons of society through a web of circumstances that commonly combined special abilities and political favour. In his *Gesta Hispaniensia*, royal chronicler and secretary Alfonso de Palencia (1998) denounced the evil effect that the participation of "new men" ("homines noui") of obscure origin had on politics (for example, 137–8). The bestowing of new titles and the creation of a "new nobility" (Mitre Fernández 1968, Moxó y Ortiz de Villajos 1969b, Sánchez Prieto 2001: esp. 23–114), or rather its renewal via a restructuring of its different branches in the fourteenth and fifteenth centuries (Binayán

Carmona 1983, Quintanilla Raso 1999), offered opportunities for individuals of low birth such as Miguel Lucas de Iranzo, who became condestable (constable),[24] or others such as Fernando Alfonso de Robles, who ascended to contador mayor and noble rank from his lowly beginnings as an escribano assistant ("mozo de escribano"). Of lowest birth and with little estate or lineage ("de muy poco estado y baxo linaje"), according to the *Crónica anónima de Enrique IV* (83), Iranzo may have been the son of a labourer. Iranzo's ascent into the office of condestable included posts such as page, *halconero* (falconer), corregidor of Baeza, *alcalde mayor* of Baeza, *alcaide* (governor) of Alcalá la Real, alcaide of Jaén, and *consejero real* (royal counselor). After arming him caballero, King Enrique IV appointed him Chanciller Mayor de la Poridad (chief chancellor of the secret seal), and later condestable. Such rapid social ascent from the lower strata was possible, but humble origins were not easily forgiven by members of noble families with longer pedigree such as Pérez de Guzmán, or by Carrillo de Huete (1946) in *Crónica del Halconero*, who made a point of noting that Robles lacked linaje ("no era omne de linaje" 71).[25] In his harsh portrayal of Robles in his *Generaciones y semblanzas*, Pérez de Guzmán (1965) decries the ascent of people such as Robles, an "obscure man of low lineage," and their eagerness to exercise power over the "prelates and caballeros" of longer and more illustrious lineages (34–5). In the words of don Álvaro de Luna, Alonso Pérez de Vivero, whose family line appears to descend from conversos (Carlé 1993: 79), went from young servant ("rapaz," also meaning rascal) to *contador mayor del rey* by virtue of Luna's protection. This is recounted in the so-called *Abreviación de la crónica del Halconero* (text published by Mata Carriazo in his preliminary study to the *Refundición* attributed to Lope de Barrientos [1946: cxcix]). Different kinds of services could merit an upward move from seemingly any social stratum, as in the case of Alfonso Álvarez de Toledo, or of Sancho González, who was most likely a *trapero* (dealer in used clothes) and therefore a mediano. He gained hidalgo status, which was extended to his sons and daughters and their descendants ("fijosdalgo notorios de solar conosçidos") by Isabel and Fernando in 1477, as recorded in the *Tumbo de los Reyes Católicos del Concejo de Sevilla* (1968, 2: 99–103). Alfonso Álvarez de Toledo, a member of a closely knit converso family network in Toledo, and a cousin of the famous *relator* (court reporter) Fernán Álvarez de Toledo, was a trapero who exemplified the type of socio-professional mobility possible at the time. Under the protection of constable don Álvaro de Luna, he joined the Castilian treasury in one of its highest offices as chief controller, and was named regidor of Toledo. In 1445, he received the village of Casasbuenas from King Juan II and became a member of the Royal Council, amassing great wealth. His second wife's sister married Baena's colleague, royal secretary Diego Romero (on whom, see next chapter) (Martz 2003: 38–43). The advantage of favouring an individual of a lesser-known family over one of noble lineage was that it ensured the loyalty of the newly empowered. Bureaucrat caballeros, such as well-known poets Juan Álvarez Gato and Gómez Manrique, had their own path to socio-professional ascendancy. This will be detailed in the next chapter.

Caballeros, Medianos, and Economic Power: Urban Interaction

As has been suggested above, changes in the role of the nobility and the growth of state government were met with other important developments in fourteenth- and fifteenth-century Spanish cities. Starting in the fourteenth century and peaking in the fifteenth was a period of agrarian expansion resulting from changing demographic patterns, a wider availability of land, and an improvement of crops. These changes gave way to commercial growth, urban development, and the establishment of a commercial oligarchy or precapitalist commercial bourgeoisie (Iradiel Murugarren 1999: esp. 614).[26] The growth of cities towards the fifteenth century thus fostered an urban middle class, notably in cities such as Burgos, Córdoba, Madrid, Segovia, and Valencia, and was strongly represented in Andalucía.[27] Breaking the three-tier feudal stratification were the medianos or *ciudadanos*, a term used by Diego de Valera and others. In his *Crónica de los Reyes Católicos,* Diego de Valera (1927) points to a civic social order made up of the groups that appeared to acclaim Isabel as the new queen and which included "nobles e ciudadanos e populares" (4). Jews and conversos were well represented in mediano professions as merchants and artisans,[28] although they were employed in a variety of others as well.[29] Historians have noted the presence of conversos in jobs with municipal power, particularly those in city councils.[30] Several waves of converso families gained access to positions of power within the concejo in Seville throughout the fifteenth century. They also showed strategies of strong group solidarity by way of marriage unions and participation in local political, economic, and commercial activities (Sánchez Saus 1998: esp. 376). As part of the state bureaucracy, conversos could hold escribano, *pregonero* (town crier), treasurer, *contino,* or *secretario* (secretary) posts.[31] In light of recent studies on the caballeros cuantiosos or de premia, it is crucial to note the role they played within urban government, as well as the move towards aristocratization on the part of the conversos from the lower ranks of the nobility or the higher ranks of the lower estates (Cabrera Muñoz 1999), while forming, as has been pointed out in the case of Montoro, part of a "commercial bourgeoisie" (Costa 2001: 13; further discussion in Gómez-Bravo 2010).

Mobility was more accessible for those individuals who were close to the higher bounds of their group and therefore closer to the next social level. Thus, the commoner elite, which has been termed "élite del común," consisting of mediano wealthy artisans and small merchants, was able to compete for power with the urban oligarchy formed by local noble families, which tried to control access to its own ranks[32] and showed a strong resistance to the inclusion of middle-class merchants and artisans in their circles (Valdeón Baruque 1990: 516–17). The relative social porousness of the borders between groups helps explain shared aspirations and economic and commercial activity as well as the creation of family ties.[33] This does not mean that the lower group necessarily sought assimilation into the higher group or the adoption of its values and ideology. Groups such as the *élite del común* sought to be on a political par with the urban oligarchy but not necessarily to become incorporated into it (Val Valdivieso 1996). Since the fourteenth century, the urban middle class had developed sociocultural and political strategies to redefine chivalric ideals and create their own definition of citizenship and therefore of the social order (Rodríguez Velasco 2010). It is also important to realize that these groups were not homogeneous and that there was internal conflict in addition to the conflict between social strata.[34]

As landowners of urban and rural properties, with interests in certain forms of taxes such as the *alcabalas* (indirect taxes on trade transactions and products that provided important revenues for the Crown) and through involvement in commerce, nobles rivalled merchants and middle-class artisans, with whom they also competed through their participation in urban government. They fit into an interloping model that has been described as the *merchant caballero*.[35] In this they had a counterpart in the escribanos, also part of the urban elite, who could combine their jobs with land ownership and commerce.[36] In cities like Valladolid, the incorporation of middle- and lower-level merchants into the urban elite further encouraged the shift (Rucquoi 1997: 173–358). The merchant caballero and his contiguity to middle-class artisans were already implicitly recognized in *Libro de buen amor*: "Las vuestras fijas amadas / véadeslas bien casadas / con maridos cavalleros / e con onrados pecheros, / con mercadores corteses / e con ricos burgeses" (Juan Ruiz 1992: 453–4).[37] In his *Espejo de verdadera nobleza*, Diego de Valera denounces these caballero activities as a declassing of the noble caballería: "Ya las costunbres de cavallería en robo e tiranía son reformadas; ya no curamos quánto virtuoso sea el cavallero, mas quánto abundoso sea de riquezas; ya su cuidado que ser solía en conplir grandes cosas es convertido en pura avaricia; ya no envergüençan de ser mercadores e usar de oficios aun más desonestos, antes piensan aquestas cosas poder convenirse; sus pensamientos que ser solían en sólo el bien público, con grant deseo de allegar riquezas por tierras e mares son esparzidos" (text in Penna 1959: 107).[38]

In addition to their military function, the caballeros were a socioeconomic group heavily dependent on monetary worth and involvement in positions of urban power. As Monsalvo Antón (2003) has pointed out, the caballeros were the socially dominant group in municipal government, while the regidores were the ruling elite (424).[39] However, the caballeros slowly lost their power to the royally sanctioned regimiento, which tended to be monopolized by the great urban oligarchies, and thus their presence at concejo meetings during the fifteenth century gradually diminished.[40] This closed political elite was supported by the monarchy for reasons similar to the ones that led to the establishment of the *corregimientos*: to exercise greater control over the cities. The itinerant nature of the court, which Cañas Gálvez (2007), Torres Fontes (1953), and Rumeu de Armas (1974) have detailed for the courts of Kings Juan II, Enrique IV, and Queen Isabel and King Fernando, respectively, would help it keep a strong presence in cities. Val Vadivieso (2006) has pointed out that the strong pillars of royal control over cities were the corregidores,[41] the urban oligarchies, and the idea of the common good ("bien e pro común") that was becoming widespread (esp. 26). This model would eventually bring about the strengthening of the Crown and the consequent loss of local power, the overall forces playing a role in the defeat of the *comunidades* by the Empire.[42]

Multitasking or *Pluriempleo*: Court and City

Men in the service of the royal court could receive significant positions in city government (Cañas Gálvez 2008). Some families established powerful escribano lineages that would climb to the upper bureaucratic positions at court and from there to the Consejo Real or the urban elites, becoming part of the higher echelons of the social ladder and gaining municipal power (Cañas Gálvez 2008). Attaining urban power, however, was not devoid of conflict. Starting in the middle of the fourteenth century, the laws requiring royal confirmation of individuals chosen by local governments for municipal positions allowed greater royal control. They also provided a source of conflict between the cities and the Crown and of city and royal government with collectives such as the escribanos, who sought to safeguard their interests by pressuring to keep limitations of their numbers or of other professional escribanos (royal, apostolic or others) in order to avoid competition in what could be a very lucrative position.[43] Narbona Vizcaíno (2003) has referred to the creation of a bureaucracy involving a transformation of hidalgos into administrators ("funcionarialización de hidalgos") through which the Crown could control urban government and dissolve the solidarities that helped local elites hold powerful hegemonies (584). This benefited both the monarchy, by gaining influence over municipal government, and the officials, for whom this provided multiple sources of income. Job compounding, *pluriempleo*, worked for both political and economic reasons, and was ultimately a source of power. Service as the royal physician, servant, carver, steward, equerry to the king, or royal secretary could open the door to jobs in city government, including the much-coveted regimiento.[44] Author Diego de Valera held many different posts during his life, including royal diplomat, member of the concejo or city council of Cuenca, corregidor of Palencia and Segovia, *maestresala*[45] of Kings Enrique IV and Fernando, member of the *Consejo Real*, and alcaide of Puerto de Santa María. A famous royal secretary (documented as such in 1488), Hernando de Zafra, probably of converso origin, was a member of the "middle urban nobility" who had begun his career employed in the Contaduría Mayor de Hacienda and as *escribano de cámara* (escribano of the royal chamber) (1468–74). Zafra held many jobs and different escribano positions. He was said to be of "obscure origin," allegedly having left his parents' house carrying just an average writing desk and materials ("escribanías medianamente aderezadas") in order to look for opportunities at court (Ladero Quesada 2005). Zafra had a stellar career, accumulating some urban government positions. He was named, for example, regidor of the town of Ronda in 1485, *alguacil mayor* (chief bailiff) of the town of Marbella in 1486, and regidor of Granada in 1498, although he unsuccessfully lobbied, through the intercession of the Archbishop of Granada fray Hernando de Talavera, to be appointed to the high-ranking position of contador mayor. Multitasking was further aided by the fluidity of competencies attached to each municipal post, as well as the kaleidoscopic variations of such competencies in each town, at times rendering useless tags such as "regidor," "corregidor," or "alcalde" (mayor), as noted by Monsalvo Antón (2003: esp. 428–9).

The possibility of increased income through the appropriation of competencies was key in the ambition of the corregidores, who were on the concejo's payroll, for increased rights over specific functions (González Alonso 1970: 101–3). A secretario could also often be an escribano, *notario* (notary), *contador* (accountant), or *oidor* (judge), as well as a member of the Consejo Real.[46] Thus, converso Fernán Díaz de Toledo, who had a law degree, served under King Juan II not only as secretario but also as escribano, royal referendary, oidor, *notario mayor de los privilegios,* and member of the Consejo Real.[47] Converso author Fernando de Pulgar called himself "escrivano" in one of his letters, written in his old age (Letra I, 35), as well as in the *Libros de acuerdos del Concejo Madrileño* (1932–87) (hereafter cited as *Libros de acuerdos*), in which on 5 September 1481 he is referred to as "escribano de cámara del Rey e Reyna." He was also identified as a "coronista" (chronicler) in the 1486 printed edition of his *Claros varones de Castilla,* and similarly in the *Libros de acuerdos* in 7 December 1487, and as a resident of Madrid at various dates during the 1480s in the *Libros de acuerdos.* In addition to positions at court and city, multiple bureaucratic posts were available at the home of important nobles or church dignitaries. In such a position was Pero Guillén de Segovia, a professional escribano by necessity, who came to work for Archbishop Carrillo and was in charge of his finances as contador, while benefiting from Carrillo's patronage as writer and poet. He states his situation in a poem addressed to Carrillo (ID1740, in Moreno Hernández 1989: 35–6, 289).

The compounding of jobs was all important for persons with aspirations to upward mobility because it brought increased wages. The salary of the regidores ranged from 500 to 5,000 maravedís per annum in the last decades of the fifteenth century (Polo Martín 1999: 165–72), while the *escribanos de concejo* earned from 1,400 to 10,600 maravedís annually, plus additional charges per document drawn (Polo Martín 1999: 345–9).[48] It is clear that access to public office came with higher rewards than the assigned yearly salary. Regidores such as Pedro González de Hoces earned much higher wages while employed by the queen or king for different commissions and special services, as is obvious from the details provided by his 1455 will (published in Ostos Salcedo 2005: 354–5). In addition, these jobs came with economic and political advantages and privileges that could increase the economic value of the post. In fact, there is evidence that these positions in towns such as Burgos were purchased for hefty sums of money by people in already good socio-professional standing because of the heightened prestige and power they brought (Guerrero Navarrete 1986: 177–93, 405–7). The need to compound jobs was also sometimes attributed to the relatively low salary that each could provide, causing poets like Juan de Mena to complain of low wages ("chico salario") in reference to posts as "coronista" and "secretario" (ID1799, v. 5, in reply to ID1798). Mena's earnings are documented to have compounded to 3,750 maravedís per annum, later adding 20,000 per annum and 15 daily, plus income amounting to several thousand more maravedís from various sources (Fuentes Guerra 1955: 87–8, 131–2; Street 1953: 163–6). Economic need – justified or not – was a malaise that several poets complained of, including Mena, Gómez Manrique, and, perhaps more famously, Villasadino and Montoro. While many poets represented themselves as lacking, in fact they were well off. Complaining about a lack is a way of expressing a proclivity towards the accumulation of economic resources. In this sense, the accumulation of titles and positions is mirrored by the accumulation of wealth.

The system in place put working low and middle nobles in close proximity to other bureaucrats, including university graduates, caballeros, escuderos, or people of lower birth but adequate connections, economic means, or perceived abilities. This contact allowed them professional access to posts in urban government, as well as wealthy noble or royal households and therefore competition. The relation of these professionals and poets could be described as one of *social contiguity*, in which persons of varying social levels interact within shared networks by virtue of their professional occupation and the social situation it facilitates, fostering internal ties of solidarity or conflict among ranks and also among equals vying for socio-economic advantage. The lower and middle nobility were by definition in a relation of social contiguity, as were the urban nobility, the medianos, and the wealthier members of the lower classes, such as the labradores ricos. This, despite the opportunities for social mobility, put many different people in potential "contiguous relation" contact. This proximity allowed for random contacts and encounters that materialized in economic and textual transactions. The ideals of caballería and nobility coexisted with those of the medianos as well as those of other classes. In addition, caballería and nobility were unstable categories that, though ideally fixed, were subject to constant negotiation by the social shifting of their constituents, opening the question of the very nature of identity and its modes of self-constitution. Textual productivity and mastery over stylized language would prove to be a key element in the processes involved in such self-constitution.

The porousness and multidirectional overlap of various social groups, the "myriad social networks" emphasized by Michael Mann (1986: 1, 19, and passim), and the possibilities for aggressive jumps across several levels suggest a model based on a continuum of irregularly interlaced porous strata more than a perfectly defined and compartmentalized social pyramid. If, as Deleuze and Guattari (1987) state, "the concern of the State is to conserve" (357), the means for the emerging absolutist state to achieve that goal included a multiplication of the competing personnel in its service, as well as the proliferation of texts and the professionals who produced and reproduced them. Textual capabilities were a clear tool for political power, but they were also key to the act of self-generation in the context of socio-professional networks of interaction.

2 Escribano Culture and Socio-professional Contiguity

The material means at hand, and the skill and value attached to paper transactions were common currency in circles that drew much of their cultural and economic worth from the written word. The transition from a "notary culture" to (pre)humanism from the fourteenth to the fifteenth centuries has been noted as key in establishing changes in books and writing (Petrucci 1992). However, this did not mean that scribal or notarial competencies were superseded by other professional occupations. Instead, they remained at the core of writerly culture. Caballeros, hidalgos, and medianos shared in what could be called escribano culture. Many of them actually fulfilled escribano duties at different points in their careers, or applied scribal practices in their trade, following the intense emphasis on record keeping and on textual production as a whole in both court and city. More than providing ground for metaphorical allusions, this insider's knowledge of escribano modes of work facilitated the transfer of methods and materials from the scribal desk and pen to the poet's, both often being one and the same. Further, the proliferation of texts and documents in cities and individual households meant a widening of the figure of the poet, beyond the earlier professional but also broader than a mere court satellite or subaltern. The multiplication of people who wrote poetry on a wide array of occasions, even if it was just a few texts, from political to everyday topics, along with the rapid dissemination of paper, made poetry a widespread social phenomenon that went beyond a court elite. Poetry as stylized discourse was the expected mode of interpersonal communication in social networks, valued as a display of mastery over language, the expression of inner worth. Poetry, like escribano deftness, could improve social standing. This would be the case through the sixteenth and seventeenth centuries.

Escribanos

Because of its implications for all textual production during the period, it is important to understand the intricacies of escribano work. As was the case with other city and state employees, escribanos were known to hold more than one post, which allowed public escribanos to place themselves in powerful positions in city government, for example, acting as jurados (Pardo Rodríguez 1994: 162–3).[1] An individual who had gained the trust of the Crown or who merely had the right family connections could hold several different positions involving document production and political responsibilities at the same time. In court, the functions of notaries (notarios), secretaries, and escribanos overlapped so much that these terms were used interchangeably or grouped together with other ones such as relator or *refrendario* (referendary) (Torres Sanz 1982: 83–124). City legislation similarly used the terms escribano and notario interchangeably, as was the case in Seville (Bono Huertas and Ungueti-Bono 1986: 44–56). Further, public and city escribanos could also become court or royal escribanos and vice versa.

Escribanos were generally in a good economic situation, and in some cities there was a strong pressure to appoint escribanos who were caballeros de premia or members of the local elite families (Corral García 1987: 22–3, 25–6; Carpio Dueñas 2000: 326–8). Even though they were not supposed to have executive power, escribanos exercised significant influence in city councils and at court.[2] Escribanos' ability to thrive economically and enjoy privileged status is clear in the case of Alfonso González, escribano in Madrid in the fifteenth century (Rábade Obradó 1994), as well as more generally that of the *escribanos mayores* studied by Pardo Rodríguez (2000), and by Guerrero Navarrete for Burgos (1986: 100–2). A similar situation is that of the escribanos of Seville, some of whom also attained the prestigious post of jurados and enjoyed a good economic situation thanks in part to their real estate properties, advantageous marriages, and participation in commercial trade, as well as tax farming (Bono Huertas and Ungueti-Bono 1986: 25–7). Escribanos rarely held a university degree and had mostly a "practical knowledge" of legal texts derived to a large degree from their work on document drafting (Bono Huertas and Ungueti-Bono 1986: 27–9; Ostos Salcedo 2005: 58), as well as the mandatory period of apprenticeship under contract in an established escribano shop or *escribanía* (Pardo Rodríguez 1994: esp. 148–64). Only a very small fraction of the large numbers of court secretarios and escribanos had a university degree during the reigns of Juan II (3.5 per cent), Enrique IV (1.1 per cent), and Isabel and Fernando (3.5 per cent) (Phillips 1986: 478–9). The period of apprenticeship included the possibility of boarding and learning ("criar") at the escribano's house, a practice that was exercised by Fernando de Pulgar, who by his own account at any given time had a number of pupils living in his house. In fact, Pulgar (1982) is remembered for wryly denouncing the fact that many "old Christians" had no qualms about sending their sons to board and apprentice in the house of converso escribanos, secretaries, and bureaucrats, while at the same time enforcing discriminatory laws against the conversos (Letra XXXI).

Escribanos carried out complex functions that included both scribal and notarial responsibilities. The *Siete Partidas* lay out the laws concerning escribanos (Partida 3, Title 19), as well as those regarding the material production of documents (Partida 3, Title 18), which formed the basis of similar legislation in Díaz de Montalvo's (1990) *Ordenanzas reales de Castilla*, and which would in turn give way to the *Ordenanzas de los escribanos públicos* issued in Alcalá de Henares in 1503, as well as other city-specific ordinances, such as those issued in 1492 for Seville.[3] Escribanos had responsibility for the books of records, which they often kept at home, a practice that presented some problems for the preservation and accessibility of records (Arribas Arranz 1964a: 247–8; Pardo Rodríguez 2002: 107, and passim). The escribanos regularly conducted business in their shops or escribanías, but also drew up documents in their homes or that of their clients, using either the escribano's book, booklet, or cuaderno, or just a scrap of paper (Bono's epilogue in Pérez-Bustamante 1985: 64–5; and Ostos Salcedo 2005: 69; see chapter 5). After he had jotted down the initial notes in the escribanía (*taula* or table in Valencia), the long task of registering and writing the documents took place either in the escribano's study or bedchamber, at the client's house, which at times meant travelling to another town, or at another location (Cruselles Gómez 1998: 119–23). The post-mortem inventories of scribes' homes in fifteenth-century Valencia show that a great deal of scribal work was done in a small study adjacent to the bedroom or in the bedroom itself, fitted with a table, chairs, writing instruments, and storage furniture, often in the form of boxes, chests, or even bags, where the loose papers and bound registers could be safely kept. The room often contained other objects conducive to sociability, such as chess sets or music instruments. Notary Bernat Costa kept the finished documents waiting to be picked up by the clients in pine and fir wood chests of drawers, whereas the loose papers ("paperots") with the initial notes were placed in a chest. After the contents of these papers were noted in the register, the loose papers were not archived with particular care and tended to disappear (Cruselles Gómez 1998: 114–19). Similar practices applied to court officials. Upon the death of royal secretary Hernando de Zafra, Isabel and Fernando ordered Zafra's widow, Leonor de Torres, to send to court Zafra's record books pertaining to his many positions. Zafra left so many of these books that in a letter to the king the widow had to ask him to specify which ones needed to be sent (Ladero Quesada 2005: 124–6). Because of the potential for loss or inaccessibility of records, there was a move towards keeping municipal books in an institutionalized archive, which initially took the form of a simple chest. It was to be permanently located in the room where the concejo members met. The escribano usually was in possession of one of the three keys that opened the chest (Corral García 1987: 66–7). Following a similar archival practice, individual household records were stored in chests in private homes. The usefulness of such a system was evident in 1440 when Santillana settled a dispute in the town of Trijueque by consulting his wife Catalina and then asking her to search for the appropriate documentation in the chests that had belonged to his mother.[4] A need for a larger repository was felt in regard to royal records, for which the fortress of Simancas was later chosen as official royal archive (Conde y Delgado de Molina 1998).[5]

Public escribanos could employ auxiliary officials simply termed "escribanos de oficio" or "escribano de" followed by the name of the city where they worked (Bono Huertas and Ungueti-Bono 1986: 30). There were also "copyist escribanos" ("escribano de letra de obra" or "escribano de mano") (documents in Bono Huertas and Ungueti-Bono 1986: 91, 142, 391), who copied books of any nature for a price; for this reason they could also be hired to teach reading and writing, as is documented in Córdoba (Ostos Salcedo 2005: 75–6). The vast need for escribanos helps explain the various specializations, as they appear for example in the accounting papers of the city of Seville, where there is mention of *escribanos de los alcaldes mayores, escribanos de la armada, escribanos de cámara, escribanos de la cárcel, escribanos del concejo, escribanos del consistorio, escribanos de latín, escribanos mayores del concejo,* and *escribanos mayores de la justicia,* in addition to *escribanos públicos, escribanos de quiebras, escribanos reales, escribanos de cámara del rey,* and *escribanos de los pleitos* (see indexes in Collantes de Terán Delorme 1972–80 for document references). Public escribanos typically had a shop in a designated public street (for Seville, see Bono Huertas and Ungueti-Bono 1986: 30–4) but could also be privately employed for work in a wealthy household. The escribanos of Córdoba, for example, had their shops in the commercial "calle de la Escribanía Pública," which stretched from the plaza de San Salvador to the commercial street of Marmolejos. The street that housed the escribanos was one of the places for public proclamations (*pregones*) (Escobar Camacho 1989: 235), showing the close link between escribanos and *pregoneros*. Both were municipal employees, although the pregoneros were not among the highest paid.[6] The *colación* (also *collación*)[7] of San Andrés in Córdoba was home to many escribanos as well as Jews and conversos, including Juan Alfonso de Baena (*Cancionero de Juan Alfonso de Baena* 1993: xvi). In Madrid, public escribanías could be found in the centrally located Plazuela de San Salvador (Vizcaíno Villanueva 1991–2: 143), where the concejo met and which was also the city's mercantile and later administrative centre. This was the location of poet and bureaucrat Juan Álvarez Gato's residence. In Toledo, the escribanías were situated close to the *alcaná* or smaller Jewish quarter, which also housed small Jewish shops, including those dedicated to garment making (León Tello 1979, 1: 187, 241–2, 367). There is evidence of guilds or professional escribano associations. In the city of Seville there is an example of a "cabildo de los escribanos" with appointed officers (see appendix in Piergiovanni 1994: 631–3); and there was a *cofradía* or confraternity of escribanos in Córdoba (Ostos Salcedo 2005: 70).[8] The powerful royal escribanos formed an exclusive confraternity, the Cofradía de Todos los Santos in Valladolid (Cañas Gálvez 2008: 395), in addition to the "Cofradía de la Concepción de la Virgen Gloriosa Señora Santa María de escribanos y procuradores de la Corte y Chancillería" mentioned in Diego López de León's will in 1452 (Carlé 1993: 69). Furthermore, escribanos seem to have followed the common tendency of members of a certain group to live in the same colación. These "group building" practices, along with a common place of work, facilitated both the conflict and solidarities associated with such close contact. Thus, the escribanos were at the heart of urban life and an indispensable everyday presence. Further, the court served as a similar common space of interaction for escribanos and other bureaucrats. Such interface transpires in many of the poems they produced. Baena, for instance, revealed this escribano fellowship in ID1584 when he asked two other royal and "very loyal" escribanos ("escrivanos muy leales"), Martín Gonçález and Sancho Romero, two of the most important royal escribanos of the time, for help with their pens ("ayuda de su péndola").[9] Each of the five whole stanzas in this poem opens with the same clamorous cry to "Mi señor Martín Gonçález, / otrosí Sancho Romero." He tells them of his misfortunes and very near poverty, and calls himself a poor squire ("pobre escudero"), a condition that only they can relieve with the help of their pens or inkwells ("con la pluma o tintero"), but without charging a fee ("acorredme con alguna / pendolada sin dinero"). Baena hoped that with their help the piece of paper with his poem and request ("carta") would reach the king as well as don Álvaro de Luna ("el grant Señor de Luna"), who would undoubtedly ("sin dubda ninguna") help him.[10]

Escribano Wares and Attributes

The preoccupation with the physical and material aspects of writing evident in Baena's poem is pervasive in escribano culture for obvious reasons. Hands, pen, paper, and ink were materials and instruments of writing and signs of the writing trade, mainly that of escribanos and also of secretarios, as well as that of the many officials whose posts involved record keeping.

The weight placed on accountability and traceability for all documents and their copies enhanced the need for the textual identificatory traits of the escribano, including his signature, signum, the use of the first person when writing, and the disclosure of the escribano's full name.[11] It was in this vein that Juan Alfonso de Baena presented his poem-document (carta) to the king: "yo, Juan Alfonso, un vuestro escrivano / con mucha mesura e grant reverençia / a vuestra persona de alta excelençia / presento esta carta besando la mano."[12] The formula used by Baena, "I, Juan Alfonso, your escribano," and the references to the material support of his text ("carta") suggest both a presentation ceremony of his poem and an escribano-authenticating signature. The transfer of escribano marks into poetry is noticeable in other cases. MN16, possibly one of the fragments of the so-called *Cancionero de Fernán Martínez de Burgos*, bears the authenticating signature: "Escriuiolo ferrnand martines de burgos fijo de juan martines de burgos escriuano publico que fue de la dicha çibdad [...]" (text in Dutton 1990–1, 2: 78).[13] Fernán Martínez de Burgos is not only one of the authors in the compilation, but also the compiler and scribe. He authenticated his book by adding his notarial signature in verse in the conclusion, taking advantage of the fact that both the notarial formula ("notary of the present document") and his name amount to two perfect octosyllables: "Notario de lo presente / Fernand Martinez de Burgos" (ID6958, MN23).[14] His cancionero is centred around a compilation of his father's work, notary and poet Juan Martínez de Burgos. Fernán and Juan were father and grandfather of Diego de Burgos, also a poet and escribano of the Marqués de Santillana. The terminology used in MN16, MN23, and other cancioneros reflects the current notarial practice that referred to the execution of the material tasks of writing, or putting pen to paper as "escriuir," and differentiated them from the commissioning of the writing by another ("fazer escriuir").[15] The closing signature functions as an authenticating mark and by way of the hand creates a chirographic notarial continuum that links the individual texts and cements the vital structure of the cancionero.

The constant contact of the hand with paper was a measure of an escribano's worth. A contract involving a public escribano of Seville taking an apprentice to work in his shop specified the need for the apprentice to write without raising his hand from his work for the duration of the time of service ("non alçar la mano fasta ser conplido el dicho tienpo" [Bono Huertas and Ungueti-Bono 1986: 347]). Far from being the exclusive mark of an escribano, scribal wares and attributes appear also to have been used by nobles with administrative duties. This is the reason why, for example, we find a writing desk ("escrivanías de asiento") in the 1490 post-mortem inventory of Gómez Manrique's belongings (Caunedo del Potro 1991: 108). Manrique's constant writing activity undoubtedly contributed to the failing eyesight that necessitated the four pairs of eyeglasses listed in the same inventory (Caunedo del Potro 1991: 103). Manrique's inventory shows similarities with those of notaries such as Antoni Pasqual in Valencia, who kept his arms, books, money, and eyeglasses in his bedroom (Cruselles Gómez 1998: 117–18), which shows the very personal space in the home where arms and letters mingled. If he did not personally write down his official letters, Manrique would always sign and write a brief farewell, declaring himself "at your service," often with the wording "a lo que la merçed de vos otros mandare" (see for example the letters published by López Nieto 1999). Often portrayed as skilled at both arms and letters, Santillana also assumed escribano tasks by writing his own business correspondence. An example is his autograph letter sent to Madrid's city council asking for permission to send cattle to pastures under the concejo's jurisdiction (text in Domingo Palacio 1907: 141–3). In turn, Santillana fostered the poetic creativity of his own secretary and also poet Diego de Burgos, mentioned above, son of notary Fernán Martínez de Burgos, allowing Santillana and Diego de Burgos to be textual collaborators.[16] A telling example is that put forth by noble Lope García de Salazar (1967) in his *Las bienandanzas e fortunas*, where he described the diligent night study and active reading and writing of Julius Caesar as amounting in output to that of three scribes ("escrebía más que tres escribanos" 2: 104). The image was undoubtedly based on García de Salazar's own experience as escribano and public notary (discussion of relevant documents in Avenoza Vera 2007). The mark of a good writer was graphic mastery as much as a tight command over language.

Heavily employed by the Church, escribanos and notaries were not only recorders of trials but also key figures in the inquisitorial process as keepers of the keys to the chests where documents were archived, just as in secular institutions. With the need for massive recording and cross-referencing of texts, the documentary and archival practices of the Inquisition used administrative procedures similar to those of municipal and state government. Inquisitorial documents, including letters and confessions, were promptly archived in *registros* (registers), which were in turn put into chests placed in special rooms.[17] The Inquisition meticulously inscribed all that it touched, not only parchment and paper but also the condemned's faces, bodies, and special garments (*sambenitos*).[18] Royal secretary Fernando de Pulgar echoed both the material nature of a similar inscription and its internal imprint, occasioned by inquisitorial persecution in a letter against the work of the Inquisition in Andalucía: "días ha muchos que en el ánimo tengo escrito, y aun con rruin tinta" [for many days I have had my soul written over with vile ink] (published by Mata Carriazo in the introductory study to Pulgar's *Crónica de los Reyes Católicos* [1943, 1: L]). The pervasiveness of escribanos and public familiarity with their professional wares explains why even popular preachers such as Vicente Ferrer could build religious analogies to the materials of writing in his sermons, and the escribano attributes of love for the pen, inkwell, and writing paper (text in Cátedra 1994: 456). Inkwell, scissors, knife, pen, and their function in the escribanía (meaning both writing desk and accessories and the escribano's place of work) are mapped to the powers of the Church. The scissors are the Church's power to absolve and condemn, the knife and awl show the Church's power over penance and forgiveness, while the paper and ink show the Church's power to "make" prelates and priests who will guide the souls and write about doctrine with their pens (text in Cátedra 1994: 655–6). Because of the perceived power that such ready access to the written word gave them, both Church and State tried to enact tight control over the escribanos. The fourteenth-century author of a treatise on confession, Martín Pérez, instructed confessors to interrogate escribanos on the potential sins specific to their trade, which included producing faulty copies, writing lies in order to finish their work in haste, or using poor handwriting and bad ink (in his *Libro de las Confesiones* [2002: 436]). Fraud on the part of the escribanos was punished with a fine or by the cutting off of the offending appendage, either hand or the thumb, in the *Fueros* of Cuenca and Béjar (Corral García 1987: 79), as well as Alfonso X's *Siete Partidas* (Partida 3, Title 19, Law 16).

Escribanos, Medianos, Conversos

The extended escribano image in Vicente Ferrer's sermon evinces the power that escribano activity conveyed. The anxiety engendered by socio-economic and ethnic difference was further grounded on the prestige of Jewish knowledge. In regards to many textual and documentary practices, Spain's cultural genealogy was often traced to Judaic practices. King Alfonso X (2001) in his *General Estoria* embraced such a genealogy for scribes and notaries, making obvious the common ground among notaries, scribes, and chancellery work: "del linage de Simeón vinieron los chanceleres de los judíos, e éstos son a los que la Santa Escritura de la nuestra eglesia de Cristo llama en el latín *scribas*, ca esto da a entender *scriba* en el lenguage de Castiella, chanceler o notario" (1: 492).[19] The prestige of Jewish knowledge would continue to be highlighted through later centuries. This is the position that Rabbi Mosé Arragel took when presenting his Bible translation to Luis de Guzmán, the impressive *Biblia de Alba* (Alba Bible; or Biblia de Arragel) (1920–2), while emphasizing the heightened knowledge of Castilian Jews in particular and the usefulness of such knowledge to the lords and kings of Castile: "Esta preheminencia ovieron los reyes e señores de Castilla, que los sus judíos súbditos, memorando de los sus señores, fueron los más sabios, los más honrados judíos que quantos fueron en todos los regnos de la su transmigraçión en quatro preheminencias: en linaje, en riqueza, en bondades, en sçiençia. E los reyes e señores de Castilla siempre fallaron que todo a lo más que oy los judíos auemos de glosa ssobre la ley e en las sus leyes e derechos e otras sçiencias, fue fallado compuesto por los sabios judíos de Castilla; e por su doctrina oy son regidos los judíos en todos los reynos de la su transmigraçión" (3).[20]

The texts show the anxiety and conflict generated by the shifting borders in a socially contiguous situation. Cancioneros MH1 and PN8 contain two poems that present two different models of *galán*, both poems appearing interlocked in PN8. The first poem, "Coplas de la gala" (ID0141) by Suero de Ribera, details the ideal young gallant's (*galán*) characteristics and behaviour, including the type of horse he needs to ride; the necessity of singing and writing poetry ("por coplas e consonantes / bien cantar e conponer" [in stanzas and rhyme / well sing and compose] vv. 43–4); the type of clothes he should wear; the food he should eat; his attitude towards women; and the apparent disdain for money, particularly that of others, which he should borrow with abandon. Only occasionally did he pay his tailors and shoemakers (vv. 89–96). According to the second poem

(ID0517), whose importance was highlighted by Vendrell de Millás (1968), conversos embodied the other style ("otro estilo") of courtly people ("gente bien cortesana") as quintessential merchants and escribanos. This converso type could be stereotypically identified because he was circumcised, obeyed the dietary laws of kashrut, and showed traits that fit a given ethnic profiling with his "nariz luenga e bermeja" (v. 18). This converso also had to earn a living. Pen behind ear ("la pluma tras el oreja"), he lived from his skilled work ("arte de que se mantenga" vv. 19–20). He also needed to be dexterous ("diestro") with all sorts of merchandise ("todas mercadorías") (v. 28), and have a meaningful converso last name, such as "Santa María." The pen was the quintessential escribano instrument and symbol. Its position behind the ear was easily recognized as an escribano mark since a pen was placed behind a new escribano's ear in the formal ceremony in which he took office (Bono Huertas 1979, 2: 254–5; Guerrero Navarrete 1986: 100; Rábade Obradó 1995a, for a specific account). All of this, along with the stereotypically converso professions of merchant and escribano that the poem assigns to him, helped identify him as a "synagogue gallant" ("galan / del solar de la sinoga" vv. 33–4). The poem presents this type as a sign of the times, a newly emerging development that follows the Torah and which will stay a growing trend, if nothing is done to improve the situation ("bien creo se fallaran / algunos tales agora / si los tienpos asi van / a serviçio de la tora / si el fecho no se mejora" vv. 37–41). Following a similar set of associations in his *Coplas de Vita Christi*, fray Íñigo de Mendoza (1968), of converso descent, further called for the disaggregation of courtier and merchant by satirically calling the offenders to a "Christian circumcision" ("çircunçisión christiana") by cutting their (socially unacceptable) practices, thereby both mimicking and mocking Jewish practices by allegedly marking not the flesh of the "señores" but their evil deeds ("las obras de maldad"). More specifically, nobles need to circumcise their proclivity to become merchants ("çircunçiden los señores / el tornarse mercaderes"), which is irreconcilable with the virtues, graces, and honours ("virtudes, graçias, honores") of a noble (ID0269, in his *Cancionero* 64, 67). Pulgar considered merchant and escribano to be typically converso professions when he denounced the "new Guipuzcoa statute" which forbade conversos to live or marry in the land. Pulgar finds it highly ironic that Guipuzcoans would have no qualms about sending their sons to apprentice in the houses of conversos, mostly secretaries and other pen bureaucrats such as Pulgar himself. These old Christian sons of the north, in Pulgar's (1982) words, filled the houses of merchants and escribanos ("fenchir las casas de los mercaderes y escrivanos de acá de los fijos de allá"). These Christian sons favour the pen over military professions and therefore gravitate towards the houses of converso bureaucrats: "Quanto yo, señor, más dellos vi en casa del relator aprendiendo escrevir, que en casa del marqués Yñigo López aprendiendo justar. Tanbién seguro a vuestra señoría que fallen agora más guipuzes en casa de Fernand Alvares e de Alfonso de Ávila, secretarios, que en vuestra casa ni del condestable, aunque soys de su tierra" (Letra XXXI, 106).[21]

The law prescribed that an escribano had to fit a certain ethnic and genealogical profile, have good morals, and be of good social standing and "linaje" (Bono Huertas 1979, 2: 219). Jews, but not conversos, were banned from the office of escribano (as were Muslims, heretics, and members of the clergy [Bono Huertas 1979, 2: 215–18; Corral García 1987: 23–4]). Theoretically at least, Jews were not allowed to hold public office, although this was not always obeyed and practices differed. In fourteenth-century pogroms the perceived association of notaries and Jews caused the burning of the notaries' shops in the central Plaça de Sant Jaume in Barcelona in 1391 before the assault on the Jewish *Call* (Durán Cañameras 1955: 135). In addition, the pressure was strong to avoid putting conversos in positions of power over old Christians (García Marín 1974: 188–91; Lalinde Abadía 1970: 85–7). There was a clear tension between the model of inequality prescribed by the law and time-sanctioned practices that in some cities placed conversos at the heart of urban government. In cities such as Burgos conversos held positions in public office and in general enjoyed the support of the monarchy (Cantera Burgos 1952, Guerrero Navarrete 1986: 405–6, and passim; Márquez Villanueva 1957).

The negative stereotype of escribanos and other public officials as conversos was constructed in the anti-Semitic "Traslado de una carta de privilegio que el rey don Juan II dio a un hijo dalgo." In this mock document, in which King Juan II allegedly gave a nobleman what were perceived as converso privileges, the king bestowed on a hidalgo the same rights that the "marranos" were accused of having attained, such as access to any royal office, including posts of alcalde, regidor, jurado, and public escribano ("oficio real, así de la alcaldía, regimiento, como de juradería y escrivanía pública"), or those related to the health professions, such as apothecary, physician, and surgeon ("boticarios, físicos y zurujanos"). It is important to note that the document also points to groups, associations, and confederations that were intended to build converso solidarities ("que os reciban en sus concilios, juntas, aiuntamientos e confederaciones e ayudas"), as well as a converso-specific "subtle and deceitful" discourse.[22] In addition to these noble and old Christian anxieties about converso agency and solidarities, which were intended, the text states, to cause the self-destruction of old Christians ("dando ocasión a que se maten los unos con los otros"), the mock document clearly betrays fear of the potential economic and professional dominance of the conversos. The document establishes an overly simplistic dichotomy between converso public officials and old Christian nobles, exposing some of the key issues that affected the intense socio-professional competition among those in a position of social contiguity. Asenjo González (2006: 170) has highlighted the anxiety over the destabilizing potential of the conversos due to their elevated numbers and high degree of professional specialization, as well as the failure of the aristocratic society to integrate them and other emergent groups. In addition, these texts reflect the very material conflicts taking place in urban centres, whose government the monarchy was increasingly trying to control. As Val Valdivieso (2005: 1041 and passim) remarks in her overview of urban conflicts, the revolt against conversos headed by Pedro Sarmiento, which resulted in the anti-Semitic *Sentencia-Estatuto* and the laws of purity of blood, sought to eliminate the access of conversos to positions in the city council or concejo, and, as Weiner noted, to exclude them from the position of public escribanos in particular (in his introduction to Horozco 1981: 10–11).

The conflict stirred in the socio-professional or political arenas was mirrored by poetic discourse. In ID0838, García de Astorga mocked the converso identity of a secretary who had insulted him by comparing the secretary to a piglet among swine or *marranos* ("lechonçillo entre marranos") and to a "sad owl" and "mouse full of wax" ("gesto de triste lechuza / rraton fagado de çera") – presumably acquired through his professional occupation – in a description that calls to mind Mena's "carnes magrescidas." If García de Astorga was of probable converso origin, as Ponce Cárdenas (2005) has claimed, this and his profession as escribano would feed the expectation for virulent poetic exchanges against other conversos as well as old Christians.[23] In ID0839, he replies in no uncertain terms to an ostensibly old Christian escudero who had allegedly criticized his writing, calling it the work of a peasant ("labrador") and pronouncing it displeasing to those who, like him, are "de palaçio." The poem is a revelatory exercise in the relation of social status, ethnicity, and writing. As a pen professional, Astorga states his superiority to the *escudero*, whose ability to write poetry is limited to some cold lines ("alguna cosilla fria" v. 13) allegedly without fault or "raza" ("vuestro trobar sin rraça" v. 16), an activity that is further hindered by the escudero's need to be vigilant about his poor, ill-fitting clothes, which he can barely keep on his body as they are so old they are falling off (vv. 23–30). All this, warns Astorga, should not be an obstacle to the fulfilment of the escudero's religious obligations. Recontextualizing a common saying, he advises that the escudero may cross himself with his foot, if unable to use his hand ("que os santigues con el pie / si no podes con la mano" vv. 44–5). The poor escudero is thus deprived of the use of the hand, the escribano's pride.

The socio-professional animosity that converso bureaucrats could inspire in poets of noble lineage such as Juan de Dueñas moved the latter to demand from King Juan II the disassociation of the royal court with the conversos (Vendrell de Millás 1958a: 189–90, 193, and 1958b). As Rábade Obradó (2008) has noted, the strong presence of conversos at court raised forceful complaints, since at least the reign of Enrique III and continuing until well after that of Isabel and Fernando. In two harsh poems to the king that seem to have gained him the royal wrath and eventually exile (ID0457, 0458), Dueñas presents the "viles," among which he notes the conversos, as dirtying the king's court and stealing the jobs from the good but poor noble hidalgos. Dueñas's goal is obviously to keep these competitors from taking the king's gifts in money or in kind, particularly the conversos, who are as enemies of the good as life is from death ("quanto mas a los conuersos / de los buenos mas aduersos / que la vida de la muerte"). And he continues to make a case for the poor hidalgos like him, who he claims will ultimately prove to be the king's support: "Antes señor deues dar / grand parte de tus algos / a muchos pobres fidalgos / que en tu Reyno veo andar / que señor deues creer / que al tiempo del menester / estos te an de aprouechar" (ID0457, vv. 50–63).[24] In the second poem, complaining about the ill reception of his first one, Dueñas blames converso slanderers for his loss of royal favour (ID0458, vv. 89–96).

Underneath social and ethnic conflict lay economic competition. The children of the labradores may play with those of the hidalgos, but, upon reaching adulthood, argues Dueñas, the first belong with their oxen, and the latter with the king (ID0457, vv. 1–28). In Dueñas's view, the peasants should respect and obey the hidalgos in order to repay for the security that the hidalgos' ancestors provided to the peasants, making a case for the primacy of birth over wealth or, perhaps more to the point, a case for (high) birth leading to wealth. The perceived importance of economic standing for the noble was decried by authors of noble lineage such as Fernán Pérez de Guzmán (1965), who famously lamented that at the time nobility was founded on wealth ("en este tiempo aquél es más noble que es más rico" *Generaciones y semblanzas* 18). Juan de Dueñas himself complained to the Condestable don Álvaro de Luna that virtue had been overshadowed by a well-stocked moneybag: "mas ya no se fazen mençion / del que virtudes mantiene / saluo señor del que tiene / bien poblado su bolson" (ID0460, vv. 33–6).[25] The practice of imbuing nobility with Christian ideals, along with a discourse of masculinity and ethnic "cleanliness," proved central to nation-building discourses in subsequent centuries (Fuchs 2003: esp. 1–4). In addition, as is clear from the underlying logic in the writings of authors such as Dueñas, wealth was in practice a noble's pursuit.

The Economy of Poetic Practice

The strife among those competing for resources such as the gifts in coin or kind from wealthy superiors shows a trade in which poetry was a key commodity that would procure other commodities and rewards. In a poem addressed to Gómez Manrique, for example, converso bureaucrat and escribano Juan Poeta (also called Juan de Valladolid) asks the former for wheat. An outraged Manrique reminds him of the large amount of wheat that the Archbishop of Toledo had given to Poeta (ID3398).[26] The poem goes on to state that Poeta has obtained the wheat not by working hard in the fields but by bartering his bad verses. Gómez Manrique refuses to provide Poeta with his patronage, claiming to be a better poet himself. The question Manrique poses early on to Poeta, enquiring why the latter comes begging from the poor caballeros ("pobres caballeros" v. 10), is as telling as the lines where he emphasizes his disdain of the trade in poetry ("que esta vuestra poesýa / saltará en mercaduría" vv. 33–4) [that your poetry / will go on the market]. Still, Manrique concedes that Juan is an elegant poet ("elegante poeta" v. 48), but makes sure to remind Poeta of his humble origins. His father was a town crier (pregonero) and his mother was an inn maid, who must be at fault because such an eloquent Homer – as he calls him – cannot claim such a lowly parentage (v. 57). Manrique could not dispense praise to Poeta without at the same time insulting his origins, making praise in this case dependent on blame. Similarly to "poor hidalgo" Juan de Dueñas, Gómez Manrique claims membership in the group of the "pobres caballeros." This group counts the iconic poet Alfonso Álvarez de Villasandino, among other notable members, in its number ("yo, un pobre cavallero" ID1299, v. 7),[27] along with others who went on to quickly gain wealth and political power and a noble title, such as Juan Pacheco, who became Marqués de Villena, or the great don Álvaro de Luna before his ascent to power. In the *Crónica de don Álvaro de Luna* (1940), as the queen is trying to negotiate the prompt marriage of the young courtier, don Álvaro complains to the grandees that he is too poor and young a caballero to be quickly and forcefully married away (26). Don Álvaro claims to have vowed not to marry on account of his being a young and poor caballero, not yet having had the chance to prove himself (24). It is important to weigh the potential role of these "poor caballeros" in the ongoing debate on nobility and in its definition of something rooted not in wealth but in deeds and virtue. A later *caballero pobre*, don Quijote, may help to explain the issue when he says that poverty leaves the caballero no other choice but that of virtue to prove himself: "Al caballero pobre no le queda otro camino para mostrar que es caballero sino el de la virtud, siendo afable, bien criado, cortés y comedido y oficioso, no soberbio, no arrogante, no murmurador, y, sobre todo, caritativo, que con dos maravedís que con ánimo alegre dé al pobre se mostrará tan liberal como el que a campana herida da limosna, y no habrá quien le vea adornado de las referidas virtudes que, aunque no le conozca, deje de juzgarle y tenerle por de buena casta, y el no serlo sería milagro; y siempre la alabanza fue premio de la virtud, y los virtuosos no pueden dejar de ser alabados" (Cervantes 1998: 676).[28] Thus, for don Quijote, low income encouraged the exercise of other merits for the caballero or, put another way, the accrual of "virtue" as a way to show good breeding and to increase personal worth. Poverty was not an obstacle to but rather an inducement towards self-improvement. Praise was but one of the possible rewards.

As poems such as the exchange between Gómez Manrique and Juan Poeta quoted above indicate, poetic worth could tip the scale in the author's favour and garner valuable rewards. Poetry held a place of privilege as the ultimate mark of social legitimation and socio-economic inclusion, as Juan Alfonso de Baena punctiliously explained in the introduction to his cancionero.[29] Santillana (1988) himself had eloquently written about the importance of poetry in his *Prohemio*, where he states that poetry springs out of gentility, a clear intellect, and an elevated spirit fuelled by a celestial zeal that is "insatiable" (439).[30] Weiss (1990) brilliantly demonstrates the complex factors that played into the formation of fifteenth-century poetic theory and its central position as privileged discourse. Poetry was the controlled aesthetic language par excellence, enabling a display of intelligence and mastery over discourse emphatically governed by rules. As the favoured discourse of aesthetic taste in court and urban settings, poetry provided a prime venue for sociability and conflict. For this reason, it proved a privileged medium for the construction of an identity of the self as well as a group identity which it developed in networks of social interaction. This role helps explain the large number of individuals, whether nobles, bureaucrats, or medianos, who wrote poetry during the period, extensively or just occasionally. It also helps explain the meteoric rise in poetic output in the period: over 7,000 extant poems circulated in many variant copies, a process further fuelled by the growth of courts and cities. Professional and political occupations encouraged the formation of clusters of authors and other notable figures. Besides the royal court, a well-known example of such a group is that of Archbishop Carrillo, who welcomed the formation of a circle of poets and intellectuals around him, including Guillén de Segovia, fray Íñigo de Mendoza, Juan Álvarez Gato, Pero Díaz de Toledo, Rodrigo Cota, Juan de Mazuela, Antón de Montoro, and Juan Poeta, all of them conversos, as well as such old Christian and nobles as Gómez Manrique, and even Santillana, who had his own intellectual circle (Cummins 1973; Moreno Hernández 1989: esp. 32–77). Sociability in urban centres, such as that exemplified by Montoro in Córdoba, or the intersection of court and city exchanges facilitated by the multiplicity of positions held by different poets and the itinerant court, all encouraged poetic practice and provided venues for its dissemination.

As Baena had eloquently explained, poetry was an "infused knowledge," a God-given gift, but it also involved learning. In this, as in much of the discussion over poetry and the poet, the arguments paralleled those that debated the nature of nobility and caballería or that revolved around the opposition of birth vs. learning. As with nobility, the ideal poet is born, but his works will ultimately establish his status. It is telling that caballeros and hidalgos such as Gómez Manrique clearly present themselves as military and poetically apt, while claiming untrained status as poets. In his introductory letter to his cancionero, dedicated to Rodrigo Pimentel, Count of Benavente, Gómez Manrique (2003) details the training he received for his two occupations ("ofiçios"): he was initiated in his military career in his brother Rodrigo Manrique's household, having been nursed in it since infancy and encouraged by classical readings on the subject such as those by Salustius. However, although he was a prolific author, he never studied poetry (98), which enables him to ask forgiveness for the quality of his poetry while launching into a long discourse on the value of learning for the caballero (97–109). Though such statements were ostensibly made as a display of (false) humility, this did not obscure the fact that they also clearly distinguished the author from paid, trained professionals or labourers of the pen, while suggesting inborn superiority. In a gloss in his translation of the *Aeneid*, Enrique de Villena (1994) clearly delineated the nourishing effect that poetry had on those of noble understanding, allowing them to become closer to those of noble birth: "ansí como los que desçienden de nobles e grandes parientes antiguados son dichos generosos, ansí los entendimientos que son habituados a la soliçitud e cura de las otras sçiençias antiguadas e aprovadas son dichos generosos e nobles. E lo que los primeros han por natura, éstos lo alcançan por doctrina. E a tales entendimientos como éstos ansí dispuestos nudre e cría la poesía" (48).[31] The discourse on nobility thus provided a clear subtext for the arguments about the nature of poetic ability. In light of the potential instability of socio-professional boundaries, such understanding of poetry was subject to sundry uses, which included affirming the ennobling power of poetry, appropriating its privileged position for conflict and slander, or using it as a tool for social legitimation. For this reason, it was also a clear marker of social exclusion, the argument being that social (and professional) marginality precluded the ability to produce good poetry. Therefore, any attempt to marginalize an individual or group would include the claim of their poetic ineptness. The Jewish mark of the converso in fifteenth-century Spain was synonymous with being a bad poet and with the practice of slander,[32] which was by definition a tainted discourse. The moral implications of discourse were highlighted when those belonging to a non-hegemonic religion were declared to be deficient in both behaviour and language, particularly the sublimated language of poetry. However, the case for associating poetic practice with nobility and with socioliterary legitimacy in turn facilitated the defence of poets of a lowly social station and helped make a case for their worth (both aesthetic and socio-economic), thwarting the intrinsic elitism with which some authors approached poetic discourse. Álvarez Gato exercised this line of reasoning when he extolled the poetic deftness of Montoro, the tailor, or Mondragón, who was a *mozo de espuelas* much like Sosia,[33] Calisto's servant in *Celestina* (ID3118, esp. initial rubric, and vv. 16–20, 81–90). If the poor and lowly are found to possess virtue, argues Álvarez Gato, those who produce greater works should be considered good above all others: "sy virtudes son halladas / en el pobre y en el chico / ... / sy mejor obra hiziere / ayanle por el mas bueno" (ID3118, vv. 81–90).[34] If the proliferation of writing, particularly poetry, among the upper and middle classes in the fifteenth century is any indication, it seems that the pressure becomes stronger for all to "do good works" and be diligent about exercising the power of the pen in order to prove their sociocultural worth. Good works thus made a good poet, which in turn could claim an elevated status. At the same time, active learning could polish natural ability. The potential circularity of the argument made the exchanges among poets all the richer and, at the same time, more conflictive. It also highlighted the value of writing and of control over poetry as "infused knowledge," signalling a way towards both self-generation and social legitimation.

Poetry and Bureaucracy: Working Poets

Poets openly acknowledged the combined role that economic and cultural wealth played in the ascent to power, often vying for the resources given by a wealthy superior via poetic competitive exchanges with other poets. Most fifteenth-century cancionero poets had one or several jobs. For example, Gómez Manrique courted several potential patrons as a way of supplementing his income. Early in his career he is known to have entered into such a relationship with King Enrique IV's chief financial officer, his contador Diego Arias, a powerful converso. In the lengthy introductory address to Arias Dávila when forwarding a poem that the contador had asked him to send, he confesses to not being able to depend economically either on his poetic toil or on the amounts allotted for him in the king's books, a situation made worse by the contador's remissness in paying (ID0093) (553–4). Manrique emphasizes his financial need in his poem, declaring his coffers to be neither full nor empty ("e nin tengo cofres llenos / nin vazíos," ID0094, vv. 44–5).

The famous offspring of one of the most important noble families, Gómez Manrique was an accomplished poet who, lacking primogeniture status in his family and any big titles, made his career holding various positions, such as that of royal contino, corregidor – first of Salamanca, then of Ávila and lastly of Toledo –alcaide of Toledo, and member of the Consejo Real.[35] Gómez Manrique, also lord of Villazopeque and Cordovilla, relied on his work for his income while paying close attention to the advantages brought by education and lineage.[36] Gómez Manrique's professional profile shows a political career that was highly favoured by the Catholic court, which placed men in its confidence in multiple important and sensitive positions. Royal continos and other collaborators were appointed to key posts in urban government such as the corregidurías (Montero Tejada 1999; Polo Martín 1999: 80–122, 146, and passim). The job of corregidor was particularly reserved for "segundones" or younger sons of noble families, who, like Manrique, made a professional career out of it, being favoured for these and other government and administrative positions over the University letrados (Montero Tejada 1996: 39–45, 211). Of particular significance is his appointment in 1477 as corregidor of Toledo, a key but particularly conflictive city where Manrique exercised his abilities in abating urban conflict, which involved the use of both violent and political means, and included the arrival of the Inquisition in 1485 (López Gómez 2004). Manrique would remain corregidor of Toledo until his death. Toledo was one of the cities with the highest rates of violence, with several raids against conversos (López Gómez 2004: 169). Isabel and Fernando started their reign granting pardons in 1475–7 (López Gómez 2004: 187), but aggressively punished those who took action against their power. The punishment could be as harsh as death and loss of properties, as happened to Juan Rótulo in Toledo in 1478, when some members of the local oligarchy were planning an uprise against their corregidor. On 30 July 1478, Isabel and Fernando wrote to the corregidor thanking him for his work, and granting him a recompense of 150,000 maravedís from the goods confiscated from Rótulo (López Gómez 2004: 190). Similarly harsh punishments went to converso regidor Juan de Córdoba, Hernando de Alarcón, and others (López Gómez 2004: 190–1). As Pulgar (1943) recounts, referring to Toledo in his *Crónica de los Reyes Católicos*, the monarchs ordered justice for many criminals and robbers "muchos onbres criminosos e robadores" who had gone unpunished for earlier crimes (1: 423). By 1485, when the Inquisition arrived in Toledo, the urban violence in the city had been abated thanks in large part to the strong political presence of the monarchs via the corregidor and regidores as well as public displays of punishment. This had paved the way for the institution of the Inquisition with only isolated attempts of resistance (López Gómez 2004: 193).

It is generally accepted that Gómez Manrique was friendly and lenient towards the conversos.[37] This view is mostly based on Alfonso de Palencia's (1998) statement in his 1484 *Guerra de Granada* that Gómez Manrique, prompted by the citizens of Toledo ("convencido por el juicio unánime de los ciudadanos" [convinced by the unanimous judgment of the citizens]), was instrumental in delaying the arrival of the Inquisition in Toledo, in order to spare the city from the ensuing impoverishment, similar to that which had afflicted Seville upon the Inquisition's arrival (116–17). The *Relación* of the establishment of the Inquisition and the *autos* celebrated in Toledo from 1485 to 1501 noted by a witness, later copied by Sebastián de Horozco (1981), similarly tell of a Gómez Manrique who was torn between punishing the conversos involved in the revolt against the Inquisition and the need to preserve the city. Gómez Manrique saw that if the guilty were punished the city would become depopulated and decided to assign monetary fines instead: "fue sabida e descubierta la dicha trayçión, e gómez manrique, que era corregidor a la sazón en la dicha çibdad por el Rey, prendió a algunos conversos que eran en la trayçión, e supo la verdad e lo que tenían ordenado. […] E porque vido gómez manrique corregidor, que faziendo justiçia de tanta gente, la çibdad se despoblaría, acordó de les dar pena de dinero" (98–9).[38] His job as corregidor, however, appears to have put him in a position to support the overall politics of the Inquisition, as befits a faithful collaborator of Isabel and Fernando's politics. In addition, Isabel and Fernando put punitive pressure on the corregidores' compliance. Pulgar (1943) recounts how in 1488 in the city of Valladolid the king and queen reviewed the work of the Inquisition against the heretics, while ordering another inquiry or inquisition ("ynquisiçión") on the corregidores. At the end of the corregidor's term, the monarchs ordered a thirty-day moratorium during which a close review of his tenure would take place. If found at fault, the corregidor could go to prison, be subject to the appropriate fines, and lose the ability to hold any future positions. What is interesting about Pulgar's narrative is that both inquisitions appear together, much like a stream of logic that united the corregidor as ultimately responsible for the city's order and the Inquisition as an instrument of control. The mention of the punishment inflicted on some Jews in Toledo, who were stoned to death, is also eloquent in stating what the city represented for the order of the whole nation (*Crónica de los Reyes Católicos* 2: 353–4).[39] Horozco's (1981) *Relaciones* show Manrique presiding over the procession of the penitents reconciled by the Inquisition, and an altarpiece (*retablo*) of the period, that was later destroyed, portraying him in the same position next to a preaching Vicente Ferrer (97), perhaps as a reminder of Ferrer's 1411 apocalyptic sermons on the coming of the Antichrist in Toledo and his instrumentality in the (violent) efforts to convert the Jewish population. Gómez Manrique's (2003) advice to Fernando and Isabel in his *Regimiento de príncipes* unequivocally recommended harsh punishment ("crudamente punir") against Averroist disbelief in an afterlife typically associated with conversos, as it brings on all kinds of evils such as war, hunger, and the destruction of cities (vv. 181–225). From the standpoint of his literary production, Manrique's role in supporting the Inquisition is more in tune with his virulent diatribes against converso poets Juan Poeta or Montoro. The benefit of supporting the Inquisition was not solely political. Gómez Manrique's wife, Juana de Mendoza, who was head of Queen Isabel's household (*camarera mayor*) and in whose court she lived starting in 1480 until her death in 1493, received large sums from the queen in recompense, including one million maravedís from the Inquisition in Toledo. She left part of this sum to the Monastery of Calabazanos, which had been designated as the family burial site, as stated in her will (text in Rivera Garretas 2007: 145, 172). As Caunedo del Potro (1991) has argued, Gómez Manrique seemed to take great care in fashioning himself as the mirror of Christian knight when in battle, if we are to judge from the number of pieces of armour and their covers ("sambenitos") bearing the anagram of Christ as well as crosses that appear in the 1490 post-mortem inventory of his belongings (107–8).

Manrique cultivated an active literary relationship with his uncle, head of the powerful house of Mendoza and prolific noble poet, Íñigo López de Mendoza, Marqués de Santillana, in whose footsteps as great poet and caballero he was hoping to follow. Santillana had a noteworthy political and literary career,[40] and his service included an early post as *copero mayor* to a young King Alfons V of Aragon (el Magnànim), obtaining the title of marquis from King Juan II as a reward for military services. He was also Conde del Real de Manzanares and lord of Hita and Buitrago. He was a friend of Juan de Mena and exchanged poems with Montoro. In addition to the Manriques, Santillana was related to other important figures such as statesman and poet Fernán Pérez de Guzmán (his uncle). On the issue of the conversos, Santillana had expressed his admiration for the work of Vicente Ferrer in the long poem celebrating his canonization (ID0306). On the other hand, Santillana is also understood to have been friendly towards conversos and to have counted them in his "cultural circle" (Gómez Moreno 2001).

Converso poet Juan Álvarez Gato, a royal protégé with a strong presence in urban government, had a polite poetic friendship with Gómez Manrique. Also upwardly mobile, Juan Álvarez Gato, urbanite and courtier, served in various royal and noble households, and more notably in the royal court of Queen Isabel and King Fernando, where he was a contino, in constant service to the Crown. In the Catholic monarchs' court, he was major-domo and escribano of the royal chamber (escribano de cámara), as well as being variously employed in special missions.[41] Early in his career he had served in the court of King Enrique IV and worked for the notable converso family of the Arias Dávila, powerful oligarchy in the city of Segovia with economic interests in Madrid. He is documented in Madrid's city government as being part of the group of caballeros and escuderos, which puts him among the city's political elite. He held various administrative posts in the local government, such as the concejo's major-domo[42] and escribano substitute (*teniente de escribano*). As a typical member of a city elite, the poet acquired urban and country properties, fixed his residence in the neighbourhood of San Salvador, and formed a *patronato* (patronage, trust).[43] He lived near the church of San Salvador in Madrid, which was situated in what today is the n. 70 of the *calle Mayor*,[44] in the Plazuela de San Salvador (currently Plaza de la Villa). This city square, where much of Madrid's oligarchy resided, was the mercantile and later administrative centre of Madrid, and its church served as the locale for concejo meetings (Gibert y Sánchez de la Vega 1949: 159, Losa Contreras 1999: 176). Further key details of his life and status have been recently unveiled (Gómez-Bravo 2011). The 1576 *Relación* concerning Pozuelo de Aravaca shows that he was not only a bureaucrat and caballero, but also a reputed hidalgo.[45] In addition, as a prominent *madrileño*, on 19 June 1490 he was part of the numerous caballeros, escuderos, and other members of the political and economic elite that were called to accept the institution of the Inquisition in Madrid under the command of the General Inquisitor fray Tomás de Torquemada and Queen Isabel and King Fernando. Further, he was put in charge of the lodging needs of the newly arrived inquisitors.[46] Towards the end of his life, Álvarez Gato had accumulated a certain amount of wealth and his life was prosperous. Márquez Villanueva (1960) termed him a "court bureaucrat" and one of the first of the madrileño bourgeois (33, 528). His membership in the urban elite as both caballero and reputed hidalgo provides a complete picture of his status.

Another important urban centre, Córdoba was the city where converso poet Antón de Montoro spent most of his life. Montoro, a tailor and used-clothes small merchant (*ropero* and *aljabibe*), belonged to the social group of the medianos who dwelled in cities and were members of an urban middle class that included artisans, merchants, and specialized professionals. This urban middle class thrived in many cities throughout Spain, and was particularly strong in Andalucía (Ladero Quesada 1992a). Middle-class poets such as Antón de Montoro were intensely familiar with personal and official record keeping.[47] Far from the image of the jocular beggar with which Montoro has traditionally been stereotyped, he was in fact a relatively affluent mediano or ciudadano, wealthy enough to leave a sizeable inheritance for his children. He was also able to keep at least one horse and the military accoutrements needed for the participation in battle that was required of – and often resisted by – Montoro as a caballero cuantioso or de premia.[48] Montoro had contacts with the court and the urban elite, enjoying a close rapport with the Aguilar family, important Cordoban oligarchy. He also wrote to King Enrique IV, Queen Isabel, and King Fernando. Although he enjoyed relative wealth, Montoro's life, particularly during his last years, was not easy. He lived through some of the worst pogroms in Andalucía, which he bitterly decried in 1473 (ID1930), and 1474 (ID1924), and witnessed the establishment of the Inquisition in Córdoba. His wife, Teresa Rodríguez, was burned at the stake.[49] Montoro's exchanges with acclaimed names such as Santillana and Mena provide an indication of the esteem in which his poetry was held. The mention of Montoro's work in Delicado's (1994) *Lozana andaluza* shows the extent of his fame and his poetry. Delicado says of the eloquent and also Cordoban *Lozana*: "Es parienta del Ropero, conterránea de Séneca, Lucano, Marcial y Avicena. La tierra lo lleva" (348).[50] The common reference to Montoro as "the tailor" ("el Ropero") points to the complex negotiation of his status in relation to poetic legitimacy. Juan Poeta's address to Montoro in ID2722 points to the chief factors that problematize and stereotype Montoro's middle status, which include his humble origin, his relation to the lineage of David (i.e., his being Jewish), his occupation sewing simple clothes, and his lack of valour in war: "Hombre de poca familia / de linaje de David / ropero de obra sencilla / mas no Roldán en la lid" (vv. 1–4).[51] In some of the poems sent to Montoro by the Comendador Román, with whom he held a virulent poetic exchange coinciding with Román's stay in Córdoba, Román seems particularly interested in Montoro taking up his needle ("tornar de nuevo a coser"), suggesting its incompatibility with poetry and its higher profitability (ID3017, vv. 22–4).[52] Conversely, Alfonso de Velasco, alderman or *veinticuatro* of Córdoba,[53] wrote to Montoro about the advisability of changing professions and choosing one more in line with his stature as a poet (ID2730, vv. 7–12), a suggestion that Montoro refused to follow. Montoro's answers leave the matter wilfully inconclusive, communicating a sense of pride in his profession and a playful resistance to any change, stating that he is bound to hasten back to his needle ("presto me cumple volver / al abuja" ID 3019, vv. 11–12). The long poems on the topic are meant to belie in practice the alleged incompatibility of Montoro's profession with the poetic craft, to which he adds the higher reliability of his profession for an income: "Pues non cresce más caudal / el trobar, ni da más puja / adorámoste, dedal / gracias fagamos, aguja" (ID2721, vv. 11–14).[54] Building a discourse of professional pride of the *medianos*, in a long poem (ID1932) he advises converso Rodrigo de Cota to stop parading around on his horse like the gentleman he has become and go back to his former profession of spice trader or apothecary (*boticario especiero*). Montoro's poetry attests to his situation as active player within the literary networks of exchange in both court and city and an intersecting course of mediano professional vindication.

Even though Juan Alfonso de Baena, royal escribano and well-known converso poet, came before King Juan II undeservingly ("indino") to present his cancionero,[55] he enjoyed a comfortable situation due not only to his income as royal escribano, but also to the generous dowry that his wife contributed to the marriage. In addition, he benefited from good family connections, belonged to the caballeros cuantiosos or de premia of upper middle status, and had properties and interests in the city of Córdoba.[56] In a fiercely stereotyping portrayal of the converso caballero and escribano, rival poet Fernán Manuel accused Baena of always tallying up the rents of the past year with very black ink and writing desk, his prototypical writing instruments, while feigning valour in arms (ID1499, vv. 5–8). The manuscript of the Bibliothèque Nationale (PN1) is a beautifully rendered though faulty copy whose execution suggests a wealthy owner. Even though the lost original was intended, as the introduction to the *Cancionero* indicates, for presentation to King Juan II and his court, it is essential to note that a large number of poems are actually linked to important urban centres, namely Seville and Córdoba (Weiss 1990: 45).

Another native of Córdoba, poet Juan de Mena appears to have been a university graduate (letrado) and also a member of the urban political elite as veinticuatro of Córdoba. Mena was said to have been such a prudent politician that he gained significant status in city government ("boz mayor en el Cabildo"), while also being in the service of King Juan II as Latin secretary and chronicler, the latter an appointment apparently gained as a result of his presenting Juan II with his magnum opus, the *Laberinto de Fortuna* or *Trescientas*. This work praised the very powerful ("muy prepotente") king, Mena's poetry earning him the highest accolades. Mena attended the University of Salamanca, and Vatican documents show that he was at the papal court in Florence in 1442–3. During his Florentine period, he was under the protection of cardinal Juan de Torquemada. Mena's father was said by some to have been a Pedrarias, a name associated with powerful converso families of the fifteenth century, of middle class and good "nation" ("de estado mediano, de buena nación"), while not much is known about his mother.[57] Much of the information on Mena's biography comes from the 1499 printing of Hernán Núñez's *Glosas* to Mena's *Laberinto* and Valerio Francisco Romero's *Epicedio*, published in 1555 after the death of Hernán Núñez, about a century after Mena's own death in 1456. No consensus has been reached on the conclusiveness of the evidence at hand, as some of this information cannot be confirmed by additional documentation. Mena's case in particular is greatly affected by the large gap in Cordoban archives for the first sixty years of the fifteenth century and the lacunae for other decades (on which see Ostos Salcedo 2005: 11–14).

The "lack of documentary evidence" indeed plagues the figures of a large number of cancionero authors, which, as Perea Rodríguez (2003) has convincingly argued, can to an extent be remedied by the wealth of historiographical information provided by the cancioneros themselves. In "Hanme dicho Juan de Mena" (ID1947), the last of a series of three poems (ID1945, 1946, 1947, in MN32, MP2, SM2), Íñigo Ortiz de Estúñiga, a *mariscal* (marshall) who was at times ambivalent about Juan II's rule and whose family would champion the introduction of the Inquisition in Seville, responded to a poetic attack by Mena by swearing to "the one that you killed." The oath was a reference to the widespread accusation of the Jews as killers of Christ. He also used terms derived from Hebrew such as "Baraha" (blessing), in a fashion very similar to taunting poems in the *Cancionero de Baena*, in which the mariscal was a willing participant, and where he also exchanged insults with Baena (see his role in the series ID1544, 1545, 1546, 1547, 1548, 1549, 1550, 1551, 1552, 1553). In "Señor, buen frontero, lengua de Sansón" (ID1546), Ortiz de Estúñiga attacks Baena's Jewishness, threatening him with homophobic violence. Fuentes Guerra has proposed that the Juan de Mena, resident of Burgos, who appears as a witness in a 1454 royal document regarding the Cartuja de Miraflores drawn up in Burgos, in the house of Íñigo Ortiz de Estúñiga in his position as alcalde mayor, juez, and corregidor of Burgos, can be identified as the poet. If the identification is correct, it would show a shared city of residence and provide a context for the political and poetic transactions between Mena and Ortiz de Estúñiga.[58] With many of Mena's biographical details still unknown, Street (1953) found no clear evidence of Mena's converso origins, which leaves the matter as inconclusive, but Asensio's (1967) act of brushing aside the accusations and taunting by Mena's contemporaries regarding his Jewishness as "choleric outbursts" are unconvincing. Asensio's lengthy and vehement attempt to dispel any doubts cast on Mena's family origins reveals much of the matters at stake in Mena's possible Jewishness. If Mena was "one of Spain's greatest national poets of the period," accepting him and other known converso authors as the makers of a grand national literature promoting empire building would be "giving what is just its last tributary the importance due to the large river" (Asensio 1967: 351). In other words, nation building cannot be the work of converso intellectuals, who are thus denied full cultural and political citizenship status. Accepting the possibility of Mena's converso roots would prove destabilizing. Whether or not a converso, as university-trained letrado, Mena was in a middle position, one that ostensibly derived much of its status from education and ability. As veinticuatro or regidor, he was an important part of Córdoba's political elite.

The complex negotiation of converso identity and the strategies connected to conscious self-generation are evident in the work of these different poets. These were neither professional poets nor solely courtly subalterns. Cancionero poets were multitasking professionals with overlapping duties who not only worked for the court, but were also deeply immersed in urban life, sharing duties and practices with pen professionals. The power over textual production wielded by secretaries, escribanos, and other bureaucrats was of utmost importance for church and state, as well as for cities. At the same time, all those involved in textual production were conscious of the role of writing as a key factor in the power of self-generation. Attempts to control this production and contain it were negotiated and resisted, but the agency of the paper that had been set in motion proved a match as formidable as the forces contributing to its creation and dissemination.

3 Pervasive Papers

Leaving paper trails was key for the competent bureaucrat, as well as the politician and poet. As Fernando de Pulgar (1982) eloquently noted in one of his *Letras*, "Muchos tenplos y hedificios fizieron algunos reyes e enperadores pasados, de los quales no queda piedra que veemos, pero queda escriptura que leemos" (62).[1] This chapter will propose that the progression towards a multiplication of official records bears strong and meaningful ties to the flourishing of a rich tradition of writing and compiling poetry, facilitated not only by the agents responsible for textual production in the period, but also by the use of shared material supports. The material aspects of production bring into focus the persons legally and physically involved in it and, by extension, authorship as textual accountability. This interest was undoubtedly related to the increase in importance of the figure of the author as well as the shift experienced by the text from simple material support of an oral transaction to artefact described by Huot (1987a). It was also clearly part of the process that produced a decrease in the oral authentication of legal transactions prevalent in earlier centuries and a growing reliance on written records as a means to preserve the private and collective memory of those transactions, as studied by Clanchy (1993). The multiplication of texts and the improved modes of production undoubtedly anticipated the introduction of the printing press, pointing not to a breach in the manuscript/print technologies and outputs, but rather to a continuity and coexistence through later centuries. This approach is in line with the arguments issued on the subject (e.g., Martos 2011; Saenger 1975, 1996).

A vast number of poets were employed in private or official record keeping, and the strategies for copying, compiling, and preserving documents were undoubtedly useful when handling the individual poetic paper, which could in turn prove to be a powerful political and cultural tool.² Thus, the crossover between literary texts, particularly short or ephemeral pieces, and other types of documents produced in or for the individual, municipal, royal, noble, or ecclesiastic household became inevitable.³ The growth in literacy, the cheaper cost of paper, and its wider availability particularly from around 1460 on (Clanchy 1981),⁴ together with the increased bureaucracy serving the fifteenth-century state, all helped propel a "paper culture" that provided an ideal support for both poetic and administrative texts, with parchment playing a secondary role.

This chapter will look at the proliferation of loose papers as material support for poetry and other types of text, mainly documents, and their similarities. The pen professionals who were involved in textual production facilitated the agency awarded to the text and the papering ("empapelar") of the social and political spheres.

Material Texts

Paper, along with pen and ink, was the medium of choice for private and public writing; it was readily available, often sold by the ream, along with other writing materials, in booksellers', spicers', or apothecaries' and other shops.⁵ To gain a notion of the sheer volume of the texts produced, it is useful to recollect that over 300,000 documents issued by the Reyes Católicos have survived, only part of what was generated under their administration (Rumeu de Armas 1974: 15). The notoriously large consumption of paper at court during the period can be easily gleaned from the account book of Gonzalo de Baeza, Queen Isabel's treasurer. There are many entries in Baeza's account book for payments for "hands" (*manos*) – fractions of reams (*resma*) of paper, as well as other writing materials. In many instances the purchase of paper is explicitly intended for writing (texts in de la Torre y del Cerro and Alsina de la Torre 1955–6, 1: 50, 380, and passim). The purchases of paper were also at times grouped with those of private use or personal care objects such as pins or combs, and could be destined to wrap objects in the private chambers (1: 267). These papers would also be inscribed with a note identifying the contents of the packet.⁶ The pervasive presence of "papers" found in inventories and other texts is witness to the rising levels of literacy and the new writing and reading public and its needs.⁷ An amount of paper could be used as a hyperbolic writing measure; thus in *Corbacho,* Martínez de Toledo (1987) states the need for ten *manos* of paper in order to detail many of the faults described therein: "Por non detener tiempo non fablo más destos perjurios, que escrivirlos bien non bastarían diez manos de papel" (113).[8] Because of the numerous papers and books she had to read, the queen eventually required reading aids for her eyes, evidenced by the presence of numerous pairs of eyeglasses ("antojos"), both regular and for travelling ("para camino") in her chests (de la Torre y del Cerro 1968: 7, 61, 370). In addition, she enlisted the help of a reading stone ("piedra para leer") set in white horn, a magnifying glass of sorts (de la Torre y del Cerro 1968: 370).

A wealth of evidence, both in the poems themselves and in their introductory titles or rubrics, points towards an understanding of the poem as a material entity, described as a piece of paper inscribed with ink. This is the case, for example, in ID0249 (MN6e), written by Juan de Mendoza in reply to a mocking composition by Fadrique Enríquez de Cabrera, Almirante de Castilla, who scoffed at the look of Mendoza's wrinkled face after he had shaved his beard. Mendoza's reply, redirecting the argument towards the Almirante's doubtful poetic ability, demonstrates the material reception of the poem as well as its relevance for the production of texts:

> Respuesta de don Joan de Mendoça echando la culpa d'esta copla a dos criados del Almirante que ell uno escriuia muy bien y ell otro trobaua mejor
>
> De la copla que me toca
> no es vuestro mas del papel
> oyo la voz de Gabriel
> siento las manos de Coca
> no es mucho que me ganes
> pues no me vale remedio
> trobando contra mi tres
> o a lo menos dos y medio [...]
> El trauajo sera en vano
> no sacareys fruto d'el
> quereys llegar con papel
> do no alcançays con la mano" (vv. 1–8, 17–20)[9]

The poem is thus described as a piece of paper on which a composition (in this case, allegedly authored by another) is handwritten in such way as to enable the authentication of the poem's origin both in content and in its material execution. Thus the poem in question, having Gabriel's style and Coca's handwriting, leaves the Almirante's only possible ownership claim to be that of the paper on which it was written. The reference to the Almirante as only half a personage is a frequent allusion to his short height in the poetry directed to him, and the reference to his hand being unable to reach Mendoza may be a further accusation of lack of dexterity with the sword. This identification of the materiality of writing with pen, ink, and paper, and with seeing, is a clear departure from the references to an oral reception more prevalent in an earlier period, and from the archetypal constructions of an idealized text in later criticism. Further, In the *Siete Partidas*, King Alfonso X (2004) had already specified the property rights over a poetic text based on the ownership of the book as its material support. If someone unknowingly uses another's parchment to write down poetry, Alfonso rules that the book of poetry ("libro de versos") belongs to the owner of the parchment and not to the writer. If the writer was acting in good faith, the owner may pay him for his work; if he was acting maliciously, the writer will lose parchment and work. If the work had been contracted, payment is due (III, 88r).

The poem is recurrently alluded to as a piece of "paper" (*papel*), pointing to the equation of text and medium. This is further evidenced through the constant references to the material aspects of poetry in many cancionero poems; for example, Montoro in ID3041 (MN19, vv. 1–4): "Don Juan de Peñafiel / vaso de noble constancia / perdonad el inorancia / de mi no sabio papel" [don Juan de Peñafiel / vessel of noble perseverance / forgive the ignorance of my unlearned paper]; in ID2847 (MN17, v. 13), where Alexandre encourages the reader: "mirad este papel" [look at this paper]; in ID0044 (MN54, v. 40), where Sancho de Villegas requests consolation for his amorous complaint, "con solo papel et tinta" [with only paper and ink]; in Juan del Encina's ID4459 (96JE, vv. 106–10, vv. 129–30), where the medium physically mimics its content, the paper empathizing with the text, shedding tears of blood: "ay cuitado que solia / escrevir devotas cosas / y ora amor con su porfia / me manda sin alegria / que escriva penas penosas […] hago de las noches dias / llora sangre mi papel";[10] in Pedro Torrellas's ID1890 (ME1, vv. 3–10), where the poem's meaning is negotiated through the materiality of paper and the pen held in hand: "queda borrado el papel / batallan los sentimientos / dentro del campo de aquel / la pluma en mano figura / entre esperança e temor / mueve y atiende y apura / sobre esta contienda amor / tiene la plaça segura";[11] in ID0045 (MN54, vv. 38–40), where Suero de Ribera proves himself to be aware of the identification of writing materials (paper and pen) with poetic composition: "se faria larga suma / pero detengo la pluma / por no tocar al papel";[12] in ID0514 (PN1, vv. 15–16), where Fernán Manuel expresses his readiness for a poetic exchange by showing his ready pen and paper: "que yo por qualquiera vía / tinta e papel tengo presto" [that I in any way / have ink and paper ready]; in ID3017 (MN19, vv. 1, 4), where the Comendador

Román starts a tirade against an accusatory stanza by Montoro: "Con pura malenconia [...] / se mueve la pluma mia" [with pure melancholy / my pen moves]. Baena refers to the poem as "esta escriptura" [this writing] in ID1568 (PN1, v. 5); and Gómez Manrique depicts the self in the physical act of starting a poem: "La peñola tengo con tinta en la mano" [I hold the pen with ink in my hand] in ID3367 (v. 1). Gómez Manrique traces a writing landscape when attempting to eloquently explain his love troubles in ID3393. He proclaims himself unable to answer the self-posed question, "¿Qué lengua recontará / o quál mano escriuirá / mi doloroso cuydado?" (vv. 3–5);[13] even if all the land were turned into white Tuscan paper and the rivers into ink, these materials ("materiales") would run out before half of the poet's cares could be told: "Si las tierras se tornasen / en blanco papel toscano, / los ríos se trasformasen / en tinta, con que pintasen / vn dolor tan ynvmano, / los dichos materïales / serían antes gastados / que la meytad de mis males / e tormentos desyguales / ser pudiesen recontados" (vv. 11–20).[14] Troubled under different circumstances, Fernán Pérez de Guzmán uses the older term for writing, "dictar," when noting that he is writing with a trembling hand: "Con coraçon muy turbado / e la mano me tenblando, / me conuiene yr dictando" (ID0105, MN6b, vv. 361-3).[15]

The use of the word "paper" or the plural "papers" (*papel* or *papeles*) with the meaning of "written piece of paper" conveys a clear identification of medium and text that marks the material nature of the poems. References to paper, ink, and writing ("escriptura") are further enriched by constant references to "scribbling" and to scrawled lines. This we find, for example, when Gómez Manrique addresses a poem to Juan Álvarez Gato: "por estos pocos renglones, / llenos de artos borrones" [through these few lines / full of many smudges] (ID2944, MN19, vv. 21-2); or when Fernando de la Torre refers to his writing as "scribbling three blank sheets of paper": "borrar lo blanco de tres pliegos" (text in Díez Garretas 1983: 188). As these citations make clear, the poem was identified by its medium and through it enjoyed a dynamic and varied material life. As far from the Lachmannian ideal archetype as from a Zumthorian pan-oral Middle Ages, the fifteenth-century poem was firmly identified with its concrete material support, a piece of paper inscribed by the hand that with keen awareness had dipped the pen in the ink.

The Poem as Discrete Entity

The examination of the uses of poetry in relation to the medium used for its support suggests that the poem is not only a record of a social transaction, but a material part of daily life and its idiosyncratic repository. It not only carries dialogic situations into paper, but is itself part of the exchange. When, for instance, a poem speaks of a gift being sent, the text does not merely constitute an abstract signifier of the material gift but is rather a part of the transaction, both object and written speech act (Gómez-Bravo 2013). The referentiality of the text should not obscure its role as an integral constituent of the situation. It is in this sense that a text of this kind is *always* a situated text rather than a mere poetic abstraction. The long process of successive copies or editions of poems in anthologies through the centuries has served to reinforce the perception of a poem as an almost abstract textual entity. In a first step, the process of copying a single text into a fifteenth-century compilation involved the erasure of the particular characteristics of the poem, which as an artefact had enjoyed an agency of its own, possessing a three-dimensional material life that it had shared with other objects. A closer look reveals that cancioneros often acknowledge the loss of meaning involved in such erasure by using textual strategies such as a sometimes-complex apparatus of rubrics that preserves the original situation of the poem via a textualization of its circumstance, just as the text becomes detached from it.

The paratext as well as the poetic text itself point to an "objectification" of the poem, capable of movement and occupying a tangible space.[16] The poem was part of daily life as an autonomous entity independent of the book archive where it might eventually be recorded and compiled along with other texts. The agency awarded by the materiality of the paper medium allowed poems to be literally thrown about, handed around, played with, defaced, or hanged. Poem ID5793, for instance, was, according to the rubric, tied to a rock and sent as a projectile to its destination: "Cantigua sua a huma dama quelhe tyrou com huuma pedra" [Poem to a lady that he threw to her with the help of a rock]. Poem ID3086 (MH2) was sent wrapped around a rod onto a roof that overlooked a window where the lady would sometimes stop and look out: "Juan aluarez vn dia que jugaron a las cañas echo estas coplas enbueltas en vna vara a vn tejado que salie a vna ventana a do se paraua algunas vezes aquella Señora."[17] In *Cárcel de amor*, a letter from Leriano reached Laureola's chamber through the same venue (42). The material nature of poetic transmission not only facilitated distribution within a stone's throw, but also provided the appropriate medium for games that relied on texts as artefacts. Such is the case of poems written on cards and made into a deck for play, as in the anonymous *Abecedario* (ID2304), Fernando de la Torre's *Juego de naipes* (ID0594), or Pinar's *Juego trobado* (ID6637), written for entertainment in Queen Isabel's court.[18] In a like manner, the rubric introducing poem ID7301 informs the reader that Garcia de Resende created one such game at the king's command. The rubric explains how the 48 stanzas were written and each made into a card: 24 ladies' cards and 24 gentlemen's. These two groups of cards were then subdivided into 12 praise cards and 12 blame cards. Once shuffled, one of the cards was to be drawn and the stanza read aloud, with those drawing a praise card faring better than those drawing a blame card (*Cancioneiro Geral* 1990–8, 4: 341–52). Similarly complex games where the poem artefact interfaced with other objects have survived. One such example is poem ID0349, where, as the rubric explains, its various stanzas were each put inside (or around, according to the version in MH1) a specially dyed flower as a present for some ladies. The flowers containing the poem were then placed inside the poet's sleeve and the ladies were asked to put their hands in it and pull out one of the flower-poems in a sort of fortune-telling game for the New Year. Not reserved exclusively for court games, the luck-drawing protocol by which written pieces of paper of equal size would be put into a container for drawing appears in the procedure known as *insaculación* or *insaculatio* for the election of officials in many cities under the rule of Isabel and Fernando.[19] Similar to the courtly game of cards are the games laid out in card-like form in Luis Milán's minute *Libro de motes*, a book containing similar heterosocial games and small enough to be carried in a pocket, each leaf resembling a playing card of sorts. The self-evident attachment of poetry to medium exemplified by these instances points towards the life of poetry as artefact in practices that would help the embeddedness of the written paper not only on special occasions, but also in daily life.

Wandering Poems

The physicality of the poem-artefact allowed for the motility of poems. Once out of the author's or owner's hands, a poem could circulate without the approval of the owner, or even the reader of the papers, as poem ID2908 attests. The rubric in MN19 explains how a poem was lent by a servant to a reader's lady friend without the owner's consent: "Vazquez de Palencia contra fray Yñigo sobre las del vita Cristi enderezadas a su amiga porque le embio pedir la de vita Cristi y no estando en casa ge las dio su mozo sin su mandado y dice asi."[20] The motility of the poetic text facilitated its appropriation by readers, editors, printers, and copyists, who did not leave it untouched, thus allowing for an overlap of the functions of all those who left their mark. Rubrics tell of poems passing from one hand to another, being requested and sent (e.g., ID1055, ID1098, ID6013, ID6636,

ID6695, ID6797), and even acting as reading primers (ID0705). Rubrics tell that a poem could go to a lady who asked to see some of the poet's work (ID6797); to a countess after the poet fruitlessly tried to excuse himself from sending some of his writings (ID6884, 14CG); to two of the poet's sisters who asked him for a sample of his works (ID1055). A poem could be lent (ID2847), or forwarded to friends and acquaintances (ID2060, ID2065), as could a compilation of a poet's works (ID2850). Poems and their rubrics ceaselessly provide deictic references to the motility of the poetic text in its material support and trace its initial itinerary. The poems themselves confirm these wanderings, as when Gómez Manrique acknowledged receipt of Guevara's poem: "Vuestra gentil escriptura / resçebí, buen cauallero" [Your genteel writing / I received, my good caballero] (ID2959, vv. 1–2). These wandering papers elicited responses, not always positive, and the back and forth of papers could grow into a more focused exchange, at times inimical, the venue to resolve personal or poetic differences, as is obvious from many poetic exchanges, such as the unending and well-known quarrels in the *Cancionero de Baena*.

Countless references tell of a poem being forwarded to the addressee in a mode that blurred the demarcations between letter and poem. Because of the physical nature that granted them agency, poems could face the addressee and make an argument, as we find in ID6116 by Cartagena: "Otras suyas porque vna señora le escriuio que le enbiasse vna carta que ella le auia escrito de antes y embiogela y otra suya y hablan las cartas a la dama diziendo."[21] Sometimes the poem itself was sent with a letter. Juan Martínez de Burgos, poet and escribano of Burgos, copied poem ID3658 at the end of a letter (ID3657) that he sent to his son, Fernán Martínez de Burgos, also an escribano, from the monastery in Lisbon where he had professed. Similarly, Álvarez Gato's poem defending poetic talent over social status discussed in chapter 2 was forwarded with a prose letter (ID3117, MH2). The material aspects of the letter-poem, for example through the inclusion of the forwarding address written on the letter ("sobreescrito"), were preserved in cancionero codices in the form of a short narrative, often by way of a rubric. Both parts, the address written on the outside and the poem-letter it enveloped, appear copied in a linear fashion in the cancioneros, where a rubric explains the original format of the letter-poem and its material support at the exact same time that the original medium is dissipated by the act of copying. The erasure of the medium and the meaning it carries is thus resisted via its textualization. One such case is ID4397 (TP2), which is actually a verse address: "vn caballero enbio vna carta a vna dama y dezia el sobrescrito": "La dama que fuere fea / no me abra ni me lea,"[22] with its obvious invitation to open the letter, which reads ("Lo que dentro se contenia es"): "No sois vos a quien yo vengo / por eso no me leais / vos señora no mirais / el sobreescrito que tengo / tornadme luego a çerrar / y no llegue nadie a mi / que a quien yo vengo a buscar / no deue de estar aqui" (ID3556, TP2).[23] The success of this playful poem relies on its material support and on the ability of the engaged reader to handle it. Similarly, ID0044 (MN54) is a letter ("carta") in verse sent by Sancho de Villegas to his lady. The first stanza is marked as the "sobre escripto," while all but the last stanza make up the "carta," which was written in the poet's own hand ("De mi mano aquesta carta," v. 10), and ends with the date ("La fecha") of 1442 in verse. This poem shows a careful compliance with the material aspects of letter writing, with its inclusion of address and date, and records the important chirographic element ("de mi mano") that served not only as authentication of document and letter production but also as a proof of literacy and intimacy. The rubrics of this poem, both the main one and the ones guiding the reading inside the text (*sobre escripto*, *carta*, *fecha*), provide a trace of the slender three-dimensional material letter, with its creases and writing on different parts of the folded paper, and attempt to thwart the unavoidable flat and linear representation of the letter in the manuscript or printed cancionero page. There are many other examples of verse addresses and poems, such as ID2848, where the first stanza is the versified address ("sobrescrito," v. 6), while the rest of the stanzas make up the letter. The first stanza and address introduce a conceptual play on short and long applied to the short height of a person, the Almirante de Castilla, to whom both ends ("estremos") of the poem refer without naming: "Al muy syn llustre señor / que de verle es maravilla / de pequeño muy mayor / grande de los de castilla."[24] Troping this structure, the poem refers to the Almirante as embodying extremes, such as his physical shortness and his political greatness. Of a different nature is poem ID1122, a letter sent by Diego López de Haro to doña Marina Manuel: "carta de amores trobada por el mismo" (LB1), or "Carta suya que embio a doña marina manuel" (11CG, 14CG). It is followed by ID1123 (LB1, 11CG, 14CG), which is a stanza commenting on the visual device of the coat of arms that sealed the letter: "La çerradura dela carta que yva cerrada con vn escudo dela merced" (LB1),[25] and which is textualized by the rubric. Because of its motile nature and embodied agency, the poetic paper, much like the letter, makes these qualities evident through textured references to its medium.

Once the poem was written, it was distributed willingly by its author or extricated from him or her through gentle or firm prodding by friends or acquaintances. The texts were recopied and re-routed per request or (in)voluntarily. When a poem was solicited and the poet refused to publish,[26] rubrics tell of the textual violence exercised by readers as acts of extortion to induce the publication of the text. Such is the case of Garcia de Resende, compiler of the great Portuguese *Cancioneiro geral*, who had to resort to distorting the order of the stanzas in Afonso de Valente's poem, of which he had obtained a non-authorial faulty copy, so as to make him agree to provide the correct version of the poem under the threat of publication of the distorted version (ID7299, ID7300): "Trouas que afonso valente fez em tomar a garçia de rresende sem lhas mandar" (ID7299),[27] followed by "Reposta de garçia de rresende polos consoantes a todas estas trouas dafonso valente que foy achar sem lhas elle mandar. *E vam forado ordem por conseguyr as suas*" (ID7300) [my emphasis].[28] Thus the materiality of poetry involved a lack of control over distribution both by the author and by the owners of copies, and inspired a desire for control over distribution rights. References to papers being sent from one poet to another that appear in cancioneros indicate the material sources mined by the compilation. For example, in ID1301, Juan Alfonso de Baena writes in the name of Archbishop don Sancho de Rojas, in answer to a petition sent by Villasandino in writing, "por escripto" (v. 3), and refers to the material support of the poem as "leaves," "vuestras fojas" (v. 56). Much like letters and other documents, poetry sent to an individual could only be retrieved with difficulty, as in ID1571 (PN1, vv. 1–2), where Rodrigo de Harana, following a long exchange with Juan Alfonso de Baena, asks that his poems be returned to him: "Señor Juan Alfonso, dezid si vos plaze / de darme las coplas e dichos que fize,"[29] and is compelled to insist again in ID1573 (PN1, vv. 9–10): "Aún por segunda vez vos amonesto / que luego me dedes aquestos dezires."[30] As in the poem, the verb "fazer" indicates authorship in a material sense, much like the term "scriptor" meant in notarial language the person who physically wrote a text, terms that increased in use as personal references to the person/s involved in the creation of a document grew in frequency towards the fifteenth century (García Valle 1999: 18, 31–2, and passim).

Although there were vast numbers of these poetic *papeles*, only a fraction have survived. Some legal collections that have been preserved provide an idea of the many formats, shapes, and forms of the wandering loose papers and what their poetic counterparts may have looked like.[31] It is helpful to examine some of these collections, such as that comprising Christopher Columbus's personal papers, which include letters and legal documents.[32] Many of the papers that were sent or handled as letters were obviously originally folded in the shape of a *billete* with the address written on the small rectangular surface that would have resulted at the front of the folding and the date often on the back of the folded document. This format has been preserved in Columbus's letters to his friend and aid fray Gaspar de Gorricio (reproductions in Varela 1987: docs. IX through XIII). Other collections of documents and their archival modes may serve, when used alongside the textual clues found in the poems and their containers, as samplers of the material configuration of the lost poetic papers.

The Poem as Publishing Agent

The poem could thus enjoy an agency otherwise reserved for animate entities. This explains the many instances in which the poet encourages it to exercise its motility as a free agent by sending the poem to find the addressee, as in ID6108, where the poet begs the poem to go and find his lady. In the rubric heading ID3099, Juan Álvarez Gato similarly talks to his stanzas and prompts them to go in search of his lady, acting as both his messenger and as his message: "habla con estas coplas y haze mensajero y enbajada con ellas para que se topen con la señora para que supiese lo quel no tenie osadia de dezille."[33] The references in the poem to the lines in the stanzas ("renglones") (also found in other compositions, such as ID3124, v. 17) assure us that the poet is not talking about an abstract poem, but rather the very paper that was being forwarded. The publication procedure is also clearly delineated here, as the poem is said to need the venue of many other hands ("de mano en mano") to go running ("corre") to cross the physical distance that separates the poet from the lady. The "sad teary stanzas" ("coplas tristes llorosas") prove to be so not only in content, but also in their material form, as they are blurred by the many tears shed by the poet ("con muchas lagrimas tristes / que borran vuestros Renglones") (vv. 1–18). The ensuing exchange with the lady, whom the poem seems to have found without problem, is equally telling. Seeing ("vido," "vistes") the poem in her house piques her into initiating a conversation by asking the poet to reveal the identity of the female subject of the stanzas received, enabling him to give the obvious answer and to reiterate his ardent (and unfulfilled) interest in her in another poem (ID3100). A third poem, ID3101 shows that the previous composition actually reached other hands, those of a caballero friend, before or at the same time as those of the female addressee, thus completing the intended publication process and opening new exchanges with individuals other than the intended (or alleged) addressee. The rubric tells how the author's friend saw the poem and asked him about the changes in his love life, providing the occasion for further writing. This poem sequence, one of the many exemplifying the intricate intertwining of text and medium, illustrates some of the key issues regarding publication and the production of meaning. As an independent agent, the poetic paper found not only its intended destination, but also others to join the poetic exchange and help redirect its meaning, placing it within wider networks of social interaction.

A similar case of textual imbrication is that in which Fernando de la Torre tells of his writing a poem for a young lady, then presenting it to some gentlemen at court, who in turn encourage him to show it to the king, all in this order. His text shows a progression marked by different forms of discourse, starting with an explicatory prose narrative, followed by an introductory poem to the king, and then finally proceeding to the *canción* written for the lady. If we are to understand de la Torre's text literally, he depicts himself writing the letter while on the road, recounting the previous week's exploits, penning a poem for a lady and then presenting it to the king after showing it to some gentlemen at court, a process which would involve producing at least five different pieces of paper or booklets for as many different writing transactions, including the copy into a book (text in Díez Garretas 1983: 116). De la Torre's narrative tells as much about the actual steps involved in the dissemination of his work as it does about the modes of compilation engaged in making it.

The intricate modes of distribution of the poetic paper, to one or several recipients, and from them yet to others or back again to the poet, all the while making its way to even further destinations, create a complex web of textual imprints that result in variegated versions of the poetic text.[34] Because of the precarious existence of loose papers and the additional loss of documentary evidence deemed of little monetary value, it is often impossible to follow the trail of every poem, but some glimpses may be gained from direct statements in the poems themselves, as well as their rubrics, as the examples above show.

Nevertheless, due to the proximity of poems to letters and other types of documentation, a relationship that will be argued in detail in the following chapters, administrative paperwork generated in the circles where the poem circulated can offer precious additional information of the poetic paper's itinerary and may provide a detailed "publication record." The documentation of court poetry in the Crown of Castile is preserved mostly in cancioneros because the royal archives to a large extent have not survived. The neighbouring Crown of Aragon shows what might have been lost. One early example is the enlightening account of the travails undergone by some poems in the court of King Pere IV of Aragon (el Cerimoniós) revealed by the official registers of royal correspondence. Several series of royal letters seem to have provided the medium for the dissemination of poems (published by Rubió i Lluch and Balcells 2000). Along with a letter dated 8 June 1355, King Pere sent a copy of a *sirventès* he had made, encouraging the recipient, his uncle Prince Pere, to show it to all those who would like to see it (text in Rubió i Lluch and Balcells 2000, 1: 168). In the same year, the king sent the members of his council a sirventès that he had written, presumably the same one, along with a letter (2: 105). On 21 November 1371, he wrote a letter to his son Joan in which he included a poem (*cobles*) that he had written (1: 237). In yet another instance of simultaneous self-publishing, the king sent on the next day, 22 November 1371, the same poem to three other people, Berenguer d'Abella, Bernat de So, and Ramón Planella, along with a letter advising of the fact that the poem had also been sent to Prince Joan (1: 237). In the register where the letter was copied, the scribe states that three identical letters were sent with the poem to the three recipients. On 2 March 1374, Prince Joan wrote from Valencia to Bernat de Bonastre sending him an (interfering) reply or *traversa* that he had made to a poem ("cobles") addressed to the king by Guerau de Queralt and for which the king had written a reply (1: 251–2).[35] Another letter followed on 7 March 1374, by which time Prince Joan sent his father all the poems generated in the exchange, along with a sirventès by Mossèn Pere March with his reply (1: 252). Yet another letter followed in which the Prince asked the queen to have the stanzas, replies, traversa, and sirventès alluded to in the previous letters shown to her. A similar letter was sent to his brother, Prince Martí (1: 252, n1). The next day, 8 March 1374, Prince Joan wrote to Bernat de So asking him to write a poem following the rhymes of the traversa and the reply alluded to in the previous letters, or using other rhymes (1: 253). Two days later, on 10 March 1374, Prince Joan wrote to Ferrer Sayol asking him to tell him about the comments that the reading of his traversa and reply inspired in the king and queen as well as in other notable persons who may have been present. He also gave him an update on the poems he had sent and the ones he had received (1: 254). The king's poem exchange went on for a number of years. Some of these poems and poetic papers have been preserved in Ms. Regto. 1265, and Ms. Caja documentos diversos III, 2 (Arxiu de la Corona d'Aragó, Barcelona) (with photos in Canellas López and Trenchs i Odena 1988: 141, 146). These acts of multiple and simultaneous publication took advantage of administrative practices that were well established in the Aragonese court. A further example may be found in the multiple copies of a letter that were sent calling for the assembly of representatives in defence of Aragonese rights in the face of the establishment of the Inquisition. There remains a copy of the 1484 letter and the long list of addressees of simultaneous copies, and similarly a copy of the 1485 letter when the king summoned armed help in defence of the Teruel Inquisition.[36] The meticulous record keeping of the Aragonese Crown, coupled with the fortunate survival of its records through the centuries, has facilitated the preservation of precious historical documentation such as this. The consideration of the intense poetic exchanges of the period invites a look at the poem as a dynamic entity whose *flexible textuality* enables it to attach itself to people, objects, and documents and to take advantage of their ability to set it in motion.

Papering

In addition to hand-to-hand distribution, as seen above, texts inscribed on paper could be disseminated more broadly by being posted on a door or other surface, being left at an inn, or being thrown in multiple copies over the walls of residences and squares. All these constituted in and of themselves acts of legal (or illegal) publication. This form of propagation encouraged a dialogic exchange in a public space, thus openly displaying the social solidarity, as well as competitiveness, involved in many forms of writing. Such public life was the particular modus vivendi of short poetic forms such as the short *motes*, which, like the phenomenon depicted by Love (1987) for England, could be published via a copy left at an inn (*mesón*), as described by Milán in his *Cortesano* (280–1) and attested by the rubric introducing the *Motes del Almirante de Castilla* (LB5): "Motes quel almirante de Castilla enbio A los galanes y damas desde vna su villa y dizen que se fallaron escriptos en vn meson."[37]

The law went to increasingly greater lengths to control the physicality of the text artefact, either legal *instrumentum*, poetic paper, or any other document or paper. Public spaces offered obvious advantages for encouraging compliance with the law by putting a message intended for one or many individuals in front of the witnessing eyes of the community. But the same spaces were easy venues where anyone could forcefully publicize any text, a practice so common and seriously disruptive that it warranted laws positively attempting to discourage it. Thus Alfonso X (2004) in his *Siete Partidas*: "Ley terçera. de la desonrra que faze vn onbre a otro por canticas o por rimos Infaman & desonrran vnos a otros non tan solamente por palabras, mas avn por escripturas faziendo cantigas o rymos. o deytados malos de los que han sabor de enfamar. Esto fazen a las vegadas paladinamente. & a las vegadas encubiertamente echando aquellos escriptos malos en las casas de los grandes señores. o en las iglesias. o en las plaças comunales de las cibdades & de las villas. por que cada vno lo podria leer" (VII, 14r).[38] It is obvious that the practice continued through the centuries. Much later, for example, under King Manuel of Portugal, in book 5, title 79, of the *Ordenações Manuelinas* (1984), first published in 1512, legislation goes one step further and punishes not only the act of authorship of these roaming poems ("escriptos de trouas e outras cartas de maldizer"), but also the act of further publication by unwitting finders. Those who find one such paper are obligated to tear it and forbidden from publishing ("pubricar") it, showing it to anyone else ("mostrar"), or even speaking of it ("falar") (235–6). Slandering letters were disseminated in a similar fashion. Fernando de Pulgar (1982) confronts the anonymous author of one such letter that was delivered at night surreptitiously ("echada de noche y tomada entre puertas"). Pulgar's *Letra* XXI, a response to the slandering text, an injurious reply to an earlier letter by Pulgar, relates the meandering ways of the texts thus disseminated. From the hands of the slanderer ("vuestras manos"), it was published throughout the city ("la andávades publicando por esa cibdad") before reaching Pulgar's hands ("las mías") (79). Wandering papers functioned as venues for slander, conveying a discourse of dissent or resistance that would continue well into the sixteenth and seventeenth centuries, as can be seen in the Spanish poetry studied by Díez Borque (1983),[39] and also in England (Marotti 1995), and in Italy's *pasquinades* (La Sorsa 1947).

Doors were favoured bulletin boards, particularly, though not exclusively, those on palaces and churches. Notices of buildings for lease were posted on the doors of churches and public places.[40] After Princess Isabel had been sworn as the legitimate heir to the Castilian throne in 1468, the biography of Cardinal Pedro González de Mendoza explains how his brother Íñigo López de Mendoza, Count of Tendilla, nailed a legal complaint on the princess's door on behalf of Isabel's rival to the throne, Juana de Castilla.[41] Chronicles such as the fifteenth-century *Crónica del halconero de Juan II* by Carrillo de Huete (1946) show that letters containing royal mandates or any other type of public announcement were posted on the doors of important residences (364), while, under the orders of Fernando de Antequera, city gates were used to post city ordinances for the benefit of travellers entering a city such as Cuenca in 1411 (text in Cabañas González 1982: 393), and the palace gate was the posting board for courtly or chivalric announcements such as the celebration of jousts.[42] Less noted by the royal and noble chronicles, but undoubtedly as common, was the practice of using the doors of private residences for the posting of papers, sometimes libels slandering the unhappy female resident who had not reciprocated an intended lover's attentions, as stated in the chapter dedicated to the first mortal sin in *Corbacho*, where the author denounces "los libellos difamatores puestos por puertas" [slandering libels posted on the doors] thus used (Martínez de Toledo 1987: 127–8). Challenge letters (*carteles*), including inculpatory and exculpatory documents venting the animosity of men preparing a one-on-one combat, were also posted on public spaces such as doors, church entrances, and squares.[43] Inn doors were used to post prices, as was done, for example, in Guadalajara (text in López Villalba 1997: 285–6), where paper copies ("cartas"), authenticated by the escribano, were ordered to be nailed ("que las hincase") on the door of seventeen inns or *mesones*. Church doors, doors in notable buildings, and other public spaces functioned as bulletin boards for edicts, laws, and similar legal texts through later centuries.[44] Church doors were also used for posting documents that summoned someone publicly to appear before the court, the Inquisition, or for any legal action. We find such use, for instance, in the 1476 letter summoning Juan de Salazar and Íñigo Cardo for trial on the death of their grandfather Lope García de Salazar, which was ordered to be posted on the doors of several churches of cities where the accused, who had escaped, were thought to be found.[45] This reflected the common earlier practice of posting in a public space official documents that needed to be publicized, such as those announcing land transactions or donations.[46]

Slanderous poetry was similarly disseminated, as can be seen in ID4943 and ID5460, two slandering poems crossed between Castilians and Portuguese and posted on the palace doors at a time of strife between the two nations: "Este rrifam escreueram huums castelhanos ha porta do paço em castela andando laa o duque dom dioguo" [Some Castilians wrote this poem at the Palace door in Castile when Duke Dom Diogo was going there] (ID4943), to which followed a Portuguese response and posting (ID5460): "E fernam da sylueyra como a uio escreueo estoutra ao pee em rreposta" [And when Fernão da Silveira saw it, he wrote this other one at its foot in reply]. More postings on different topics are found in Portugal's (1943) *Arte de galantaria* (61), using the palace doors as a sort of billboard. Slanderous poems could be left lying about in a public space, or in private or semi-private places such as a palace; thus the famous slandering poem ("porquês") that was found in the Palace of Setúbal in Portugal (ID5980, *Cancioneiro Geral* 1990–8, 3: 343–7).

Further postings were done by means of hanging *tablas* (boards or simply pieces of paper or parchment) in churches, which provided the space for the public display of prayers and religious texts for both didactic and proselytizing purposes (Castillo Gómez 2006: 212–20). These were likely close to the tables studied by Reynolds (2003), which were made out of paper or parchment glued, nailed, or tacked through a ribbon or a strip of leather to boards for display purposes of religious or legal texts and which could also be illuminated (24–30), which would explain, for example, the references in Madrid's city council to the destruction of the tablas by tearing ("rasgar") and to the production of multiple copies.[47] Reynolds (2003) has underscored the artistic importance of loose, independent leaves intended to be hanged in public or private places or inserted in books in the fifteenth century, as well as their highly perishable nature. Mounting them on boards would be a way to avoid their loss or destruction. Similar tablas could be seen hanging from the walls of the city hall displaying a copy of city ordinances and other public information such as prices or taxes; the tablas could also display the prices of wine or other wares in the appropriate establishments.[48]

The oral publication of the legal document often accompanied its written recording (Castillo Gómez 2006: 206–9), which shows the tight connection between the work of escribanos and town criers and explains the queen's personal interest in choosing the right crier for the city. Laws, ordinances, and other texts of public interest were mandated to be publicized by the city crier. This was the case with the important document issuing from the Cortes in Toledo in 1480. The written document states at the end that the cuaderno must be disseminated both in writing by providing copies to all towns and also orally by *pregón* in all public spaces, including town squares and markets. An escribano was needed as witness to ensure the pregón was communicated as instructed.[49]

Other notices, and a variety of short texts, appear likewise affixed to the walls inside private residences in visual art. One example is the small piece of paper affixed to the wall in the shop of Saints Justa and Rufina in the altarpiece depicting their lives (photo in Gudiol and Alcolea i Blanch 1986: 269, pl. 79). Small papers with writing are also often found attached to a wall or the front of a desk in depictions of private studies and/or writing desks, as in the case of the panel of the Annunciation in the altarpiece of the Condestable (ca 1465) by Jaume Huguet, housed in the Chapel of Saint Agatha in the Palau Reial Major (Royal Palace) in Barcelona (photos in Riu-Barrera et al. 1999: 23–4). Another is seen tacked on the wall right above Mary's desk in Jaume Huguet's Annunciation altarpiece panel (photo in Gudiol and Alcolea i Blanch 1986: 415, pl. 780). The casual pinning of papers on the wall of private chambers was, as these images attest, widespread enough to warrant the identification of the writing self with its medium. Converso author Alfonso de Cartagena saw the papers that made up his correspondence with Pier Candido Decembrio as the means to endure in the latter's memory. Cartagena used a metaphor that made him a piece of paper pinned in Decembrio's memory cell: "tanta tamen fuit dignatio tua tantaque benignitas, ut memore communicationum nostrarum ... me in memoriae tuae cellula clauis dilectionis fixum haberes" (text in González Rolán et al. 2000: 414).[50] In a similar vein, realistic depictions of scraps of parchment or paper were used to mark authorship. In the central panel of a fifteenth-century altarpiece, a piece of paper affixed to the platform on which Virgin and Child are sitting attended by angels contains the authorial signature and the date of the painting: "Pere Garcia de Benavarre ma pintat any 14" (photo in Gudiol and Alcolea i Blanch 1986: 440, pl. 925). Whether anonymous or carefully marked by the authorial signature, pieces of paper entered collective or individual memory through blank walls, doors, and squares, fitting venues for the text's restless and endless journey.

The pervasiveness of the written word that papered buildings and public spaces helped to inscribe daily life, thriving in a continuum that went from lapidary inscriptions to the loose piece of paper. Modes of dissemination reveal much about the spectrum of textual practices and raise the important issue of textual agency. The punishment of the publishing reader as author in the *Ordenações Manuelinas* only confirms the overlapping action of author, reader, and compiler/editor warranted by the compelling materiality of the text and the effect that the various hands may exercise over it. The concrete nature of the papers created a tension between the motility it conveyed and authorial desire for control. This tension would help shape the very nature of authorship, as other human agents – editors, readers, printers – also vied for control.

4 The Hands Have It

On 19 October 1419, King Juan II issued a royal provision ordering all public escribanos to present themselves at court to take an exam in order to ascertain their professional skills. The document was copied, authenticated, and widely distributed. The original letter was received and its authenticity established in a ceremony that involved the material examination of the document. Before the escribano could proceed with the production of copies, law expert and royal servant Juan Sánchez Peralta held the letter in his hands ("tomo la dicha carta oreginal en sus manos"), saw it ("viola"), touched it ("tocola"), and certified that the letter was in good state and that he did not perceive any tears or marks crossing over the text ("sana e no rota ni cançellada") that would invalidate any part of it. He declared the letter to be legitimate and therefore ready to be reproduced through one or more copies ("un treslado o dos o mas") (doc. in Pascual Martínez 1981: 169).

As the act of hand-to-hand dissemination such as posting or throwing attests, paper and parchment were destined not only to be read but also to be touched. The importance of physical proximity was underscored by its validating effect. Religious and legal books as well as images lent their sacred, binding, and perhaps magical value to those whose hands came into contact with them and to their words. The book of poetry was also part of this net of artefacts with a tactile value, co-opting its ability to connect text and medium and to establish an enduring liaison with the surrounding world, both physical and transcendental. A close look at the practices involving physical contact with the book in legal and religious ceremonies will shed some light on analogous uses of the poetic anthology.

Contact with a book often replaced human touch in legal ceremonies. The worth of the text in these types of transfer could be buttressed iconographically by means of a religious symbol or an illumination, which designated the point of contact of the hand with the text. The crosses painted on books, law codes, and binding documents were to be touched by officials or other persons swearing office or acquiescing to the agreements or laws specified by the text (Silva y Verástegui 1989). The law stipulated this as the protocol for being sworn into office. Touching the Gospels was an approved stand-in procedure for that of the *inmixtio manuum* by which the official who was being sworn in put his hands into those of the King.[1] The touch of the person was transferred to the touch of the written and illuminated page. Beltrán de Salcedo's swearing into office in September 1489, for instance, used a common formulaic expression that highlighted the act of "bodily" ("corporalmente") putting the right hand on a drawing of a cross ("esta señal de crus").[2] Swearing an oath in front of God usually involved touching a book, and as such it is represented in the Alba Bible, as the illuminations depicting Abraham and Abimelech's oath show. Jacob is likewise depicted touching a book while swearing his primogeniture.[3] Many manuscripts containing copies of a town's laws or *fueros* have an illumination of the crucifixion, which is often worn out, particularly the surface area around Christ's feet, as it was customarily touched when swearing the fueros and is apparent in such manuscripts as the fueros of Uclés, Soria, Navarra, or Teruel.[4] Such physical touch could at times be used destructively on a page containing a miniature with a loathed image, such as the one depicting a devil in the Bible of Ávila (Biblioteca Nacional, Madrid, Ms. Vit. 15–1, fol. 323). The tactile connection to material support is also obvious from similar marks in medieval missals, such as the one in the Cathedral of Girona, or the *Misal de los Templarios* in the Biblioteca Apostolica Vaticana (reproductions in Delclaux 219, 381). King Enrique IV's 1462 *Cuaderno de las alcabalas* contains *formulae* and a sign of the cross to be used when swearing the oath to fulfil the tax-collecting duties owed to the king (text in Moxó y Ortiz de Villajos 1969a: 425, 445). Escribanos underwent a similar ceremony when being sworn into office (Pardo Rodríguez 1994: 153), as did anyone presenting a sworn statement in front of an escribano for the purposes of document production (see, for example, docs. 43 app., 270, in Ostos Salcedo 2005: 156, 359), as was done in the 1408 negotiation leading to the marriage of Santillana to Catalina Suárez de Figueroa.[5] In addition, touching the cross and the Gospels was part of the sworn testimony of a witness in any legal proceedings.[6] Touching the Gospels in front of a crucifix was required of those accused of heresy as part of the confession and retraction.[7] Some of the depictions of hands within the text in this type of document indeed evoke the act of swearing a document while touching it and grasping it in order to hand it over to its recipient (Sáez 2002b: 224; also Pacheco Sampedro 1997). For all purposes, Jews were instructed to swear by touching the Torah in the synagogue, in the presence of both Jewish and Christian witnesses (Alfonso X [2004] *Siete Partidas* III, 34r). Contact with the material support conveyed acquiescence with the text. Because of its legally binding implications, the material consequence of the artefact was more than equal to that of its textual content.

Both the large format and the tactile connection with the text also served the colonial efforts of Portuguese explorers at a later date. João de Barros (1988) explained how in the time of King Manuel, António Correa used Garcia de Resende's print *Cancioneiro geral* during the Portuguese expeditions through Asia. In the Terceira decada of his *Asia*, printed in 1563, he tells of the peace oath proceedings with the King of Pegú (Burma or Myanmar) in 1519. After this king had finished his part of the peace ceremony, António Correa picked the grand poetic compilation of the time, the *Cancioneiro geral*, over the breviary of the ship's chaplain or any other book of prayer because of the *Cancioneiro*'s large format (it was a folio). The other books did not represent so well the magnitude of the Christian faith, so the book chosen was intended to make the oath more impressive. As he opened the *Cancioneiro*, Correa's eyes fell on a poem by Luis da Silveira on the topic of *Vanitas vanitatum*, which filled Correa with awe and resolved him to be bound by the oath on account of the solemn words read from the *Cancioneiro* (Barros livro 3, cap. 4, fol. 67r–v). In this case, the large format of the *Cancioneiro* did not disappoint, and its content lived up to the grandiose expectations of its size. The physical contact required for the oath ceremony both imbued the cancioneiro with transcendence and was in turn validated by its contents.

Inscribing Touch

The twelfth-century cartulary of Leyre (*Becerro Antiguo de Leyre*) is a good example of the use of hands in manuscript illumination to convey physical contact and handling of the text. For instance, an illumination shows the abbot holding the roll open with both hands, while an open hand holds what may be the catchword with the thumb, middle, and index fingers in what appears to be an invitation to turn the page in fol. 50 (in Silva y Verástegui 1988: 16, 19). The hand with which a monk points to a place in the manuscript of the fourteenth-century Breviary of Pamplona (in Silva y Verástegui 1988: 125–7), as well as that of Gabriel pointing to the roll containing the words of the Annunciation in the illumination of another Pamplona breviary (in Silva y Verástegui 1988: 136) bear strong iconographic resemblance to the *maniculae* that are abundant in manuscripts. These visual representations show an iconographic path that goes from the human hand that handles the written page to its graphic extension as textual marker with varied applications.[8] The angel in the illumination depicting Ahasuerus's restless night points to a passage in the book held open by one of the boy attendants with a similar hand, in fol. 393r of the Alba Bible (photo in Nordström 199). Large hands appear in cancioneros as well, such as the anatomically accurate sixteenth-century hand pointing to a stanza in Santillana's cancionero (SA8, fol. 14v), complete with nails and outlines of the joints. There are similar appearances of disembodied hands in other types of art. The hands that help when reading a book are compellingly represented in Gil de Siloe's funerary sculpture of Prince Alfonso, brother of Queen Isabel, in the Church of the Monastery of Miraflores in Burgos, which shows the prince kneeling with his hands joined together in prayer in front of an open book propped up on a richly covered stand and helped by a disembodied hand coming out of the wall and shown in the act of turning the page (photo and discussion in Yarza Luaces 2005: 61–3). The illuminations of saints in the codices of San Martino (Saint Martin of Leon) (ca 1130–1203), the pilgrim, theologian, author, as well as canon at the church of San Isidoro in León, similarly use the hands to engage the reader (in Viñayo González and Fernández González 1985: 56), while grotesque figures use their tongues to point to paragraphs in the text. As Pacheco Sampedro (1997) has shown, hand depictions in the cartulary of the monastery of San Juan de Caaveiro constitute a detailed record of the role of the hand in the act of preparing, swearing, and handing over the document, which is often depicted as a roll. That the inscribed rolls (also referred to as phylacteries, ribbons, banderoles, or scrolls) depicted in, for example, manuscript illuminations are signs of a residual orality that would work much like a speech balloon is an interpretation challenged not only by codicological and diplomatic evidence, however scant, but further by the use of hands in the iconography of the period. The emphasis on the hands and on the text as roll deserves to be considered in relation to the passage from voice and mouth to an increased emphasis on visual and tactile contact with the text. In this vein, it must be noted that the illuminations that show a scroll coming out of a figure's mouth are answered by scores of others that show the scroll originating in the figure's hand, positioned as if it were being held by a corner. The depiction of the banner and the roll seem to blend in miniatures of the

Annunciation such as that of the 1486 Book of Hours by Jan Spierinck.[9] In this miniature, Gabriel holds the roll with the words of the Annunciation with his hand, which is mirrored in a similar roll held by Mary with her response. There are also instances of scrolls that appear with the disembodied hand holding a pen in the act of writing its contents, such as the one found in the illumination depicting Belshazzar's feast in the Alba Bible, which shows the writing on the wall from Daniel 5:1–31. The same Bible (fol. 11r) contains a miniature of Rabbi Mosé Arragel as he kneels in front of a seated King Juan II of Castile, both of them identified not only by their visual depiction but also through the flying rolls that each holds in his hands and which remain partially furled for an easier grasp; the rolls state the identity of each of the figures (photo in Nordström 1967: 14). Both types of depiction, the mouth- and the hand-originated scroll, could indeed represent authorship, but the hand would be poignantly and literally pointing towards the text as artefact and to the contact not only of author but also of user, reader, or otherwise, through his or her own hands. Although the roll format and its depiction will be studied at greater length in the following chapter, it is useful to note here its function in relation to hands, because the iconography conveys a clear reference to modes of textual handling.

Therapeutic Touch

The need of physical contact for some authors was so great that there are narratives of some of the most prolific going to bewildering extremes to be able to write. An early example is that of the twelfth-century San Martino. The book of his life explains his peculiar *modus scribendi*, when, crippled by terrible headaches and debilitating osteoarthritis, he was unable to sit on a chair and work at his desk. He found a solution by having cords hung from a beam by his desk so that he could write suspended from them, while they supported the weight of his body.[10] For San Martino, being able to put hand (and stylus) to the wax tablets (*tabulis ceratis*) when he wrote meant negotiating his body's ability to work despite a crippling disease. Counteracting the physical exertion involved in writing, contact with paper could have a beneficial effect on individual health if it was inscribed with incantations or similarly powerful texts. Texts of such nature have survived, such as the amulet in Grenoble, Bibliothèque Municipale, MS 4376, France, fourteenth century (photo in Sirat 2002: 100). The use of these extremely common amulets containing prayers or incantations was widely advised in medical treatises, such as Fernando de Córdoba's (2002)

Suma de la flor de cirugía, which recommends writing some words referring to the story of Job on a piece of paper ("cartilla") and tying it around the neck as treatment in conjunction with herbal remedies (109–10). The effectiveness of such pieces of paper relied heavily on the credibility of the explicatory titles heading the texts they contained, as is obvious in Vicente Ferrer's attempts to discredit rubrics that headed widely circulated copies of prayers: "Mas non pensedes, buena gent, que aquellos títulos que ponen en las oraçiones primero que sean verdat, nin son ordenados por los santos, sinon por algunos ypócritas mentyrosos o por algunos echacuervos, diziendo que si omne trae aquella oraçión consigo que nin morrá en fuego nin se afogará en agua, e otras cosas muchas vanas e mentirosas" (text in Cátedra 1994: 355).[11]

The physical contact with the text suggested by these instances, joined with its religious and legal bearing, helps explain similar uses of books of religious poetry. When King Alfonso X took to his bed so ill that he feared for his life, he refused the remedies recommended by his doctors and asked instead to have his book of *cantigas* in honour of Mary brought to him. Upon contact with the book, the king was instantly and miraculously cured, as he tells in a first-person narrative in his cantiga "Muito faz grand' erro, e en torto jaz" (Cantiga 209), illustrated by the miniatures in the Florence manuscript (Florence, Biblioteca Nazionale Centrale, Banco Rari 20, fol. 119v; Keller and Kinkade 1983: plate 32). Writing thus documented and ratified the value of touch, as did the visual depictions that accompany King Alfonso's cantigas. Even in the case of intrinsically miraculous objects or their fragments, as is the case with relics, papers served as the means for tactile contact and as conveyors of meaning. A case in point is that of the relics wrapped in inscribed papers identifying their contents in Queen Isabel's chest, or a surviving example of the same, the fifteenth-century reliquary, the Relicário dos Santos Mártires de Marrocos, now at the Museu Nacional de Arte Antiga in Lisbon. Here, the inscribed paper is the means of access and repository of both relic and its meaning, the potential separation of paper and relic resulting in the loss of ascribed value for the object (further discussion in Gómez-Bravo 2013).

Chirography

Material access to textual production was promoted by an increasingly powerful monarchy, which relied ever more heavily on the advantages of literacy. In addition, other developments from previous centuries helped propagate authorship and the personal production of texts. The emphasis on hands and the tactile connection with texts had to have a strong bearing upon the material production of texts. Increased literacy and chirographic ability in larger sectors of the population brought about a rise in some forms of textual production, much in the same way as the access to personal computers spurred the wide availability of desktop publishing.[12] Scribes and authors became closely related figures as writing became expected of the monarchy, nobility, and the officials employed in their service. Aristocrats, appropriating the prestige that came with the skills of trained professionals, learned good penmanship and drawing. As a result, handwriting grew in importance as a mode of authenticating a document or a poem.[13] Many of the documents of the period emphasize this personalization of writing, which was also encouraged by the formation of private cultural spaces (Gómez-Bravo 2005). These skills were also useful to the monarchy; they gave the ruler personal control over document production, which manifested itself in securing secrecy for the writing of politically sensitive documents. In the fourteenth century, King Pere IV of Aragon (el Cerimoniós) took full advantage of a tight system that gave him full control over the professionals employed in document production while also allowing himself direct participation in the composition and writing of particular documents (Gimeno Blay 2006). The techniques used for personal writing could be as idiosyncratic as that described by Garcia de Resende (1994) in his *Chronica* of Portuguese King João II, who would never dip his pen when writing, instead requiring that his chamber assistant provide a freshly prepared pen when the one he was using ran out of ink. This provided a strong intimacy with the king and his affairs, giving Resende the opportunity to see what the king wrote ("tudo o que elle escrevia"), until one day, seeing that the matter at hand was highly sensitive, Resende turned his face so as to give the king some privacy. It was a proof of the king's strong regard for Resende that the king told him to turn his face back towards him, saying that if he did not trust him he would not have employed him as a close assistant. This mark of confidence was to make Resende not proud of himself but more faithful to the king (434). The narrative in his *Chronica* shows that he was both.

Royal penmanship was so important that chronicles of Enrique IV's reign emphasized the fact that King Enrique IV was a great escribano of all sorts of script ("muy grande escrivano de toda letra"), a statement that was repeated in other chronicles of Enrique IV's reign (e.g., the *Crónica anónima* 1991, 2: 478). Queen Isabel granted the same importance to personal handwriting in letters, because it would give a greater sense of intimacy. The amount of autograph correspondence between Isabel and Fernando and the personal tone of their letters are witness to this effect (letters in Prieto Cantero 1970). A similar use of handwriting may be found in the correspondence with other people of particular esteem. In a cédula written to the Countess of Feria in response to a letter announcing that the countess had successfully born a child, the queen added a note in her own hand. The note gives the usual excuse for not writing the whole document in her hand – her many occupations – and ends by explicitly stating that the short note at least is indeed in her own hand, with the common expression: "De my mano. Yo la rreyna" (facsimile in Millares Carlo 1983, 3: doc. 294). Much earlier, in 1152, another important woman, doña María, wife of Count Poncio, had confirmed a charter of the town of Castrocalbón "propria manu" adding her own signum containing the letters of her name (photo in Romero Tallafigo et al. 1995: 108–11, plate 9). If the author him or herself could not write, the scribal hand, often that of a trusted collaborator or secretary, was identified. When King João II of Portugal was short of time and could not write in his own hand, he begged forgiveness of Princess D. Beatriz, a key political figure until her death in 1506, with whom he kept an intense correspondence on political matters. He made sure to note that he had left the physical writing of the letter to his trustworthy scribe and begged D. Beatriz to continue using her trusted secretaries, such as Isabel de Sousa, or the king's envoy Rodrigo Afonso, both of whose handwriting he was able to recognize, for her own letters (documents in Chaves 1983: 284–93). The distrust of a personal or sensitive letter that had gone through (too) many hands is patent, as is obvious from his warnings ("que nom passem por mujtas mãos," in Chaves 1983: 290).

Personal and critically important correspondence was expected to come directly from the pen of the sender, as many documents of the period demonstrate. Isabel and Fernando exchanged letters with King João II of Portugal "de sua mão" on Prince Afonso's impending marriage to Isabel and Fernando's daughter (documents extracted in Chaves 1983: 151–2). Documents written in Queen Isabel's hand appear in the post-mortem inventory of her possessions. In one of the chests, there were five such memoriales containing provisions for different people (see chapter 7).

Ecclesiastical dignitaries had similar concerns. In his instructions for the administration of his household, Archbishop Hernando de Talavera (1930), who had served as confessor and key adviser to Queen Isabel, placed utmost importance on the responsibilities borne by those with access to his private chamber or library, and therefore to all his documents, but more importantly those written in his own hand (33). Trusted secretaries boasted of this proximity to their masters' hands and, much like

Resende in his narrative of the peculiarities of King João II's writing, similarly used it for self-promoting purposes. Juan de Vallejo (1913) tells in his account of the life of Cardinal Cisneros of his participation in the production of some of the cardinal's more sensitive documents. He disclosed that, as trusted page of the cardinal, he was not only with him in his private chamber when the cardinal had to write important letters to King Fernando, but that his was in fact the hand that wrote the documents, imitating the cardinal's handwriting on the latter's orders (93, 112–13). The authenticating value of chirography could thus be both preserved and thwarted: a complex operation that revealed the importance of textual cooperation.

Official and public documents made similar uses of the authenticating power of chirography. In city registers the recognition of the handwriting of the various officials involved in the city council was part of the authenticating process. Alfonso X's (2004) *Siete Partidas* strongly emphasized the authenticating nature of the escribano's hand and even provisioned for potential shifts in the chirographical makeup of the document, caused by the different inks and pens used ("mudamiento de la tinta o de la peñola") as well as individual escribano variations, including those occasioned by illness and age: "E otrosi se podria desuiar la forma dela letra por enfermedad o por vegez del escriuano. ca de vna manera escriue onbre quando es mançebo & sano & de otra quando es vieio & enfermo" (Partida 3, Title 18, Law 118).[14]

Document production necessitated modes of authentication, particularly as entries were noted in a regular fashion with the participation of many hands. Indications such as "están en letra del dotor de Madrid" served not only as authentication but also as a finding aid of the appropriate entry in municipal records (in *Libros de acuerdos* 1932–87, 1: 88). The fifteenth-century Catalan personal letters, many of them by *notaris*, published by Martorell (1926) contain references to the hand(s) involved in writing and delivering the text ("Escrit de mà mia" [Written in my hand] 86) ("Un memorial he rebut per mans de Lopiz, escrit de vostra mà" [I have received a memorandum from the hands of Lopiz that was written in your hand] 87). The material aspects of the letters are further marked by copying the address and the date of receipt following the text of the letter, and including references to any other documents or objects accompanying the letter; for example, some keys (60), a receipt (86), money (77–8), a bundle of letters ("plec de lletres intercluses en la present") (83–6), or books for binding (117–18). Other professionally produced documents such as post-mortem inventories of books list texts copied "in the deceased's own hand" (Iglesias 2002) and went so far as to note those in which the deceased's hand was one of several that produced a particular text (Iglesias 2002: 275). The emphasis on attesting that a document had been written, signed, or marked by the sender is also noticeable in other texts such as the challenge letters (*carteles*), which cross the line between the literary and the legal text.[15] These letters were typically signed in the challenger's own hand and bore his seal (Riquer 1963, 1: 13). In the famous exchange of carteles between Don Diego Fernández de Córdoba and Don Alonso de Aguilar (the elder brother of the *Gran Capitán*, Gonzalo Fernández de Córdoba) in 1470, the former sent a cartel signed in his own hand that denounced the waste of paper and ink that so delayed them in using their hands to fight instead: "E çesen los carteles que cosa vergonçosa es entre caualleros gastar tinta y papel sin venir a las manos" (text in Raulston 1993: 406).[16]

Royalty and members of the Church hierarchy were not the only ones sensitive to the advantages and responsibilities related to the production of personal documents. Writing one's own correspondence was a sign not only of authentication but also of the close intimacy that was expected from family and friends. Santillana's grandson, Íñigo López de Mendoza y Quiñones (1440–1515), second Count of Tendilla, first alcaide of the Alhambra, and captain general of the kingdom of Granada (also regidor of Guadalajara, as documented in 1485 in the city council cuadernos [López Villalba 1997: 286]),[17] was not only a cancionero poet but also the author of a large amount (over 5,500 pieces) of correspondence that has been preserved in four volumes. The registers are fair copies of the intense letter and document production by Tendilla. The registers are mostly in the hands of Tendilla's secretaries, namely, in the first decades, Íñigo López, later sent to court as the count's major-domo, and, for the later period, royal escribano Juan de Luz, both of them conversos, as well as a number of unidentified escribanos (Moreno Trujillo et al. 2007: 50–3). The registers show the stages of co-production of documents by Tendilla and his secretaries, as Tendilla corrected or added to the text copied.

All letters were officially authored by Tendilla, as they were by necessity signed by him. Moreno Trujillo has hypothesized that Tendilla dictated his letters to Luz (in the introduction to Moreno Trujillo et al. 2007: 53–7). This poses the well-known question of authorship in such situations, as Luz (or another secretary), whose hand is pervasive in the documents dated in the last decades of the count's activity, could conceivably have worded a good deal of the colourfully composed documents, some bordering on the literary. The loss of most of the single papers impedes the assessment of the degree to which Tendilla may have used written outlines or drafts of the documents for his secretary's use. However, the Archivo Histórico Nacional manuscripts point to the flurry of papers that accompanied the registers, as they contain, preserved between their leaves, loose papers in various shapes and sizes that correspond to drafts, notes, lists of documents, or original letters in the hand of Tendilla ready to be sent (Moreno Trujillo et al. 2007: 31–2, 69–70, and passim). This evidence is further enriched by references to Tendilla's notebooks or *cartapacios* (on which see chapter 5), where he wrote and vented his frustrations in the same manner as he does in many of his letters, pointing to a possible cross-referencing among various documents. The few letters in Tendilla's hand with more personalized wording and those that are mainly formulaic are at both ends of a spectrum that points to the gradated practices of authorship surmised from the consideration of the material production of texts. This kind of textual cooperation for the purposes of improvement is reminiscent of that evidenced in cancionero poetry, in which Tendilla was an active participant. The different shades of authorship that this modus operandi suggests need to be taken into serious account and deserve a more in-depth study. While Moreno Trujillo has called Tendilla's an "intellectual authorship" (in the introduction to Moreno Trujillo et al. 2007: 53), the evidence of the papers, registers, and active literary activity by the count suggests what could be termed rather an "authorship by degrees," a participatory authorship in partnership with his secretaries and escribanos. This is visually evident in some of the texts, written in successive sections by the count, Luz, and the register's escribano (see photo example in Moreno Trujillo et al. 2007: pl. 5). In the registers, the copyist specifically marks the documents that were written in the count's hand ("De su mano," e.g., fol. 19r). The number of letters that were indeed in the count's own twisted handwriting point to the chirographic expectations of the recipients. He offered profuse apologies when physical ailment prevented him from fulfilling his personal writerly duty, as in his letter to Micer Girónimo Vianelo: "Perdonad, magnífico miçer Girónimo, porque no va ésta de mi mano, que estoy malo de mi romadizo, y quedo vuestro como hermano" (1996, 1: 157).[18] This same ailment prevented him from writing for a few days, and further apologies appear in different letters, particularly to those who knew his handwriting well (1996, 1: 157–8). A sad disposition could have a similar adverse effect on his ability to write (e.g., 1996, 1: 302), while a stomach illness later prevented his writing to the king, as did old age (1996, 1: 247, 250, 251), until Tendilla must have become aware of it sounding too much like a bad excuse to warrant credibility and decided to shift his argument: "Yo, señor, no pido perdón a vuestra merçed porque ésta no va de mi mano porque lo que yo escrivo es todo hojarascas y quien quiera lo puede ver y tanbién porque me corre mucho de la cabeça" (1996, 1: 252).[19] Tendilla's handwriting was particularly gnarled and hard to read, a characteristic that seems to be shared by other noble or royal writers (Queen Isabel's contorted handwriting and the complaints it inspired are a famous example), and it appears to have been cultivated thus so to some degree by nobles looking to distance themselves from pen professionals (Cortés Alonso 1986: 6–7). Tendilla often uses his bad handwriting as a valid excuse for having someone else care for the material preparation of his letters: "Yo he dexado de escreuir y no de seruir, y la cavsa es porque hago mala letra y peno en haserla."[20] The physical nature of writing and its reliance on material support explains the constant references to *seeing* letters or documents in Tendilla's correspondence (e.g., 1996, 1: 243, 250); and letters are acknowledged as received and memoranda as *seen* (1996, 1: 83, 339). The material letter, as opposed to the oral message, allowed for repeated readings of the same text by the same person. Thus Tendilla asked one of his captains to reread a letter of his as an aid to help him remember Tendilla's instructions (1996, 1: 288). This use of writing to dispense orders led Tendilla to muse on the power vested in the mastery over paper and ink, which may at times, as he complains, be the only power he has in his administration: "pues no puede onbre proveer syno de papel y tinta" (1996, 1: 343). This is different from other cases in which the letter seems to function as a credential (*carta de creencia*) for a messenger, who is to relate the details orally. Both types of transmission could go together as options. One such case is that of two letters sent to Francisco Ortiz, his secretary at court and nephew of secretary and major-domo Íñigo López. One of the letters contains a long and potentially compromising narrative to give to the king's secretary, while the other bears just the *creençia* so that Ortiz could decide on what was appropriate to relate (in Moreno Trujillo et al. 2007: no. 277). The different hands in Tendilla's registers, mainly those belonging to his secretaries and escribanos, in addition to his own, display a rich process of textual cooperation. This process is further evinced by the paper trail that has been partially preserved, as well as the acute self-awareness of the materiality of writing present in the letters.

Textual Cooperation and Authorial Authentication

Tendilla's grandfather, Íñigo López de Mendoza, Marqués de Santillana (1988), had posted notice early on in his *Infierno de los enamorados*

(ID0028) to anyone finding fault or shortcomings with his work to take the matter and the pen in his own hand: "el que defectos fallare, / tome la pluma en la mano" (136). Similarly, Álvarez Gato's sent his piece (ID3123) to Jorge Manrique for "improvement": "A don Jorge manrrique Rogandole que fauoreçiese vna obra suya que le enbiaua a ver": "noble varon escojido / a quien sirue mi desseo / dad a mi tienpo perdido / fauor asy fauorido / que ponga afeyte a lo feo / y doliendos de mi daño / muy notable cauallero / engañad con tal engaño / que dores sobrell estaño / lo que no harie el platero."[21] Building on the well-worn humility topos, the calls of Álvarez Gato on Jorge Manrique to "adorn the ugly" much in the manner of a "silversmith gold-plating tin" plainly point to his request for material poetic intervention on the part of a worthier poet, along the lines of what Ezell described as "social authorship" for a later period for a print setting. Álvarez Gato's call to Jorge Manrique as a "noble" and "caballero" also point to the continuous link between poetic worth and social standing. His address to Manrique also makes apparent the nature of the text as collective work, one in continuous transformation by the textual community. This participatory poetics is based on the practice that treats the text as something not so much open, but rather in a continuous state of becoming. Participatory poetics is built upon the wilful cooperation of hands of various graphic abilities in the making of the text.

Tendilla's case is unusual in that, though much documentation has survived, the processes of document production he employed are so complex that they are difficult to reconstruct in full detail. Although the trail of the small scraps of paper used for first drafts was by definition highly evanescent, there is evidence that small pieces of paper were of similar service in other administrations such as that of King Pere IV el Cerimoniós in the fourteenth century (Gimeno Blay 2006: 108). As Tendilla's practices show, when preparing a fair, official, or presentation copy of his literary works, a wealthy nobleman might feel himself lacking the professional proficiency in penmanship or the inclination for long writing hours that was the mark of a good scribe. These were considerations that weighed in Don Juan Manuel's (1994) tight control of the production of a copy of his collected works in the fourteenth century ("fizi fazer este volumen ... que yo mesmo concerté"), as he explains in his *Conde Lucanor* (5–6). Similarly, in his letter accompanying a copy of his works ("copilaçión de mis obras") sent to the Count of Benavente, Gómez Manrique (2003) declared that the production had been entrusted to professional copyists and decorators ("ministrales") more skilled in writing ("escritura") and ornamenting ("ornamento") than he was as author ("conponedor") responsible for the arranged compilation ("ordenada" and "conpuesta") (108). Still, the importance of chirography is clear from the value placed on the author's own hand and in his authenticating signature in legal documents. In the document dated 15 January 1454, by which Santillana gives the town of Vallehermoso de las Sogas to his son don Pedro Laso de la Vega, Santillana's signature ("el marques"), with authenticating strokes, appears centred on the page with the rest of the signatures and text written around it framing it like a gloss (photo in Romero Tallafigo et al. 1995: 196–203, plate 44).[22] In contrast, intellectuals who were also involved in the professional production of documents and were therefore skilled with the pen could dispense with scribes and control the material production of their own writings. It was not unusual for early Spanish intellectuals, bureaucrats, and humanists such as Juan de Lucena and Alfonso de Palencia to produce fair copies of their own works.[23] The references to one's own hand present in fifteenth-century poems discussed in chapter 3 bear witness to the physical mechanics involved not only in the act of writing a poem, but also in the production of a fair copy by the same poet. A poem by Juan Boscán in MN6 (though of unidentified authorship in the manuscript) presents the poetic compilation that the poet sent to his lady at the beginning of their courtship, noting that the copy had been made in his own hand, "de mi mano trasladadas" (ID0228, MN6d, v. 5), the expression "de mi mano" bearing the same authenticating value that it did in legal documents. Sancho de Villegas's poem was likewise written "de mi mano": "De mi mano aquesta carta / te faze saber, sennora" (ID004, MN54, vv. 10–11). The lady might respond also in her own hand, and Diego de León quotes his lady writing to him with a reassuring "escripta fue de mi mano" (ID0579, MN54, v. 16). The apologies that poor handwriting entailed in fact conveyed, in its intended artlessness, the guarantee of authenticity and of unmediated personal communication, as implied by Fernando de la Torre: "nin tanpoco te marauilles por yr escripto esto de tan mala letra, por ser de mi mano" (text in Díez Garretas 1983: 201).[24] As in the case of Tendilla, these authors evince the tension between textual authentication and control on the one hand, and graphic ability and cooperation on the other. Scribal and authorial competencies grew increasingly intertwined as the material production of texts became central to both authorial consolidation and to bureaucratic needs. The exercise of graphic competencies conveyed a social meaning that had to be negotiated in light of the value attached to professional skill and to the worth embedded in the pursuit of letters. The need to exert control over the production of legal texts further provided a context for textual cooperation and for the negotiation of social salience. The role of hands in the production and dissemination of texts and the tension between authorial control and textual scattering continued well into the sixteenth and seventeenth centuries (Infantes de Miguel 1993; Rodríguez-Moñino 1968: esp. 26–34). The printing press gave authors one more venue in their attempts to control production and dissemination and, above all, ownership of their texts. These reasons were weighed by Juan del Encina and Pedro Manuel Jiménez de Urrea in their decision to put their cancioneros into print. In the introduction to one of his *Églogas*, Encina (1991) promised to issue a compilation of his works ("sacaría la copilación de todas sus obras") because they were being usurped and corrupted ("porque se las usurpavan y corrompían") and so that the whole extent of his work and knowledge would be known and he would no longer be typecast as a mere writer of shepherd plays (97). Urrea's explanation of his decision reveals more vividly the tension between the need for authorial control and the self-consciousness of publishing. Although self-publication could ensure authorial control, it also sent the text far from the author. Urrea's animadversion to publishing came from his unwillingness to stumble around taverns, kitchens, and in the hands of youngsters ("en bodegones y cozinas y en poder de rapazes") or to be subject to being judged by slanderers ("que me juzguen maldicientes") through his works (BNM Ms. 3763, fols. 12v–13r). Limiting access to culture via obscure language did not prove to be an effective enough method to control access to culture by the non-elite. Control of the means of production and reproduction was also necessary.[25] Whether or not it worked to the advantage of the author, textual motility, much like textual cooperation and chirography, actually functioned as a key factor in the tightening of the bonds among social networks. The shared touch and inscription of the paper put the material support of the text at the heart of such networks.

5 Papers Unite

The freewheeling existence that made loose papers active agents had the drawback of a potential evanescence. Papers led a precarious life because, with the exception of legal documents, they were often deemed to lack the monetary value of other household belongings. To this day, archives have relegated loose papers to sections labelled "useless papers" ("Papeles inútiles"). The instability of the material life of single papers was also due to their fragile nature as medium. The constant danger of loss, destruction, or defacement spurred the use of different strategies of preservation. In addition, the flexibility afforded by the single paper needed at times to be sacrificed for the sake of greater ease of use. Thus, specific modes of support were devised with the purpose of gathering papers together. The nature of the medium and the needs of the users inspired various modes of paper compilation. This chapter will examine some of the more relevant ones, including booklets, rolls, and envoltorios.[1]

Precarious Papers

The massive loss of documents and books can only be surmised, as may be seen in the reference to the large stock of copies of *Cárcel de amor* in the shop of a Seville bookseller: fifty printed copies of *Cárcel de amor* with woodcuts ("çinquenta libros de molde estoriados de Cárçel de amor") that a bookseller in Seville named Juan Lorenzo left in pawn in the hands of a creditor named Diego de la Fuente in 1498 (document extracted in Bono Huertas and Ungueti-Bono 1986: 97). Papers could be easily destroyed either by any of the hands that touched them or by willing action of the author, as in the declared case of poet Juan Rodríguez del Padrón when he burnt his, thus stated by the rubric of poem ID0775 (MN65): "Juan Rodriguez del Padron quando quemo sus papeles" [Juan Rodríguez del Padrón when he burnt his papers]. Similarly, the lady who exchanged letters with Fernando de la Torre assumed that her missives would end up in the fire after being read, instead of being published (text in Díez Garretas 1983: 137). Some of the preferred means of paper destruction appear to have been tearing and burning.[2] When Hernán Mejía de Jaén read an anonymous poem slandering women in an unspecified compilation, he wrote to the person he thought most likely to be the author, his friend Juan Álvarez Gato, admonishing him to destroy the infamous stanzas thus: "Rasguense quemense en brasa" [Let them be torn, burnt in hot coals] (ID3125, MH2, v. 94). An earlier poet, Alfonso Álvarez de Villasandino, wrote to the constable don Álvaro de Luna in what has come down to us as a perfectly isosyllabic poem, asking him to tear it ("rronped este original") if he found a breach in the syllable count: "E sy alguna vocal / aqui mengua en las vocales" (ID1330, vv. 9–11). In a letter written in Zaragoza on 4 December 1493, Queen Isabel asked her confessor and adviser fray Hernando de Talavera to avoid having her letters read by someone else, entrusting to him their custody or their destruction: "ruegoos questa mi carta y todas las otras que os e escripto, o las queméys o las tengáys en un cofre debaxo de vuestra llave, que persona nunca las vea, para volvérmelas a mí quando pluguiere a Dios que os vea" (in Clemencín 1821: 382).[3] She of course signed it "De mi mano" (382). Besides burning, a single paper or document could be ensured of destruction by simply not getting copied into a book or register, as practised by Tendilla (Moreno Trujillo et al. 2007: 46). As a slightly less perishable medium, a book could be destroyed less easily by tearing, but burning would of course remain, as is well known, a choice means of destruction of both books and people. In a similar treatment, books could, like people, be drowned instead, as stated later in the administrative journal kept by Antonio Gracián (1962), secretary to King Felipe II: "Di cuenta a su Majestad de lo que era el libro de Pedro Treviño y mandóme que le quemase; echéle atado en un canto al Tajo" (28).[4] When a single text inside a book or in freestanding format was the target of destruction, erasure was sought through tearing of the single paper or by crossing out the text in an otherwise acceptable book. This was the course of action mandated in regards to a prayer that Jews were suspected of reciting every day against the Christians and the Church (perhaps referring to the Birkat HaMinim, which was intended for heretics). Destruction of the prayer in any written format by tearing, discarding, or crossing out ("las rronpan tiren e chançellen") so that the text would be rendered illegible ("en manera que se non puedan leer") was repeated in Díaz de Montalvo's (1990) *Ordenanzas reales* (fol. 238v).

The obvious perils of the loose paper had strong implications for the establishment of both authorship and readership, because it was clear that the medium would have to be controlled in order to allow for the exercise of power over textual production, circulation, and ownership. The willing destruction of papers, along with the wear and tear that came from active use, meant the complete loss of a text unless it was moved to a more permanent material support.

Envoltorios

Less durable but in some ways more versatile than a book, the wrapper, folder, or bundle (envoltorio) was a widely used archival model. Papers were inserted between two pieces of hardy material, such as parchment, or sometimes just pressed paper, and the contents secured with a string tied around it. This folder provided an easily assembled conveyance when loose papers were compiled for circulation or safekeeping. Papers dealing with a common subject or being sent to one recipient were often grouped in the same bundle. The practice of sending envoltorios or bundles of letters and documents to one recipient who would then see to the distribution of the individual letters and documents to their individual addressees is fully illustrated in the correspondence of Íñigo López de Mendoza, Count of Tendilla.[5] The envoltorio was often used to store legal documents, such as those listed in the inventory of the archive of Mondragón, which clearly describes the envoltorio repository as two paper sheets ("dos pliegos de papel") covering the documents ("escripturas ... enbueltas en un enboltorio") and tied with string. The front cover of the bundle listed its contents, serving a similar paratextual purpose as that of a rubric (text in Crespo Rico et al. 1992: 143). The resemblance of the envoltorio format to that of the archival file folder or *legajo*, still used today, is clearly stated in Fernández de Oviedo's (1870) *Libro de la Cámara*, where they are used interchangeably when referring to marked bundles of loose papers referenced in the registers (39–40).

Serving not only as repository, the envoltorio was also intended to secure safe circulation of groups of heterogeneous papers, for example, Christopher Columbus's memoranda and accounts, as he himself states in the narrative of his voyage to Cuba and Jamaica: "como de todo ya escreví largo a V. Al., cuando yo partí para descubrir, y dexé el emboltorio en la

Ysavela, porque, si viniesen caravelas o algunos navíos de los que se esperavan y se despachasen antes que yo bolviese, porque V. Al. fuese de todo bien informado ... muchas cosas particulares las cuales heran neçesarias, de las cuales todas y de la instruçión enbié el treslado a V. Al. en el mesmo enboltorio" (Colón 1992: 290).[6] The inventories of his papers deposited for safe keeping in the monastery of the Cartuja in Seville also list many such envoltorios. The 1544 inventory lists, among many such "envoltorios," one with ten letters, with the paper in which they are wrapped rubricated ("Va rubricado el papel en que van enbueltas") (Gil 1989: 158; Serrano y Sanz 1930).

The intense bureaucracy put in place by the Inquisition also made use of the convenience of the envoltorio in its correspondence, which frequently included bundles with bulls, reports, and assorted letters, among other documents. A 1487 letter from King Fernando mentions one such envoltorio and its contents, among which were two letters, a papal bull for the establishment of the Inquisition in Barcelona, and some *processos* (in Torre y del Cerro 1950: 412). The administrative practices of poets when engaged in political transactions would prove useful when dealing with any piece of writing. In a letter to King Juan II discussing his plans to fortify the cities of Úbeda and Baeza, Santillana described the posting of documents on church doors at night by Rodrigo Manrique ("fizo poner sus cartas de noche en las puertas de las yglesias"). After receiving copies of the posted letters, Santillana replied and forwarded copies of all documents to the king along with an explanatory letter in an envoltorio ("los traslados de sus cartas et de la mia los quales a V. M. embio en este imboltorio") (text in Rubio García 1983: 195).

A versatile support such as the envoltorio proved useful in literary circles, and poet Fernando de la Torre instructed that his poetic card game (ID0594) be presented in such a folder: "El emboltorio de los naypes ha de ser en esta manera: Una piel de pargamino, del grandor de un pliego de papel, en el qual uaya escripto lo seguiente. E las espaldas del dicho emboltorio, de la color de las espaldas de los dichos naypes."[7] This wrapper/folder was also writing material, as the poet makes clear at the end of the stanzas when referring to the cover as envoltorio or memorandum (*memorial*) where instructions for the making of the card game were written (text in Díez Garretas 1983: 231). Any aptly sized piece of paper or parchment could be made into an envoltorio, as a way to salvage useless writings by turning them into covers. For this reason, Fernando de la Torre preemptively pleads with his female correspondent to read his writing before she even thinks of turning it into a cover or wrapper ("capa ni enboltorio") of worthier documents (text in Díez Garretas 1983: 189). In the same vein, he humbly described his *Libro de las veynte cartas e quistiones* as a coarse cover ("aforro grosero") or "scribbled wrapper" ("enboltorio borrado") to other writings that he forwarded to the lady (text in Díez Garretas 1983: 103).

Poet and royal secretary Juan Alfonso de Baena mentions the practice of grouping papers containing poems by various authors in such an "emboltorio" (in poem ID1584, vv. 17–24), pointing to a mode of material circulation of papers before their inclusion in the poetic compilation. Baena's poem exposes the materiality of the culture surrounding the written word and the traffic with ink, pen, and lines recompensed with the gold of the royal treasure, or perhaps just as pressingly, with a mule (as is the case in ID1582, ID1503, or ID1584). Baena's calls to put the poem in an "emboltero" or "emboltorio," together with his references to the value of the texts, reveal not only his archival preferences, but also the writing practices of secretaries and escribanos. In the poem, Baena requests the help of two fellow "escrivanos," further asking that his "lines" be put in a wrapper or folder ("mis renglones tales quales / ponedlos en enboltorio"[8] vv. 19–20) where his composition may not come last ("do non sea el postrimero," v. 21), and requests their help with pen and ink ("con la pluma o tintero," v. 29), begging that his work (which he characterizes as a free-standing "carta," v. 38) may not be the object of mockery. Baena sent poem ID1593 to Juan Carrillo along with an envoltorio ("emboltero," v. 5), with an artless gift ("vos embío un emboltero / con presente desdonado," vv. 5–6), which Baena hoped would encourage Carrillo to ask the prince for a cash gift for Baena, as the last verse makes clear ("en me dar de su dinero") [in giving me money].

Rolls

Baena's envoltorio summons up the etymological relation of the word to *liber involutus*, and to the *rotulus* or roll,[9] made since earlier centuries out of stitched strips of parchment or even paper. The relation between *rótulo* and envoltorio/emboltero is further suggested by other sources, such as the Alba Bible. In the brief lexicon preceding his translation of the Hebrew Bible in the *Biblia de Alba* (1920–22), Mosé Arragel explains that a "Rotulo. Es o enboltero o proçeso de libro, carta" (26). The Alba Bible further illustrates the close relation between rótulo in the sense of roll and the envoltorio format. In gloss 126 penned by Arragel as commentary to Reyes, libro IV, cap. XXII, the Torah is described as being wrapped or rolled ("enbuelta") in a scroll ("rotulo"): "vna atora enbuelta asy como oy la tienen los judios en rotulo" (822); and further, "el libro de la ley en rotulo estaua, estaua la enboltura del rotulo enbuelto a dos partes" [the book of the Law was in a scroll rolled into two parts] (844). The miniature, reproduced in Paz y Meliá's edition between pp. 822 and 823, shows the Bible open at the page in Deuteronomy (28, 36) cited in the passage, clearly being intended to portray a Bible. The iconography does not follow the clear statement of the Bible format. This is undoubtedly due to the fact that Arragel refused to execute the illuminations (10), which were charged to the master illuminators from Toledo ("maestros pintores" 15) hired by Arias de Encinas, the Franciscan friar who was to oversee Arragel's work, and instructed by him in writing as to the content of the illuminations, further guided by those found in the large illuminated Christian Bible that was supposed to serve as model (15). However, some of the illuminations, such as the menorah in fols. 77r, 88v, or 236v, that fall squarely in the tradition of Hebrew illuminated manuscripts, may have benefited from Arragel's input.[10] Similar uses of the term rótulo are found elsewhere. In 1484 a hamlet dweller brought a list of 300 men who were volunteering to support the establishment of the Aragonese Inquisition in a "rótulo or piece of paper" ("en hun rotulo o paper"). Given the context and the content of the document being a long list of 300 names, rótulo appears to refer to a long strip of paper that could be rolled up and perhaps posted (text in Sesma Muñoz 1987: 51).

The value of envoltorio and the related *proceso* as scrolls can be further inferred from their use in different translations of the Jewish Bible, themselves in codex form. Throughout the 1553 Ferrara Bible the word envoltorio is used as a translation of *megillah* or scroll. One meaningful example is that of the scene where God gives Ezekiel his book ("emboltorio de libro") and commands him to eat it, as told in chapter 3 of the Book of Ezekiel (*Biblia de Ferrara* 1996: 787).[11] In contrast, earlier translations of the Hebrew Bible render megillah as *proceso de libro* or *rollo de libro*,[12] whereas the Latin Vulgate has Ezekiel consume a *volumen*. The eleventh-century Bible of Sant Pere de Rodes (also known as Biblia de Rodas), a product of the monastic scriptorium of Ripoll, and currently at the Bibliothèque Nationale in Paris, bears an illumination of this passage, with God's hand introducing a scroll into Ezekiel's mouth (photo in Gudiol i Cunill 1955: plate 22). This may have provided the inspiration for the account of Saint Isidore of León forcefully feeding a small book with the Scriptures to a hesitant Saint Martin (San Martino) of León, as recounted in the latter's life (ch. 52, text in Viñayo González and Fernández González 1985: 25).

Although the surrounding evidence invites the understanding of Baena's envoltorio as an archival term referring to a bundle of documents protected by a cover and tied securely, the fact that texts in roll format continued to be produced under certain conditions should not be overlooked. In addition, any envoltorio could also be rolled for convenience, particularly, though not exclusively, when containing large or oddly sized documents that needed to be guarded from creasing, either because they were destined to be posted or otherwise displayed or simply to protect their legibility. Traditionally identified with pre-Christian writing, scrolls were not automatically superseded by the codex, as much evidence originating from codicological and visual art sources patently shows.[13] Although the Christian Bible came to be identified with the codex format, and the Jewish Bible preserved the scroll form,[14] different cultures and religions used either format as support for various texts. Some types of Jewish texts, such as the Haggadot, were written in codex form, as may be seen in manuscripts such as the beautiful fourteenth-century Barcelona Haggadah (British Library, Ms. Add. 14761), and the illuminations contained in them.[15] The categorization of the scroll as Jewish and the codex as Christian as the privileged format for the Bible is further questioned by the existence of Jewish Bibles in codex form, and the dichotomy of scroll versus codex and Jewish versus Christian Bible needs to be reviewed for this period in light of them. One relevant example is the stunning fifteenth-century Lisbon Bible (dated 1482) at the British Library (Ms. Or. 2626). The roll continued to provide a convenient support in Christian Spain for genealogical rolls such as the so-called *Rotlle de Poblet* (or Genealogia dels Reis d'Aragó), the well-known illuminated roll of the Crown of Aragon kept at the Monastery of Poblet. Other types of texts used a similarly large roll format for display purposes, as exemplified by foundational confraternity documents, such as the one pertaining to the establishment of the "Confradía de Santo Domingo," dated 1268 in Tárrega.[16] Devotional and religious rolls were also kept in private homes. In the last years of the fifteenth century, Elisabet, wife of March de Parets, denounced to the inquisitor that she had from her husband, among other items, a roll made out of eight thin pieces of parchment containing some writings from the Bible: "Un rotllo de vuit pergamins prims en que son scrites de ploma les generacions e altres coses sumades de la Biblia" (in Carreres Valls 1936: 110). This description recalls other rolls with similar content, such as the Genealogy of Christ that may have been the work of Peter of Poitiers (Newberry Library Ms. 22.1).[17] Other types of religious rolls include devotional rolls aimed at attaining indulgences.[18] Rolls such as Newberry

Library Ms. 122, containing Battista da Montefeltro's poem "O glorioso padre, almo doctore" praising Saint Jerome (Triggiano 1999), may have been intended for reading and display during a communal religious ceremony, in a use akin to that of liturgical rolls such as the Exultet Rolls of Milan and Southern Italy.[19] As Daly (1973) argued, the format of the liturgy rolls was chosen by the Church because of the prestige it conveyed for imperial documents in Rome, and for the power it communicated, both civil and religious (336). In addition, oversized documents such as papal bulls could be conveyed or stored at some point as rolls throughout different centuries,[20] even if they were later folded and archived as two-dimensional artefacts.[21] If the oversized document was of particular importance because of its being key to, for example, the property claims of a family, it could later be bound with the document folded in such a way that it could be open to full size for reading purposes. The importance of the document necessitated a fittingly precious binding. The Biblioteca Lázaro Galdiano in Madrid holds several such documents (e.g., Ms. 706, the Papeles de Sancho de Paredes, Registro 14426, M35,13; as well as Ms. 656). Royal archivist Pere Miquel Carbonell differentiated between rolled documents ("cartes canonades") and folded documents ("cartes planes quadrades") kept in different sacks in the royal archive in Barcelona (in Martínez Ferrando 1961: 88). A further erasure of their life as rolls comes from the practice of copying many large-sized documents into booklet form. The position of seals hanging at the bottom of many documents makes it all the more difficult to envision something other than a rolled document.[22]

Further, the roll format was used for chronicles (Fossier 1980–1), the listing of contents of early libraries, and for legal disputes, as may be seen in the roll held at Western Michigan University's Waldo Library (Ms. 142). It was additionally used for administrative purposes in the making of cartularies (Paden 2004–5), or the fifteenth-century Catalan *cabreo* held at the Cathedral Archive in Lleida (Ms. LP_00938),[23] and for accounting purposes as in the Navarrese *roldes* (e.g., the one studied by Carrasco Pérez 2000, measuring 127.2 × 30 cm), and in the well-known English exchequer and local pipe rolls, and also used in household accounting, or even for cookbooks, such as the Viandier de Sion's (photo in Flandrin and Lambert 1998: 9). *Rotuli* were the support used for the copies of donations or acts in Europe (Guyotjeannin 1989: 131). The roll was a convenient format for a potentially "never-ending book," to which more textual units could be stitched when creating a cartulary, when sharing news that travelled from one monastery to the next in earlier centuries. The early monastic roll functioned as a means to carry news and necrology notices from one community to another through addition of new parchment to the roll by each community. These rolls could also contain poems or other texts. This practice may actually be traced back to much earlier centuries, as Turner (1978) found that the vertical roll format was always used for documentary purposes in Greco-Roman antiquity and earlier (esp. 34–51). In Spain, the roll had a long history as a favoured support for legal texts. The twelfth-century *Fuero de Estella* and its thirteenth-century revision are both written on rolls (the first one, which is actually missing two of its parchment leaves, measures approximately 422 × 35 cm, while the latter is a longer 752 × 36 cm) (Lacarra de Miguel 1927: 404–5; Arribas Arranz 1948: 7). In spite of Church and royal regulations from the fourteenth century onward encouraging the substitution of the roll in favour of the cuaderno in legal documents, rolls in fact continued to be used during the fifteenth century in the compilation of legal processes (*proçesos*) (Arribas Arranz 1948: esp. 8–13).[24] Several documents belonging to Christopher Columbus, for instance, use this format. In the 1544 inventory of Columbus's documents are several folders with writings on long, narrow pieces of paper written vertically ("de escrituras escritas en papel a la larga" Gil 1989: 159). There are similarly shaped folders tied with string (Gil 1989: 160).[25] Covarrubias (1998) later attests to the continued use of long documents made by stitching folios together containing edicts and posted on the doors of churches, tribunals, and public places. The entry "carta" in his *Tesoro de la lengua* states: "Cartapel, la escritura larga, que junta pliego con pliego, y no buelve hoja, como los editos que se fixan a las puertas de las yglesias, tribunales y lugares públicos" (313),[26] a roll that joins sheet to sheet and end to end and does not turn as a page.

Tiras and *Procesos*

Escribanos were in charge of recording and producing copies of legal processes (proçesos), and such activity was regulated in legislation such as the ordinances for the escribanos of Seville issued by Queen Isabel in 1503 (Rodríguez Adrados 1988: 781–2; also Bono Huertas and Ungueti-Bono 1986: esp. 54). This is of particular interest when considering the omnipresent poetic legal battles (often referred to as proçeso or *processo*, with further references to *lo procesado*), found in cancioneros such as Baena's.

A "proceso," his very long poem (ID0285, MH1) that appears to have been intended as an introduction to MH1, is what Baena handed to the king ("resçibit en vuestra mano"). In these poetic processes, the king acted as judge in charge of the sentence or *sentencia,* an expected outcome identical to that of a legal process. In ID1549 (PN1), the king reviewed all documents involved in the legal proceso, including initial poem, responses, and replies ("todo lo proçesado ante d'esto, assí la reqüesta como las respuestas e replicaçiones"), and then deferred to Pedro López de Ayala (in rubric). Baena goes on to encourage Ayala in his role, asserting that he is knowledgeable in legal matters and extolling the virtues of good notaries in any well-ordered proceso (ID1549, vv. 9–16). The references to the proceso and its being anchored on materials such as quality white paper appears in many of the poems in PN1, as the king is asked to have the whole arranged in smooth, white Tuscan paper ("mandat ordenar lo que es proçesado / en blanco papel, broñido, toscano" ID1505, vv. 19–20). A short poem-document termed cartilla (v. 24), consisting of thirty-six lines, conveyed a formal petition to don Álvaro de Luna and Count don Fadrique to act as judges in a poetic dispute (ID1483). In contrast, when Baena asked King Juan II to take with his own hand ("resçibit en vuestra mano") a very long poem ("este escrito" ID0285, vv. 7–8), which comprised 200 stanzas in the MH1 version, it was presented as a "proçeso" (v. 12). The poem was addressed and submitted as an independent document, and referred to within the text of the poem as a revolving document ("al propósito de ençima" [to the purpose stated above], v. 1196; "mi proçeso bien deyuso" [my proceso well below], v. 1204). At the end Baena asks for a definitive sentence ("sentençia definitiva," v. 1580), crowned and written in gold letters (vv. 1583–4). The similarities with documentation prepared for legal processes of the period are staggering, as exemplified by the "proçeso de pleito" in a 1466 document from the town of Madrigal (complete text in Ser Quijano 1998: 259–60). The use of the term "lo proçesado" leaves open the possibility that it could refer to a compilation of papers geared for legal action and thus in roll format. The constant references to proceso, *pleito*, sentencia, notario, *escriptura*, carta, juez, secretario, escribano, *libramiento*, *albalá*, and other legal terms strewn through the poems in Baena's cancionero make evident the crossover between literary and legal texts and their formats.

Since Baena's poems and those of his opponents were in his words presented in an envoltorio, the possibility yet exists that the word may refer to a roll, or a rolled folder. Unlike earlier rolls, which were made out of parchment, fifteenth-century rolls containing the documents relating to legal proceedings were made out of paper sheets and gatherings (cuadernos) (Arribas Arranz 1948: esp. 6–7). The potential implications of poetic processes being read and written at least at first in roll form would be supported by mentions in the documentation of the period of *tiras* or strips, which were the textual unit by which Isabel's public scribes were paid, and which could be integrated (stitched or glued) into a larger legal roll.[27] In fact, there is evidence that already in the fourteenth century the *palmo* (approx. 20 cm) or handspan was used as the measure to calculate payment to scribes, pointing to the use of strips (tiras) in scribal work.[28] The constant reference in the ordinances to the "tiras de lo proçesado" in relation to legal testimonies makes clear the use of tiras in the compilation of legal processes, as well as their use as a basic medium by the escribanos. The physical features of the tira and the number of obligatorily dense lines were carefully regulated.[29] A tira was obtained from the folding of a full *pliego* (445 × 310 mm) lengthwise, resulting in two long and narrow pieces of paper of 445 × 155 mm each (Arribas Arranz 1948: 21). The extensive regulation on the tiras in respect to the prices that the escribanos were allowed to charge includes Díaz de Montalvo's (1990) *Ordenanzas reales* (fols. 43v–45v, and passim). The royal letter given by the Catholic queen and king to the city of Valladolid regulated the rates charged by public scribes, the tira being the unit used for payment to the escribanos (text in Arribas Arranz 1953: 22–8, esp. 23–4). The tira was the unit used for the copies of the *proçesos de escrivano* given to the implicated parties in a legal process after its "publication" ("e desque fuere fecha publicaçión e se diere traslado a la parte, de cada tira un maravedí") [after it is published and the copy given to the parties, payment for each tira is a maravedí]. This was stipulated in the law specifying the rights of the escribanos (Ley XIII) issued by Isabel and Fernando in Seville in 1491 in a cuaderno on alcabalas. The *Ley* goes on to further regulate the fees that may be charged by the escribanos using the tira as the unit (text in Ser Quijano 1998: 290). Petitions to the pope were also made in strips, which were stitched together to form a rotulus of group petitions (Linehan and Zutshi 2007: 999). These rotuli lost importance once the letter responding to the petition had been produced. The contents of the rolls survived in the registers known as *Registra Supplicationum* (Linehan and Zutshi 2007: esp. 998). Few such Spanish rolls are extant; they include the 1305 Barcelona roll housed at the Archivo de la Corona de Aragón, and the 1307 Madrid rotulus kept in the Archivo Histórico Nacional (Linehan and Zutshi 2007). The use of the tira as draft paper or as the medium for a loose *nota* may be seen in one extant strip attached to the bottom of fol. 3r in the notarial register of Dueñas preserved in the Archivo Ducal de Alba in Madrid. The strip was used to note an inventory taken on site and is written as a roll, the strip having to be turned over lengthwise in order for the back side to be read, rather than from right to left as a page.[30] Also in tira form are some of the fragments preserved from fifteenth-century Navarrese roldes (photos in Serrano Larráyoz and Velasco Garro 2010). The tira format was used for other accounting purposes, as exemplified by the alphabetical index measuring roughly 95 × 310 mm, cross-referencing the names of the clients in "court banker" Ochoa Pérez de Salinas's (1980) books (facsimile published with Pérez de Salinas's *Libro Mayor*). Merchant accounts such as Andreu Conill's as well as church accounts used a similar format for record keeping (examples of documents and photos of docs XX, dated 1420, and XXII, dated 1442, which measure 401 × 141 mm and 298 × 110 mm respectively, in Gimeno Blay 1985).

The Mutual Referentiality of Text and Image

Diplomatic evidence on the roll as material support is complemented by that supplied by visual art. This is a valuable source of information, given that relatively few rolls have survived[31] and can be used in combination with diplomatic and textual sources. In visual art, these rolls, at times identified as phylacteries, rótulos, or even banderoles or banners, often bear an unmistakably close visual link to extant parchment or paper rolls.[32] Images from manuscript illuminations, woodcuts, paintings, and sculptures help interpret textual references to the medium, adding their own layer of meaning to narratives of textual transactions. Depictions of scholars and professional scribes and secretaries, as well as private residences, show vividly the physical characteristics of the various types of material support used for writing.

The gathering process that would help transpose vertical and horizontal tiras as well as scrolls to codex form may be seen in the depiction of the compilatory work of Ramon de Caldes in the twelfth-century cartulary *Liber feudorum maior* in the Archivo de la Corona de Aragón (fol. 1) published by Miquel Rosell (1945). In this full-page illumination, Ramon de Caldes is depicted reviewing horizontal and vertical documents in the form of strips with King Alfons II of Aragon (el Cast or el Trobador), a noted troubadour. The documents are neatly stored in a space ostensibly intended for safekeeping purposes situated behind Caldes and the king and clearly visible between the two. They are represented extended in their full form in rectangular shape with writing laid out horizontally or vertically, depending on the document. To the right of Caldes, we see a scribe copying on a longer strip that hangs down from the desk. This may be intended to represent a first step in the compilation of the documents into the cartulary or it may be the document witnessing the king's charge to Caldes to compile the cartulary, perhaps the very copy kept at the Archivo de la Corona de Aragón. Other twelfth-century cartularies, such as the *Libro de los testamentos* of the Cathedral of Oviedo, present a similar depiction of individual documents in their illuminations.[33]

The roll as the visual representation of a single legal record appears elsewhere, as for example in the late twelfth-century cartulary known as *Libro de las Estampas* or *Testamentos de los Reyes de León*,[34] which contains copies of donations made to the cathedral by seven kings and one countess. A depiction of each donor appears at the beginning of the appropriate copy holding what is ostensibly intended to represent the original document issued by him or her portrayed as a roll with a pendant seal. On each document we can read the legal formula of the donation with the name of the donor (e.g., "Ego Rex Adefonsvs confirmo"). The illuminations portraying the act of donation through the instrumentum are intended to serve as both witness and authentication of the act.[35]

The lavishly illustrated *Vidal Mayor*, commissioned by King Jaume I of Aragon (el Conqueridor), is a compilation of Aragonese laws that the large illumination in fol. 1r portrays as the result of the labours of Vidal de Canellas, Bishop of Huesca (1236–52).[36] It is important to note that the numerous miniatures in this manuscript illustrate the laws described in the text and emphasize the legal processes followed in compliance with the laws. For this reason, there are a great number of legal documents portrayed, many of which are rolls.[37] These rolls are often seen on the laps (or other supporting surfaces) of dutiful secretaries, as documents are drawn up to record the sale of a house, donations, wills, testimonies, and other similar transactions. Rolls appear throughout the manuscript in the hands of individuals or their lawyers: taking them to court and preparing litigation or other types of legal documents such as wills, privileges, or contracts, with the judge and the lawyer being identified by their characteristic attire.[38] The rolls always have lines drawn on them so as to represent writing, the illuminations being too small to allow for actual text to be inscribed. A grotesque marginal figure that stands between the two columns of text on fol. 165r appears to represent a judge, albeit with red animal legs. He holds an unfurled roll with one hand while pointing to the text with the other, hinting at the relationship between the particular text and its material support, and therefore between roll and codex, the one serving as a source for the other. If the iconography of the roll is studied vis-à-vis that of other types of material support, such as the booklet and the codex in the illuminated manuscript, it appears quite clear that the roll often portrays the earliest form of written text in its metamorphosis into a cuaderno or booklet and/or book. This is particularly evident, for example, in the miniatures that show lawyers holding a roll containing a will, a witness account, or any other such preliminary document, along with the sentence in cuaderno form, where the judge has presumably incorporated the materials from the rolls presented to him. The illumination in fol. 222r, which opens the section on legal sentences, holds the representation of a booklet with the written sentence ("sententia") presented by the judge to the two litigants;[39] and fol. 184r holds the miniature of what may be a sale document in the form of a booklet read at the time of sale. Other equally telling examples are found in the section that deals with the damage to or loss of a legal document (Book 3, fols. 125r–126r). Finally, a book (the Gospels) held open by the judge or sometimes by another figure appears often in the miniatures as the object on which the witnesses or parts implicated in the legal process swear their testimony by touching the book with the hand (e.g., fols. 92v, 102v, 106v, 112r, 125r, 126v, 152v, 174r, 207v).

In the manuscript copy of the Consulate of the Sea (Consolat del Mar) in the Archivo Municipal of Valencia, fol. 15 bears an illumination representing the king hearing the petitions of citizens and caballeros, some of whom are shown holding unrolled sheets on which one can read the beginning lines of the individual petition to the king (photo in Bohigas 1960–7, 3: 13–14, plate 142). Similar iconography appears in a representation of the female dispensation of justice: the panel with the Enthroned Virgin and Child, possibly part of an altarpiece from the church of Santa María de Monzón in Barcelona, executed by Bernat Martorell in 1437, now in the Philadelphia Museum of Art.[40] The Child is sitting on Mary's lap while both hold a long roll inscribed lengthwise with what is meant to appear as writing, while the female incarnations of the four cardinal virtues surround them. At Mary's feet, on both sides of the panel, appear two female figures holding rolls also inscribed with writing but this time to be read across the width of the roll. Since the altarpiece belonged to the church where the Catalan parliament, presided over by Queen María, sat at the time, it seems obvious that the connection between Mary and María is underscored and that the rolls point to legal documentation. Like Mary, María could not review legal cases and dispense justice without the aid of the cardinal virtues.

The exemplary nature of the depictions of Mary underscores their acute referentiality to the material life of the period, as is clearly the case with depictions of the Annunciation. The iconography of the Annunciation typically shows Gabriel holding a roll while Mary holds another. Depending on the rendition, the corresponding words of the Annunciation may be written on the roll(s). In the Book of Hours that belonged to don Alonso de Zuñiga, Gabriel hands Mary such an inscribed roll (Domínguez Bordona 1969: plate 148). There are similarities between this and the iconography of the story of the handing of the Mandylion and Christ's letter to King Abgar. In the altarpiece from the convent of Santa Clara, today in the Museu Episcopal in Vic, the letter is portrayed as a fully written, unfurled scroll that looks much more like a horizontally inscribed tira than a banderole.[41] Depictions of the Crucifixion such as that found in Queen Isabel's Book of Hours now in the Royal Chapel in Granada often present a figure with a realistic roll representing the (legal) written record of the wondrous event.[42] In the miniature depicting the Crucifixion in the Missal of Bertran de Casals, the centurion points to an unfurled roll that declares Jesus to be the Son of God (photo in Planas 1998: colour plate 1). A similar depiction of the centurion (identified by an inscription) with a roll pronouncing Jesus to be the Son of God is found in a crucifixion mural in the old Cathedral in Lleida (ca 1350) (Gudiol and Alcolea i Blanch 1986: 216, plate 8). Legal manuscripts such as the Decretals of Nicolò di Giacomo in Milan's Biblioteca Ambrosiana also illustrate various forms of material support, including rolls, large pieces of parchment or paper unfolded for reading, and books (photo in D'Ancona and Aeschlimann 1949: plate between pp. 158 and 159).

The different choices of material support available for the execution of texts are regularly depicted in visual art. A catalogue of sorts of the available options in the fifteenth century is displayed in portrayals of the four Evangelists. The different writing surfaces and furniture pictured further exemplify the repertoire of writing resources at the disposal of each writer. Supported by lap, flat board, inclined surface, or by desk with lectern, the writer is seen at work on a horizontal scroll, a vertical roll, a single leaf, a booklet, a quire, or a book.[43] An example may be found in the illuminations of the Evangelists at the beginning of the section of the Gospels on fol. 32r of King Alfons V of Aragon's Book of Hours (Villalba Dávalos 1964: 109–10, and plates 47, 48, 49, 50). There is a similar depiction of varied writing instruments and material supports in the illumination portraying the Evangelists in the 1409 Missal housed in the Biblioteca Nacional in Madrid (fol. 154v) (description and plate in López de Toro 1964: 36, and plate 26). But these are not the only religious figures depicted in the act of writing. Zachary and John appear in a ca 1365 altarpiece writing with pen and inkwell on rolls (one lengthwise and the other along its width) complete with red initials, in Joan de Tarragona's *Retablo de los santos juanes* (photo in Gudiol and Alcolea i Blanch 1986: 230, pl. 24). The material referentiality of the roll is evidenced by manuscript illumination such as that of the *Llevador del plat dels pobres vergonyants* from the church of Santa María del Mar in Barcelona, now supposedly lost but known through a reproduction in the Arxiu Mas. In this miniature, the infant Jesus, seated on Mary's lap, is found, pen in hand, in the act of writing on a roll-phylactery that is draped horizontally on his own lap and whose end flies off to a side like so many other phylacteries, while an angel holds his inkwell.[44] Also illustrative of the many different kinds of material support, as well as venues, used in the period are the illuminations that depict a private study, such as that of Boethius in the manuscripts of his *De Consolatione Philosophiae*, e.g., Harley Ms. 4335, fol. 1, where the author is writing in a kind of scriptorium in one depiction and reading in bed in another when Lady Philosophy appears to him, almost in a kind of inverted Annunciation scene (photo in Kren 1983: 158). In an illumination of Boethius's book in Paris, Bibliothèque Nationale, Ms. lat. 6643, fol. 1, the author's private study is brimming over not only with books, but also with rolled strips of paper or parchment that hang out of shelves and desk (photo in Kren 1983: 161). Petrarch is portrayed sitting at his desk writing on a blank book, with the verso of the left page finished and just the first line written on the recto of the right page, in fol. 6 of the fifteenth-century manuscript Aix-en-Provence, Bibl. mun. 1800.[45] A fifteenth-century manuscript of the encyclopedic *Visión deleytable* by Alfonso de la Torre (Bibliothèque Nationale, Paris, Ms. Esp. 39) bears an illumination showing the author writing on the left side of what appears as a blank bound book, with the right side and the space below the author's pen left blank, and the page only marked by the rulings (plate in Domínguez Bordona 1929: 138–9) (see chapter 6 for the use of blank books). Indeed, we may notice depictions of similar types of material support in paintings such as Domenico Ghirlandaio's well-known *Saint Jerome in His Study* (1480, Ognissanti, Florence). Here Jerome appears at work with a folded piece of paper in his hand and others hanging from the side of his table and the top of his shelf. The representations of Jerome provide some of the more illustrative depictions of men's private studies, much as the scenes of the Annunciation do for women. The fifteenth-century painting of Jerome in his study belonging to the Valencian school (Niccolò Antonio, Colantonio, Museo Nazionale di Capodimonte, Naples) depicts the scholar with sundry papers, and books strewn anarchically through the study, papers slipped behind a thin metal bar affixed to the side of the desk, rolls, pieces of papers hanging from various locations, strips of paper tucked inside books, or papers still partially folded – possibly letters or bills – tacked on the side of the desk or the edge of the shelf, as well as many objects.[46] Jaume Ferrer II provides a similar depiction of Jerome in the altarpiece dedicated to Jerome, Martin, and Sebastian (Museu Nacional d'Art de Catalunya, Barcelona). The portrait of Jean Miélot, scribe and translator to the Dukes of Burgundy, which appears in a manuscript copied around 1450 by Miélot himself for the library of Philip the Good (Brussels, Royal Library, MS 9278, fol. 10r), is a well-known miniature displayed on the cover of Christopher de Hamel's *Scribes and Illuminators*. In it the author can be seen sitting at his desk with his bed behind him and a fireplace in front of him, suggesting that he is at work in his private chamber; at his side is an open chest containing books and written paper in small roll format. Equally telling are fifteenth-century depictions of the Annunciation that portray Mary as an active reader and writer often seated at her desk with scissors, pen, penknife, ink, books, and unfurling pieces of paper around her. She can be seen thus in a panel of the altarpiece in the Paeria of Lleida (Jaume Ferrer II and collaborators, Retaule de la Paeria, Paeria de Lleida, Spain) (on which see Puig 2005). The bed in the background, half disclosed behind a curtain, also suggests the female scholar working in the privacy of her room. Authorial composition emerges from fifteenth-century texts, images, and material culture as an activity located in the solitude of the (bed)chamber. These visual representations of the author present a picture of the material processes of writing that differs greatly from that of the dictating master of the scriptorium and point to the changing circumstances surrounding the multiplication of the hands involved in textual production.[47]

All these depictions clearly question generalizing interpretations of the roll/scroll as an idealized representation of a mere "speech balloon," even if it may at times carry those connotations. Rather, they point to the varied material supports that professional (and amateur) scribes could use, preserving a testimony of the medium that time has largely erased. By further providing a narrative contextualization of the medium, images encourage a cross-examination of the texts themselves.

Poetic Rolls

Extant literary artefacts such as the fragments of thirteenth-century German rolls studied by Rouse and Rouse (1991), or the French *Song of the Barons* with *Interludium de Clerico et Puella*, the St Quentin roll studied by Branner (1967), or the Lambeth Palace rolls studied by Wallensköld (1917) point to the roll as support for poetic texts.[48] The clear referentiality of iconographic representations of documentary artefacts begs, as Rouse and Rouse (1991) have cogently pointed out, for a reformulation of the interpretation of the rolls that unfurl in manuscript illuminations, paintings, sculptures, and architectural pieces. The use of the so-called rótulos, rotuli, or *rolos* in earlier Provençal or Galician-Portuguese poetry is unclear and controversial, their presence being harder to document. Sirat (2006) observes that the rotulus form has been continuously used through the centuries, but because few of them have survived their importance has not been given due notice (278). In spite of the early assessment of the Vindel parchment, containing some poems and music by the late thirteenth-century poet Martin Codax, as a roll, it is not; rather, it is a large piece of parchment folded in two with writing exclusively on the inside, a "volant leaf," "folha volante," or "hoja volante."[49] The use of single leaves in the circulation of poetic texts has been well established by Avalle (1993: 61–2). However, there are eight references to "rolos" or to "rotulus" in the rubrics of the *Cancioneiro da Biblioteca Nacional* (B) to indicate that the poems were copied in roll form before being transferred to the cancioneiro, the term *folhas* referring to book leaves.[50]

Rolls do appear in the stunning miniature vignettes (six for each poem) of the thirteenth-century *Cantigas de Santa María* codices of King Alfonso X. In Cantiga 3, Teófilo abjures his faith via a signed document ("carta") carried by a devil (Biblioteca del Real Monasterio de San Lorenzo del Escorial, Ms. T.I.1 [Escorial], fol. 8r). The carta is represented in its legally binding format, recognizable as such, as a roll, which appears in various degrees of unfurling in the different vignettes. The transfer of poetic texts from one medium to another is illustrated in Cantiga 56 in the same manuscript (Escorial, fol. 83r). This process is described as an *ordinatio* and *compilatio* of sorts. The first six lines of the third stanza read: "Este sabia leer / pouco, com' o'y contar, / mas sabia ben querer / a Virgen que non á par; / e poren foi compõer / cinque salmos e juntar."[51] As the cantiga relates, a monk set out to choose five psalms in honour of Mary. In the second vignette, he is seated at his desk, with an *armarium* in the background, copying the psalms on what is obviously a roll, whereas the third vignette shows the same monk at Mary's feet reading his choice of psalms to her now in the form of a booklet, in what iconographically also resembles a book presentation ceremony, the official publication of the work. Cantiga 307 ("Toller pod'a Madre de Nostro Sennor") shows a man writing his poem ("cantar") on a long, narrow scroll, while the following vignette portrays him holding the roll in front of Mary and singing in the presence of a crowd (Florence, Biblioteca Nazionale Centrale, Banco Rari 20 [Florence], fol. 101r) (Keller and Cash 1998: plate 12), coinciding with the narrative of the text of the cantiga. The cantiga's narrative and miniatures capture the entire process of poetic composition. First, the patron suggests the creation of the work and its topic. In this case the patroness is Mary and the topic is herself and her praises. The author sits down to a professional-looking writing desk, complete with a steep writing board and two horn inkwells inserted on its side. While the right hand holds the pen with an elevated wrist, the left hand supports the pensive head of the writer. The narrative explains that after the task of writing had been confided to the good man ("bon ome"), he set out to write the lyrics, and only after that did he compose music that would suit such lyrics ("E segund' as paravlas lle fez o son" v. 45). The iconography of the scene also reminds one forceably of a book presentation ceremony, though in this case a single poem. In Cantiga 210 (Florence, fol. 120v) ("Muito, Foi Noss' Amigo"), the archangel Gabriel holds a vertical roll in his hand, though a more complex iconography is that of the illumination accompanying Cantiga 80 (Escorial, fol. 104r). As the rubric announces, this cantiga is basically a versified version of Gabriel's salutation to Mary, the Hail Mary prayer ("Esta é de loor de Santa Maria, / de como a saudou o angeo"). However, in what appears to be an iconographic merge of two referential values of the roll, the first and fifth vignettes show the king himself holding a long roll while kneeling in front of Mary, perhaps claiming thus the role of Gabriel as well as that of the devout poet, additionally signifying the value of the poem as prayer. Further depictions of rolls exist in cantigas miniatures, where the roll is the writing support and/or the means for the dissemination or presentation of the poem to Mary, and is used when writing not only on desks but also in a chamber or even in prison (Cantigas 29, 202, 291, 307, 316: where it is the support for a satirical poem, 363: where it is the support for slanderous poetry; and Escorial, fols. 194r, 212r; Florence, fols. 24v, 25r, 40r, 45v, 77v, 109r).

There are many representations of the king and his team writing on rolls in the illuminations of Alfonso X's works. At the beginning of the Escorial manuscript (fol. 4v), the illumination shows the king imparting instruction on how to write poetry (cantigas). Both the rubric and the first stanza of the prologue state as much: "Este é o prólogo das Cantigas de Santa Maria, ementando as cousas que há mester eno trobar." "Porque trobar é cousa em que jaz / entendimento, porém que-no faz / há-o d' haver, e de razom assaz, / per que entenda e sábia dizer / o que entend' e de dizer lhe praz, / ca bem trobar assi s' há de fazer."[52] In the illumination, Alfonso is pointing with one hand to the vertical roll that he is showing with his other hand to six figures (three on each side), who sit by him holding their own unfurled vertical rolls in their hands, some of which spill outside of the square frame of the illumination. The roll shown by the king has the first lines of the prologue, stopping at the end of the roll because it is not long enough to hold the whole text. Represented thus, the roll held up by the king seems to show the poem in its original support. Similarly, in the illumination heading Cantiga 1 in the Escorial manuscript (fol. 5r), the king appears holding a book open with one hand, while with the other he points towards the scribe, who is seated at his feet writing on a roll supported on his knees. On the left, another scribe is shown at work on a similar roll, this one containing music. Behind each scribe a matching group of professionals is shown at work, perhaps depicting Alfonso's team and giving a visual clue about the modus operandi of Alfonso's scriptorium. A group of scholars hold open a book while standing behind the scribe with the roll containing the text. A group of musicians playing their instruments stand behind the scribe, who is writing the music. Alfonso's miniatures show the currency of the use of the roll as support of legal, religious, and poetic texts, while at the same time depicting the transfer of roll to booklet and book. Further miniatures exemplify Alfonso's keen interest in visually documenting acts of authorship. In the first miniature in Alfonso's *Libros de ajedrez, dados y tablas* (1283) (Escorial, Ms. T.I.6, fol. 65r) the king appears imparting his wisdom to a scribe who is sitting at his feet with crossed legs, which support what appears to be a single piece of paper or parchment. The hand of the king, iconographically nearly identical to the maniculae that aid active reading, points to the parchment on which the scribe notes the king's degree of involvement in authorship: "Aqueste libro que mando fazer el rey Alfonso" [This book that King Alfonso ordered to be made].[53] Similar depictions of kingly/scribal action include that found in the *Llibre Verd de Barcelona*. King Jaume II of Aragon (el Just) points an index finger to a horizontal roll supported on the knee of the scribe, who appears to be writing on it. The king is dictating a letter and is depicted in the act of speaking, his hand representing his command and his authorship of the text, which the scribe is merely copying (photo in Yarza Luaces 2004: 287, plate 23). The roll is also representative of the whole work in author portraits. Such is the case in a twelfth-century manuscript at the Biblioteca Laurenziana in Florence (Ms. Plut. 82), where atop a miniature depicting the presentation scene of a book is a human figure perched on a tree holding a roll that contains the name of the author, Petrus de Slagosia, indicated by a "fecit" (photo in D'Ancona and Aeschlimann plate CVII).

Although the majority of fifteenth-century manuscript cancioneros are not as lavishly illuminated as Alfonso's codices, one such source is the *Cancionero de Pedro Marcuello* (CH1), which offers much evidence for our topic.[54] The manuscript is filled with miniatures depicting the author and his daughter Isabel, for whom Marcuello was trying to gain a position in court. Throughout the manuscript, author and daughter appear holding rolls with verses corresponding to poems on the facing page. These rolls, obviously tangible, are often held in the hand like any other object. For instance, on fol. 36v Marcuello and his daughter hold a roll inscribed with poetry with one hand while with the other they hold up a bunch of fennel, an herb that symbolizes the union of the Catholic monarchs because of its spelling (with Isabel's "y" in Castile and with Fernando's "f" in Aragon: "Ynojo," "Finojo"), as the poem explains. In the miniature on fol. 45v, Marcuello and his daughter are portrayed, as they are elsewhere in this manuscript, outside of the main scene, praying, with a roll unfurling from their hands. On fol. 69v, the roll with the first line of a poem peeps out of the sleeve of Marcuello's daughter, calling to mind the practice of tucking notes in one's sleeve, as practised by Queen Isabel herself (see chapter 7). Young Isabel and Saints Peter and Andrew exchange verses via rolls stemming from or grasped by their hands in the miniature on fol. 114v. This depiction invites further conjectures on the intended uses of cancioneros as prayer books. In this regard, the roll appears linked to religious poetry, including versified versions of basic prayers. We see rolls held in young Isabel's hand, as she prays in front of the appropriate saint or religious image, on fols. 98v, 101v, 102v, 103v, 104v, 105v, 106v, 107v, 108v, 109v, 110v, 111v, 112v, 114v. Conversely, the saints in these miniatures also hold rolls, and their index finger is usually extended, suggesting their role as teachers. There is an intriguing use of the rolls in the double depiction of Santa Fe on fols. 52v and 53v, the rolls being held by the four Evangelists that surround the scene of Marcuello's kneeling daughter and the Santa Fe. On fol. 52v, each one of the four symbols of the Evangelists holds a roll with the corresponding name. However, on fol. 53v the tetramorph's rolls are inscribed with the four lines (one per roll) of the poem written on that page, which responds to young Isabel's prayer on the previous page that they aid and protect her queen and king. This appears to be a daring association of poem and Gospel for the benefit of the royal persons, thus contributing to their messianic portrayals current during the period. In most cases, the rolls in the illuminations, which bear the first line or lines of the facing poem, directly refer the reader to that poem, thus creating a visual association of roll and book, the one literally inscribed in the other. The transfer from roll to book would thus be suggested through this interplay of miniature and text.

Another significant instance of poetry copied (and posted) on rolls is that which is portrayed in the altarpiece executed by Jorge Inglés, the

Retablo de los Ángeles commissioned by Íñigo López de Mendoza, Marqués de Santillana, for the chapel of the Hospital de San Salvador in the town of Buitrago built by Santillana.⁵⁵ Because it is an altarpiece, the question is open as to the possible uses of the text for liturgical or other purposes, much like the existing liturgical rolls. Two panels of this altarpiece, the middle section of the retable, are well known for their widely disseminated portraits of Santillana and his wife, Catalina Suárez de Figueroa, kneeling on either side of the crowned sculpture of Mary.⁵⁶ More interesting for our purposes is the top section of the altarpiece, two large panels with twelve angels, arranged symmetrically with six on each panel. Each of the angels holds an unfurled roll containing twelve stanzas (one per roll) of Santillana's poem on the twelve (not seven) Joys of the Virgin. This is Santillana's "Gózate, gozosa Madre" (ID0322, included in MH1, MN8, MT1, SA8, 11CG, 14CG), which has thirteen stanzas. The roll with the last stanza appears in the panel bearing Santillana's kneeling portrait; affixed to the wall just behind the poet, it is clearly visible on the same level as Santillana's face. There is no roll or any other type of writing in the opposing panel with the kneeling image of Catalina Suárez de Figueroa. The rolls are intended to look like parchment or quality paper rolls written in a gothic hand, complete with red initials, that is commonly used in the codices of the period (though in SA8 only the main initial is red and blue, while the other ones are fancy but uncoloured).⁵⁷ It is no coincidence that this last stanza appears in the painting very close to the poet. While the first twelve stanzas all begin with the invitation to rejoice ("Gózate"), proper for a poem dedicated to the joys of Mary, and are written in the second person singular, through which the voice of the poet addresses Mary, the thirteenth stanza alludes to the twelve joys in sum ("Por los quales gozos doze") and shifts the focus to the poet. In this last stanza the preceding emphasis on the joys gives way to the poet's appeal to Mary. When asking Mary's help to attain the joys reserved for those blessed with heaven, Santillana introduces the only verb in the first person singular form, the appeal of the first person subjunctive ("que goze"), which redirects the subject of the poem to the person of the author. Thus, Mary's "gózate" yields to the poet's own joy ("que goze"), and a poem of praise turns into one of petition. Santillana's portrait next to those words leaves no doubt as to the identity of the recipient of such entreaties for joy or the subject of the subjunctive "goze." SA8, which is directly related to Santillana's circle, also places the thirteenth stanza in a privileged position. While the other twelve stanzas are copied three to a page (fols. 189r–190v), the thirteenth heads a fresh page (fol. 191r) and is separated from the first stanza of the next composition by a wide space in the centre of the page and the red rubric that introduces the next poem, thus sharing the page with only one other stanza. The painting helps make clear the close relationship between author and medium and the referential value of a support that stood for the authorial self.

The representation of the material support of texts in different art forms draws attention not only to their visual referentiality but also to the tactile nature of the texts. The trompe l'oeil in Jorge Inglés' altarpiece and its inter-medium relation to Santillana's manuscript cancionero further evokes the material grounding of poetry and the complex dynamic of medium transfer. As the illuminations in the manuscripts of Alfonso X's also make clear, this relationship is far more complex than that of poem to book. Instead, they imply a physical notion of poetic composition as well as modes of transfer from one medium to another that include the passage from roll to booklet and to codex.

Cuadernos

As the roll format continued to thrive, the "leaf" medium had also been used to full advantage through slate, lead, wax, ink, and other types of writing tablets.[58] Multi-leaf support in the shape of cuadernos, *qüerns/coerns*, or *cadernos* makes a constant appearance in inventories. In these pages, the term "booklet" is used as a translation of the terminology found in medieval Spanish (cuaderno, *quaderno*), Catalan (*qüern, coern*), and Portuguese (*caderno*) at the time. The complex nature of the book models examined here defies a rigid distinction between quires or gatherings, understood as a codicological unit intended to join similar units as part of a book, and booklets that have a more independent nature. The Spanish term "cuaderno," like the Catalan "qüern" and the Portuguese "caderno," may also mean "quire." When appearing as a bundle of booklets or quires, sometimes tied together, these terms differentiate bound and unbound books, as is the case, for example, of the *Crónica de España* owned by King Duarte (Nascimento 1993: 285). Similarly, the inventory taken after Queen Isabel's death differentiates among material supports such as book, cuaderno, and cuadernos, listing for example a "quaderno escripto de mano en papel"[59] and loose booklets tied together with the title written on top: "Vnos cuadernos atados con vna cuerda de cañamo que tienen escripto ençima. Revelaçiones de Sta. Brigida."[60] The post-mortem inventories detailing the contents of the chests with the possessions of Queen Isabel painstakingly record other identificatory traits in the material aspects of books and documents. This precision was needed in order to differentiate specific copies of a given text. The binding of the books was always specified as to the materials and colour, even if they were just bound in parchment,[61] and more particularly if they were made out of precious metals or stones.[62] Recording the type of material support was so important when drawing up inventories that these also specified details such as whether a letter was written on a booklet or the number of cuadernos that made up a book that had become unbound (e.g., Torre y del Cerro 1968: 34). A similar distinction among books, cadernos, and other forms of material support or formats can be found in the will of Queen Leonor of Portugal (Cepeda 1987). Works in booklets are customarily mentioned in inventories (e.g., in Madurell i Marimón 1974: 65, 66). The chest with the belongings of Princess Isabel-Queen of Portugal, Queen Isabel's daughter, which remained in Spain after the former's death, held, for example: "Catorze quadernos de papel de quarto de pliego escriptos en papel de molde ques CONFISIONAL del Tostado" (Torre y del Cerro 1968: 380).[63] The 1423 partial inventory of the books left by merchant Guillem de Cabanyelles included, among others, two booklets with Solomon's Proverbs: "Item, .II. coerns de paper en què eren scrits los Proverbis de Salamó" (in Madurell i Marimón 1974: 46). The loss of a cuaderno that was part of a book resulted in the sometimes irreparable loss of text, as thirteenth-century poet and notary Gonzalo de Berceo (1992) famously complained in his *Vida de Santo Domingo de Silos*. He could not finish his story because the book from which he took it ended; a cuaderno was missing, which, he hastened to say, was not his fault ("perdióse un quaderno, mas non por culpa mía"); and venturing an ending would be a foolish thing ("escrivir a ventura serié grande folía") (stanza 751). In a similar vein, a cuaderno was the measure and the unit used to assess progress in the copying of a codex, and the unit of payment for such a job. A 1432 document details payment for a manuscript done by commission, which specifies that the finished work consisted of thirteen and a half cuadernos, each of eight leaves. Payment for each cuaderno was one hundred maravedís (in Zarco del Valle 1870: 281–2). A copyist was often contracted to work on booklets, which were provided for him, as in the production of a copy of a *Vita Christi* or *Cartojá*, with the stipulation that the copyist should hand in each booklet as it was finished (text of the contract in Sanchis y Sivera 1930: 37–8). Likewise, royal chronicler Alonso de Santa Cruz 1951 refers to the process of writing draft versions of historical accounts in cuadernos (1: 20). Similarly, booksellers measured the length of the blank books they sold in cuadernos or even "hands" (manos, which amounted to about five *cuadernillos*).[64] The booklet's flexibility and variety of functions show an overlap with the well-known university *pecia* (also a cuaderno).[65] Printers would sell works in booklets to be combined and later bound, or in ready-bound books. In addition, as a means of authentication and a measure to secure their integrity, many early printed books often state the number of quires and the number of leaves per quire that make up that particular book. For example, statements such as the one found at the end of Fernández de Santaella's (1992) *Vocabulario eclesiástico,* where cuaderno is used in its original sense of *quaternion,* are common: "Tabla delos quadernos que estan en este libro. a. b. c. d. e. femenino género h. i. k. l. m. n. o. p. q. r. s. t. u. x. y. z. y. todos son quadernos excepto .y. que es quinterno" [Table of the cuadernos in this book. a.b.c.d.e. feminine gender h.i.k.l.m.n.o.p.q.r.s.t.u.x.y.z. and they are all cuadernos, except for y, which is a quintern].

Cuadernos facilitated the recording of several texts or entries of various lengths during a certain period of time. This also allowed for the constitution of a "book in progress," its units compounding meaning when joined together in the ultimate archive that was the bound book. Although many never reached the binder's hands, cuadernos may have been intended for later binding in most cases, however precariously. During the time of its independent existence, however, the cuaderno was a highly portable and flexible repository of sundry texts. One such example is the account book of Pedro de Toledo, Bishop of Málaga, converso and Queen Isabel's almoner,[66] known as the *Libro del limosnero,* now in the Instituto de Valencia de don Juan. The manuscript is made up of four booklets, which show different watermarks and seem to have been separately written by at least three different hands. It was later bound, leaving some pages blank in order to accommodate later additions. A similar composition is found in the "Formulario de cartas y mercedes del reinado de don Juan el II," a compilation of letters and other documents issued during the reign of Juan II that was most likely ordered by his son Enrique IV to provide model documents for use in his own administration. This formulary, held in the Biblioteca de Palacio in Madrid, is composed of booklets and single sheets that were added as needed, with the whole bound together later. Several hands were involved in the copy. Some of the rubrics were copied consecutively, while others were written by a different hand, perhaps as the documents were identified and used in the chancery.[67] A similar formulary exists in the Biblioteca Nacional, Madrid (Ms. 6711), which, though sharing many similarities with the one previously mentioned, was most likely intended for private use, perhaps by a notary, a scribe, or a judge.[68] These constitutive models show the flexibility of the cuaderno used with the purpose of aiding textual cooperation.

Cuadernos were a common unit in legal proceedings, and the statement of the number of pages, hand, and signature included at the end were obligatory marks of authenticity.[69] Some laws and procedures were stipulated independently via separate cuadernos, referred to as such in legal terminology (e.g., *Cuadernos de las alcabalas*), the cuaderno format allowing constant updating and change (Ladero Quesada 1999b).[70] The breadth of cuadernos as support may be seen in their variable length; the *cuadernos de alcabalas* ranged from 6 to 49-plus folios (Moxó y Ortiz de Villajos 1969a). Cuadernos proved very useful for the documentary needs of both state and local administrations. Notaries found them a convenient medium for recording the notes of the documents to be drawn (*cuadernos de protocolos*) (Durán Cañameras 1955: 156–63), as well as city council meetings and business. The extant cuadernos of the concejo of Guadalajara have a varying number of pages, with no cover or binding and just a single thread running each cuaderno's full length holding the pages together, much like the cuadernos kept by escribanos in other cities, such as Cuenca (Chacón Gómez-Monedero 2005). In many cases, single free-standing pages inscribed by many different participating hands have been preserved along with the cuadernos, thus combining two flexible supports (description for the concejo of Guadalajara in López Villalba 1997: 32–4). Cuadernos constituted a basic support favoured by escribanos because of their flexibility. One year's work or that belonging to a set chronological unit could go into a cuaderno and then all of the cuadernos would be bound together into a volume for safekeeping purposes, although not all necessarily made it to the binder. Valencian notary Vicent Saera could take note of between ten and fifteen transactions on any given day, which he jotted in abbreviated form in a loose piece of paper or a small notebook. These "quadernets" were always written in his own hand and bound at the year's end. He was prosperous enough that he could employ two scribes for the tedious job of writing the official copy of the document in full and entering it into the registers (Cruselles Gómez 1998: 138). The notarial custom of using cuadernos and the wide availability of blank books and booklets at booksellers' shops undoubtedly encouraged a similar use of cuadernos for private accounting. Alfonso X's (2004) *Siete Partidas* cautioned against entering the wrong amount owed by a debtor in one's private cuadernos either deliberately or through forgetfulness.[71] In a similar manner, official notarial records of individual households, when issued by public escribanos, followed the cuaderno system, with each cuaderno being dedicated to a particular type of transaction, in chronological order. They contained blank spaces and a blank last page used for later additions by different hands; the cuadernos were likely to be later bound together. The books of Pedro González de Hoces, veinticuatro of Córdoba, are good examples of these practices (Ostos Salcedo 2005: 111–16, 121–3, 131).[72] The Inquisition used the cuaderno support, as well as loose papers, letters, and memoranda, profusely, as detailed for example in the description of the archived documents related to high inquisitor Lucero in Córdoba (Gracia Boix 1982: 80–95).

Even allowing for some looseness in the terminology of the period,[73] it is clear that there was a widespread use of a flexible codex model embodied by the cuadernos as a writing medium. The flexibility afforded by loose papers was extended through the use of booklets, which provided more protection for the text than a single paper and also enabled the joint circulation of several texts. The use of cuadernos had many other advantages, not the least being their low price and the ability to make them at home. It also facilitated the portability of reading and writing. Thus we see Queen Julia pull from her breast a small cuaderno that she always carried with her ("un pequeño quaderno que traía continamente en los pechos") in Diego Ortúñez de Calahorra's (1975) *Espejo de príncipes y cavalleros* (168). From the early fourteenth century has survived a tiny paper cuaderno (measuring 63/65 × 50/55 mm) containing a poetic text, the so-called *Disputa de Elena y María*, small enough to be tucked into one's purse or, as in the previous example, on one's body.[74] Equally handy is the format of the small book or booklet ("librete o cuadernillo de pargamino") that could be carried around for astrological observation, according to Enrique de Villena's (1994) *Tratado de Astrología* (1: 545). Whether destined to become part of a larger unit or to enjoy an independent existence, the cuaderno was favoured for its portability and flexibility.

Cartapacios

Of a similarly flexible nature, the *cartapacio*, a blank personal notebook in which one may jot down anything worthy of note, shared many of the material traits of the cuaderno. In the inventory of the books that belonged to Queen Isabel and that were in the charge of her *camarero* (chamberlain) Sancho de Paredes we find five such notebooks with parchment covers that were used in the teaching of Latin to Prince Juan: "Cinco cartapacios borrados de cuando al Príncipe se mostraba latín e las cubiertas de pargamino," along with several manuscript and print cuadernos and cuadernitos (Torre y del Cerro 1968: 38). These cartapacios were recommended as an important tool in children's education. Nebrija (1981 and 1979) described the cartapacio as basically a blank notebook ("Albioli carthe non scriptae") in his *Introductiones latinae* and in his *Diccionario* ("Albiolus. i. por cartapacio en blanco"). Covarrubias (1998) specified its nature as a miscellaneous book or booklet in a constant process of formation, used for the accumulation of knowledge as it was acquired: "Cartapacio, el libro de mano en que se escriven diversas materias y propósitos, y el quaderno en que uno va escribiendo lo que dicta su maestro desde la cátedra" (313).[75] Humanists such as Palmireno and Vives were strong advocates of the use of cartapacios, which they described as notebooks or blank paper books divided into different sections. The student would write down notes during instruction in the appropriate section, and at least one section would be dedicated to jotting down interesting thoughts or sayings that could be used later at appropriate times.[76] In addition, Vives (1971) also advises the student to have a larger book in which to further develop the notes taken during class with the help of the teacher, as well as notes copied from his readings (108).

Similar items are found in the possession of members of other professions. In the inventory of Jaume Callís, doctor in law or *jurisconsult (legum doctorem)* (Barcelona, 1434) there is mention of a *cartipas* with parchment covers belonging to him, large enough to comprise eight paper coerns (in Carreres Valls 1936: 58). As this example shows, in Catalan inventories the *cartapàs* or cartipas can point to a personal book used for business purposes. In a 1440 inventory, we find another "libre o cartapàs" in small format ("scrit en paper de forma menor") (Madurell i Marimón 1974: 65), and in a 1461 inventory a "libre o cartapàs" in quarto format with parchment covers containing some accounts ("Comptes de pesos") (Madurell i Marimón 1974: 84). References to a *"cartapàs* or book" *(un libre o cartapàs)* point to the often blurred lines dividing various kinds of material support for the codex format.[77] The use of a personal cartipas at home for professional purposes invited an extension of its utility for other private writings. The inventory of a jurisconsult in 1485 lists a cartipas entitled "Repertory" written in his hand and containing some texts *(dictats)*, probably poems, which the author of the inventory considered of little value: "Un appellat repertorium de sa ma ab alguns dictats de pocha valua a manera de cartipas" (in Carreres Valls 1936: 90).[78] Tendilla found his cartapacios the ideal place in which to write and vent his frustrations: "en esto y en escreuir mis cartapaçios echo toda la ponçoña y descanso de lo con que rebentaría sy asy no lo hiziese" (in Moreno Trujillo et al. 2007: no. 1708).[79] These handy cartapacios, much like the all-purpose booklets, were ideal repositories for miscellaneous entries, particularly excerpts of great works, useful tidbits of information, memory logs, and poems.

Gathering Poems

Various modes of gathering poetic papers are documented for earlier centuries. Loose sheets or booklets (*libelli*) containing the work of the Archipoeta form the basis of the larger compilations, one of the most famous being the *Carmina burana* (Bourgain 1991: esp. 73–80). In the thirteenth century, a similar practice has been pointed for Peire Cardenal (Brunetti 1993; Holmes 2000: 9), and Miquel de la Tor (Holmes 2000: 9–10). Machaut would count the length of the finished compilation of his works ("livre ou je met toutes mes choses") in terms of booklets ("cahiers"), rather than folios, pages, or number of verse lines (S. Williams 1969: 441). Later instances include the manuscript with the poems emanating from the circle of Blois (e.g., Arn 2008). Boffey and Thompson (1989) have shown that in England in the latter fifteenth century texts were produced and circulated in booklet form, often by a "home-based scribe," making for idiosyncratic compilations (289–96, and passim).

Of the well over 200 manuscript cancioneros listed in Dutton (1990–1), most are either copies of earlier composite codices or edited volumes containing poems that were previously scattered in loose leaves or in cuaderno format. Few of these single leaves have been preserved. One surviving example may be the single page containing two cancionero poems covering the recto and verso that separates two historical works in BNM Ms. 9268 (Gómez Moreno 1985; Gómez Pérez 1969–70). Cuadernos may have contained the works of a single author, as in the case of poet Santa Fe (Tato 2005a, 2005b), or the exchanges of several authors, as was probably the case with some of the *Cancionero de Baena*'s thematic threads, or even the poems stemming from a particular circle or environment, as in the case of the poets gravitating around the household of Archbishop Carrillo. Alvar García de Santa María's will shows that he possessed "otro cuaderno de Trobas de Fernan Perez de Guzman, en papel."[80] This cuaderno may have contained Pérez de Guzmán's most popular composition on "virtudes e viçios," which in the cancioneros was headed by a rubric that identified García de Santa María as the recipient (ID0072). Given the length of the composition, this could have been a thick cuaderno.[81]

The poetic compilation known as the *Còdex de Cambridge* was written in booklets, some individually wrapped in parchment for protective purposes, in order to facilitate additions and was intended for eventual binding (Martos 1999: 457–8). Many cancioneros are made up of various booklets designed as independent thematic and stylistic units, with the end of the booklet often left blank in order to facilitate later additions of like compositions. The accumulated booklets would be later bound into a volume, and compositions continued to be copied in the appropriate sections or, when the blank spaces were filled, in the margins or in other available spaces. This is the case, for example, of BU1 (*Jardinet d'orats*) (Beltrán 1995b: 257–9), copied by a notary in just such booklets; or similarly the *Cartapacio poético del Colegio Mayor de Cuenca* published by Forradellas (1986). A similar system may have been used to copy PN1, the *Cancionero de Baena* (Blecua 1974–9), the so-called *Pequeño Cancionero* (MN15) (Elia 2003), and the Galician-Portuguese *Colocci-Brancuti*; it was used by Italian humanists in the copy of thirteenth-century French chansonniers, such as U, as well as in the production of Petrarch's manuscripts (Beltrán 1995b: esp. 262–3). The *Cancionero de Gallardo o San Román* (MH1) is composed of booklets with differing numbers of leaves, different types of hands, of paper, and of watermarks, grouping the works of various authors with no apparent order but with spaces intentionally left blank for further additions. It produces the distinct impression that some of the booklets may have been inserted into others before being bound together.[82] Many of the traits present in this cancionero match the requirements for a booklet-based codex identified by Robinson (1980) and Hanna (1986). Further evidence of such use of booklets exists in SA1, which contains an independent booklet (fols. 149–81) that was later bound with other materials.[83] In addition to having its own watermark, this booklet bears the signs of wear associated with an independent existence before being bound into the manuscript. It also contains accounting figures and many maniculae that direct the eye to various points in the list of maxims copied towards the end. The booklet ends with two poems copied as prose (ID0159, 0160), undoubtedly to economize the precious little space left. Pérez Priego (2005) has likewise identified two booklets in NN2, associated with the circle around Lucrezia Borgia and the court in Ferrara. The intense traffic of single pieces and of poetic clusters in Ferrara is further attested to by the three single pieces of paper (measuring 156 × 93 mm) that were later bound in to Ms. CL. II. 69 of the Biblioteca Ariostea in Ferrara, each containing a different sonnet written in Spanish by Pietro

Bembo to Lucrezia Borgia (Lara Garrido 2005). The work of Catalan poet Ausias March appears in the form of unbound booklets in the inventory of his belongings, as "dos llibres en paper, de forma de full, desqüernats, ab cobles" (Martos 2005: 410). Ausias March's poetry may have circulated primarily in booklet form until his works were compiled for printing.[84] Of a similar composite nature is Biblioteca de Catalunya Ms. 1744, which contains a "plec poetic" among other independent units, which were later bound together; the volume showing the action of both chancellery and accounting work along with that of a learned court official (Beltrán 2007). Beltrán (1995a) has likewise identified two booklets, perhaps autographs by Juan del Encina, and has posited an extensive use of the booklet by medieval poets (Beltrán 2009).

The dissemination of loose papers and booklets later archived in a book is a practice shared by literate communities that were themselves subject to stronger administrative and political, as well as cultural, forces. These formats show a continuum with the codex form stretching across a range of single leaves, rolls, envoltorios, cuadernos and, ultimately, books. The various formats produced by the printing press, from single documents such as bulls to chapbooks (*pliegos sueltos*)[85] to books built on cuaderno units, replicated scribal practices and were also part of this continuum. Each means of support conveyed intrinsic meaning made richer by the overlapping uses of paper that suited a combination of cultural, social, and political needs. Visual links between poet and medium point to the consciousness of the material aspects of authorship. As important as chirographic authentication was the choice of support in the complex constitution of authorship.

6 Paper Politics

The female correspondent in Fernando de la Torre's *Libro de las veynte cartas e quistiones* deemed book pages a durable and effective archival method. She saw Fernando de la Torre's efforts in dealing with her writings as an archival process of transfer into a book whose aim was to preserve writing as memory of "things past" and "forgotten deeds," a common topos for the writing of history, a process that had the power to transform her "unpolished admonishments" into a book of glory and example: "y qué juzgaré del fin vuestro que fue en tresladar mis grosseros amonestamientos e sinples consolaçiones por orden durable en fojas de libro donde las cosas pasadas se fazen presentes y los fechos oluidados se reduzen a memoria, de lo qual se sigue gloria o pena a los pasados y a los presentes enxiemplo?" (text in Díez Garretas 1983: 137).[1] Fernando de la Torre necessarily agreed with this belief in the staying power of the written word: "las letras son más memorables que las palabras que pasan como ayre presuroso" [Letters are more memorable than words, which pass like rushing wind] (text in Díez Garretas 1983: 188).

Ultimately, books functioned so well as a repository of single texts that their use encouraged the disappearance of the loose paper. The need to gather and organize assorted papers would foster different archival models, as flexible and varied as the literate communities engaged in their production. The following pages will look at some key textual practices of the late medieval bureaucracy that address the pressing need for both a flexible as well as an enduring system of fair copying. The use of documentary forms such as the registro (register), albalá (payment voucher, and more generally a document),[2] and memorial (memorandum), favoured by late medieval administration provides key insight into the material implications of such practices. Intersecting with these models are others such as memory and personal books, all of which further fostered the processes of textual transmission as both a personal and an institutional practice with strong implications for the production of literary works. The bureaucratic posts filled by many authors encouraged the cross-pollination of administrative, historical, and literary archival practices. Poetic and legal documents in particular shared a relatively short format that facilitated their copying and, when needed, their quick dissemination. They also shared the inevitable evanescence caused by their circulation in single leaves, booklets, or a similarly flexible support. The archival solutions to this predicament would benefit both types of documents as textual models continued to develop.

Poems and documents were part of the daily transactions at court, in the cities, and elsewhere. That the handling of different pieces of paper would be affected by the intersections of different kinds of compilatory formats was undoubtedly facilitated by the poetic practice of many bureaucrats and other literate professionals. Active documentary and poetic production took full advantage of the versatility and pervasiveness of the piece of paper (and, to a lesser degree, parchment), resulting in a textual hybridity that bloomed into new modes of writing. Although monastic and university textual models still bore their mark, the shift to a bureaucratic model carried its strong imprint in the fifteenth century, fuelled by the material changes brought by an increasingly powerful and complex bureaucracy.

Paper Pushing

In order to trace the passage of the individual paper to the bound book, it is essential to understand how the late medieval administration handled documents.

In 1371, the *Cortes* of Toro established a tight protocol for the production and fair copying of loose documents, ratified and expanded by the 1447 *Cortes* of Valladolid and those of Toledo in 1462. This protocol emphasized the importance of textual integrity and of marking individual responsibility in the production of a text (Maravall 1986: 473; see also Martín Postigo 1959). With the proliferation of loose documents issued by the administration came specific archival methods. Much paperwork was generated in relation to the bestowing of posts, the granting of rewards, or the assignment of yearly amounts to be paid to a specific individual, all of which were commonly made by a document issued by order of the Crown.[3] These documents were issued in the form of royal letters (*cartas reales*), royal decrees (*cédulas reales*), missive letters (*cartas misivas*),[4] and payment vouchers (albalaes). They were most commonly written on paper, dated and signed by all those involved, either materially or nominally, in their preparation.[5] One of the more widely used types of documents under Queen Isabel was the albalá. The protocol followed for drawing up, processing, and archiving the albalá offers a clear illustration of the path that the loose paper followed until it found its way to the book, a course that was familiar to all those immersed in the paper-handling culture of the period. After the queen ordered the issuing of an albalá, her secretary had it drawn up by one of the escribanos working for him. Before the original was given to the recipient, the document was made official once it was copied into the appropriate book(s), including the *libro de asientos* (an account book),[6] and the *libro de traslados* (fair-copy book) (see below). The verso of the original given to the recipient was marked ("sobrescrito") with information such as the date of copy into the official book and with the signatures of all who had been involved in its issuing. Only then was the amount specified in the document made available for payment ("librada," "libramiento")[7] to the grantee, either in one lump sum or in yearly or other specified instalments. In order to receive the payment, the recipient would take the original document to the officials in charge of issuing the payment (the contadores), who after payment would enter the amount in the appropriate books. While the beneficiary kept the original albalá, additional copies could be made for other parties involved. If at any point there was a problem or confusion regarding the grant, the recipient could present the albalá thus marked with the officials' names and the date on which it had been recorded in the royal books.[8] The privileges bestowed through a *carta de privilegio* (letter granting or confirming a previous royal document) were processed through the office of the *Escribanía Mayor* in a similar manner.

A similar system functioned in local city governments, and although there is evidence that at times some shortcuts were taken, the overall process of producing, authenticating, ratifying, and archiving documents in a book register was the standard. City escribanos issued documents in a similar manner and typically held ownership of their "books."[9] Medieval legislation concerning the escribanos suggests the use of multiple *instrumenta* when preparing a document, including a rough draft containing a brief description of the transaction, a copy of the draft in its early abbreviated form in a book or cuaderno, a copy of the full text of the document in its final form, and the document itself as it was extended to the interested parties. These documents needed to be authenticated by the appropriate signatures and the escribano's signum. The system that prevailed in the fifteenth century seems to have involved mostly three stages: first draft, inscription in the escribano's book or cuaderno, and final document. These were often referred to as *nota*, registro, and carta. However, the evidence suggests that practices varied, and the early draft could be noted down in a book or a cuaderno instead of, or in, a piece of paper. Drafts have been found, alongside the fully written document, between the leaves of a register, jotted down in small pieces of paper in a hard-to-read cursive hand with a simple outline of the main points to be included in the document, mostly the names of the person granting the document and the name of the person on the receiving end, as well as the nature of the transaction. In cases such as the drawing up of a will, the draft may include a questionnaire for the testator on one side of the draft and the answers given to the escribano on the other; this draft was filed in the register alongside the full and official text of the final document (Ostos Salcedo 1994: 195–6). Other possibilities included omitting the draft in favour of the full text in cases involving a short document and stitching or otherwise saving the full text of the final document between the pages of the draft book, becoming then "single page registers" (*hojas de registro*). The process could potentially produce a double inscription of the document in two books/cuadernos in different stages of development, as well as on two or more single pieces of paper.[10] Further, when supporting documentation was presented in a particular case, the paper or papers could be inserted into the cuaderno or book by stitching or otherwise.[11] At times original documents were archived right in the book along with a fresh copy of their content, thus combining two types of material support. Loose documents could be inserted into the book via a special pocket or "bag" (*bursa*) attached to the book cover.[12] If the book covers were thin, the integrity of the bag could be compromised, thus forcing the papers once again to be interspersed among the pages of the notary's book, by stitching or other means. Records would thus be doubly archived in the book, the repository of both draft and loose original. The book register could hold between its pages full-size documents as well as smaller scraps, as in the *Manuale* of fifteenth-century notary Pere Pau Solanelles, of Igualada (Torras i Cortina 2003, 1: esp. 19–20), or personal letters and notes (Ferrer i Mallol 1980). Notaries in Granada used either a blank book, cuadernos that would be stitched together at the year's end, or loose copies of documents stitched together with or without additional documentation (Moreno Trujillo 1995: 92–3). Earlier, in 1369, King Pere IV el Cerimoniós points to the inner workings of the system in place. In the letter written to the people of Valencia, the king forwarded a proposed bill ("proposiciones de corte") stating that it was written on two leaves in his own hand ("de nostra ma"). He asked that it be copied in the appropriate book, and that the original not be destroyed in case it was needed later on (Rubió i Lluch and Balcells 2000, 1: 221–2). A pre-stitched blank book (*libro cosido*) would eliminate some of the dangers of losing single documents, and institutions such as the Inquisition attempted to enforce their use in its massive efforts at creating an enduring textual repository of its transactions and overall activity. This was the support used, for example, for the record of the sale of expropriated Jewish belongings in Toledo in 1494 (León Tello 1979: 612).

A similar system also proved useful in fulfilling the needs of individual household administration. Books such as Fernández de Oviedo's (1870) *Libro de la camara real del Prinçipe Don Juan,* or Hernando de Talavera's (1930) *Instrucción para el régimen interior de su palacio*, deal with record keeping in the internal administration of large households, such as Prince Juan's, heir to Isabel and Fernando, in the case of Oviedo (34–42), and Talavera's own in the latter case. Oviedo and Talavera detail a complex system of books and single pieces of paper or writing material (*folios*) that various household officials, such as the mayordomo, camarero, *botiller* (marshal of the buttery), and *despensero* (steward), among others, used to take note of the objects and foodstuffs under their direct responsibility, as well as the related account keeping. These books and single leaves were to be cross-referenced and the entries in the various books and loose pieces of paper made by the official directly responsible for the item(s) recorded. The book, alternatively called "borrador," "manual," "diurnal," or "registro," was the keystone of the complex system of books generated in the administration of Prince Juan's house.[13] Queen Isabel's belongings necessitated the use of eleven of these books, whose content appears detailed in the documents published by Torre y del Cerro (1968: 394–400, 414–21). This system obviously encouraged individual textual competencies, as well as intertextual cooperation, inasmuch as the various hands of scribes, secretaries, and chamber personnel (such as the "moço de camara") would go into various books and these would be cross-referenced. As in royal administration, or perhaps even more so, given the more self-contained nature of these household bookkeeping practices, textual clusters emerged as a common practice encouraged by the common locale of the texts as well as the enforced solidarity of the hands involved in their creation. As a matter of course, skill with language and numbers became part of the daily practice of textuality, bearing obvious similarities with merchant bookkeeping. This is explicitly acknowledged by Oviedo (1870), who likens the "libro mayor" or master household book to merchants' and bankers' ledgers: "e llaman a este libro los mercaderes e banqueros libro de caxa" (39). The parallels between court or city registers and merchant registers are obvious in books such as the book of privileges given to Genoese merchants living in Seville (*Libro de los privilegios concedidos a los mercaderes genoveses* 1992). Moneylending operations, such as the double-entry one used by "court banker" Pérez de Salinas give a good idea of the complexity of the booking system used. His books are organized chronologically and by city. Even though only one of Pérez de Salinas's (1980) books, the *Libro Mayor*, has survived, he refers to other books and to the more personal "librillo" or "librico." In addition, a long and narrow separate booklet dated 13 March 1498 in the city of Alcalá de Henares, containing a cross-referencing alphabetical index with the names of clients, has been preserved, its ledger format seemingly constituted by tiras (see facsimile included in the edition). The complexities of systems based on multiple kinds of supports that were tightly interconnected brings into relief the physical makeup of each different document even when it may have vanished as an independent unit.

Poetic Bills

The proximity of cancionero poets to document production systems may be deduced from many of the poems' texts, which provide a textured reminder of the material aspects of both medium and content. There are cases in which the coincidences of legal and poetic paper became so close as to produce a legal poem or a poetic bill. The rubric that heads Antón de Montoro's ID3035 (MN19) presents the poem as an albalá, outfitted with the format and language required of such records: "Montoro al Obispo de Salamanca que le mandó dar diez doblas, e fizose Montoro el Alvala" [Montoro to the Bishop of Salamanca, who ordered ten pieces of gold given to him, and Montoro himself made his albalá]. The text of the poem gives the date of 1461. Montoro reused the more or less formulaic text of this poetic bill on at least one other occasion, following current bureaucratic practices. His familiarity with court, city, and merchant bookkeeping provides a clear example of the overlap between court and merchant accounting. It also helps explain the textual practices of middle-class merchants and artisan poets such as Montoro with strong connections to court and city government.[14] As poem ID3009 (MN19) attests, in the following year, 1462, Montoro made an almost identically worded verse albalá, this time not for pieces of gold, but for a shirt.[15] In fact, Montoro seems to have become something of an expert at poetic bill drawing, as there are records of further such albalás (or albalaes) penned by the poet. Poem ID2729, also an albalá, is dated 1447 in two of the extant manuscript witnesses, but which, according to a third, might have been reused ten years later. The texts in LB3 and SV2 both agree in the year and the amount of payment (300 maravedís), but the text in the MN19 version dates the document to 1457, increases the amount to 500 maravedís, and has other significant textual variants. In view of the poet's apparent knack for reissuing poetic bills, it is possible that the variants are due to the author's pen rather than to a change by a copyist or a reader. Montoro also wrote other types of bills, such as a versified payment or libramiento (order of payment) to don Juan Peñafiel in which he specified different amounts to be paid by different people for a cape that don Juan had asked him to sell (ID3041).[16] Further use of the poetic paper as legal document appears in ID0179 (MN19), which is ostensibly a formal letter requesting the release of a friend of Montoro's from prison, for which action a payment was due.[17]

In addition to the albalás, there are other forms of poetic billing. One such case is ID1926, in which Montoro sent a price list to a prior either to encourage an order or to coax payment for some articles of clothing already delivered. The prior, addressed as "Serenísimo señor" [Most Serene Lord], is told: "capa, sayo y jubón, / cuestan mil maravedís / zarahueles, borceguís, / bonete cuesta docientos. / Gran señor, no son dos cuentos, / por eso ved qué decís."[18] But Montoro was by no means the only one who courted albalaes and libramientos along with the favour of the doormen of noble residences. Juan Alfonso de Baena wrote a fair amount of just such compositions and included them in his cancionero. Such is the nature, for example, of poems ID1580, 1581, 1582, 1503, 1584, 1585, 1586, 1587, 1588, 1589, 1590, 1591, 1592, and 1593. ID1586 seems to be a text much like Montoro's self-issued payment, written in the third person and instructing immediate payment to Baena. Other poets, such as Alfonso Álvarez de Villasandino, also wrote poems requesting an albalá: ID1197, 1293, 1321, 1347, and 1360. We know that ID1293 had its intended effect, because the favourable reply by Pedro de Luna, archbishop of Toledo, follows Villasandino's petition in PN1. In ID1294, the archbishop notified Villasandino that the albalá honouring the poet's rhymed petition ("petiçión rimada") had already been issued ("librada") (vv. 21–32). Many others of Villasandino's poems seeking money use a similar legal language. In another poem, Villasandino complains to the king about a payment slip ("carta" in the poem, "nómina" in the rubric) that five (court) men tore after attacking him (ID1203). In his petition to Diego Gómez de Sandoval, Adelantado Mayor de Castilla, Baena asked the Adelantado to sign another one in his favour ("me firmad otra cartilla" ID1591, v. 54). In the poem that follows in his cancionero (ID1592), Baena flatters Gómez de Sandoval's wife, Beatriz de Avellaneda, Countess of Castro, and begs her for the money that she had granted to him through a signed carta ("quando mi carta librastes" v. 14). In cases such as these, the poem does not become a payment or a bill or a legal petition, but rather poem and payment are both part of a transaction that is both economic and textual and that may be as strong as the literary value of the poem or as weak as its (paper) medium.

Working nobles such as Gómez Manrique also penned poems whose main goal was to coax substantial libramientos out of those in the power to grant them. ID0094 tries to extract unspecified amounts from Arias Dávila, contador mayor of Enrique IV and powerful converso, acknowledging that a bad poem would suffer the same rigours as the ill-fated libramientos, which the contador had torn: "mas si fuere desabrido / el quemante fuego pido / sea su deuido premio, / o roto con los ronpidos / libramientos" (vv. 57–61).[19] The poem's rubric in 11CG openly reveals the economic nature of the transactions between administrative and poetic production, the poetic paper being produced as a condition for the issue of payment (libramiento): "Otra obra del mismo a diegarias contador del Rey don juan porque no le quiso acceptar vn libramiento suyo antes le hizo pedaços diziendo al portador que si no le hazia algunas coplas que no le libraria jamas."[20] Libramientos, orders of payments, or gifts that were slow to come or were never realized were also a problem for poets like Montoro, Fernán Sánchez Calavera, and Alfonso Álvarez de Villasandino, as they declare in their respective compositions ID1784 (MN19), ID1657, ID1215. The inability to cash a libramiento was a significant enough problem that the law had to protect the plaintiff by allowing him access to a higher official (the contador mayor or his lieutenant) to personally see to its payment.[21] In those cases, a poem could help push the paper bill or perish with it.

Juan de Tapia further identified the poetic bill as a hybrid document. Tapia's poetic albalá to the daughter of the Countess of Arenas (ID0554, RC1), presented as a document and sent as a letter, bore the marks of its legal, poetic, and epistolary status: "Yo te enbio esta cancion / escripta como aluala" (vv. 53–4) [I am sending you this song / written as an albalá]. The address written on the paper was also written in verse: "El sobre escripto del aluala: A ti madama maria / Carachula el sobrenombre / iohan de tapia el hombre / que aquesta aluala te enbia" (vv. 86–90).[22] This format took advantage of the similarities with legal documents in that some letters were intended to act as albalaes, in acknowledgment of the receipt of goods or another such document, the matter specified in the letter along with the appropriate legal formulas.[23] Because of their material and contextual affinities, loose papers – legal, literary, and personal – went through similar material modes of handling and imprinting from production to dissemination. Once the paper was written, the sender would unite the name and perhaps address of the recipient on the outside of the folded leaf.[24] A servant or a friend of the two parties often delivered it by hand. After it had arrived at its destination, the recipient could write on the back of the letter the date of receipt, the name of the sender, a few lines detailing the content of the letter or document, its cost, and the date of the reply. These were what could be called "archival marks" because they textualized the extratext that surrounded the paper from its inception, through its transit and its receipt, fulfilling the function of a docket for documents or a rubric for literary texts. These archival marks would be carried over to the book as archive and in a way "smoothed out" the textured paper and its situation. These loose documents were eventually filed and recorded, their presence in poetic compilations showing the overlap of cancionero and official register.

Open Books

The need to contain the texts proliferating in potentially precarious material supports encouraged archival practices that counted the book as a pivotal step. The book as archive and the fair copying system coexisted with other archival methods such as that of the "open book," created by means of piercing and threading single documents together so that pages could be easily added or removed, like a modern-day ring binder. The fourteenth-century *Ordenacions* of King Pere IV el Cerimoniós carefully established how documents should be archived, and included reference to the existing method of "filing" documents by threading (*fer enfilar*).[25] This system was particularly useful when dealing with ongoing business that necessitated the constant updating of the textual record. Account books and rosters benefited greatly from this model. Some of the better-known account books of Isabel's reign are, in fact, in threaded format, kept by her treasurer Gonzalo de Baeza (cited in chapter 3), and housed today in the Archivo General de Simancas (Contaduría Mayor de Cuentas, Primera Época, legajos 6 and 15). Held together by a cord going through a large hole punched in the upper-left corner of all pages, the books are organized chronologically, grouping the individually listed payments by the document type (cédula or *nómina*) issued at the queen's command. The sum of all individual amounts is tallied at the end of the entry and is intended to match the amount allotted by the document at the beginning of the entry. The margins are used to record any changes or to note the receipt of the payment.[26] The books provide invaluable information on the daily life of Isabel's court, and many of the names they mention may be identified with cancionero poets who stayed in close proximity to the court. Fernández de Oviedo (1870) recommended the use of just such a book for household accounting purposes. This book, composed of punched pages ("de pliegos horadados"), would be used alongside the *libro mayor* and the *borrador*. It had the obvious advantages of arranging the information more clearly so that it was easier to visualize. The threaded model also facilitated the easy consultation of single documents without the inconvenience of dealing with a bound volume. The escribano de cámara in Prince Juan's household was advised to keep a threaded companion of the master chamber book. The threaded book would contain separate entries for different categories so that it would be easier to bring a single page with the specific entry ("pliego horadado") to the Prince (68).

The *Cortes* of Madrigal in 1476 stipulate the use of the threaded book model ("rregistro foradado") for registering letters and provisions while also requiring that the escribanos write their full names on the documents they register, their simple signatures being insufficient (*Cortes* 1861–1903, 3: 31). This meaningful insistence on writing the full name on the documents by all involved in their production was reiterated in the *Ordenamiento de las Cortes de Toledo* in 1480 with regard to the documents issued by the royal *Consejo* (*Cortes* 1861–1903, 3: 119). The threaded format was also used for archiving city council documents, as is the case of the aptly named *Libro horadado del concejo madrileño* published by Millares Carlo (1927).[27] Municipal books were by definition flexible, open, and of a miscellaneous nature not only in terms of their content but also in their compilatory processes (as emphasized by Pardo Rodríguez 2002: 90–1, 98, 100–1, and passim). This model was applied in other instances. Christopher Columbus, who offers a good illustration of the many textual practices at play at the end of the Middle Ages, left a threaded book as well: the list of crew members who embarked on his second voyage to the Indies (1497) and the complex system of payment of their salaries.[28]

The establishment of the Inquisition fostered the proliferation of cross-referencing texts by creating a system that demanded a clear paper trail from denunciation to trial to sentence. Supported by well-trained administrative personnel and particularly anxious for its own purposes to establish a secure way to produce and archive text, the Inquisition also took advantage of the threaded book model, as is the case of the records of pecuniary penances assigned in Córdoba between 1533 and 1538. This threaded book is twenty-seven folios long and is kept at the Biblioteca Colombina in Seville (Ms. 59-6-20 (1)) (Porras Arboledas 2003). Other texts in the web of documents used by the Inquisition included voluntary confessions to be presented in writing in front of the inquisitors and a notary during the grace edict (*edicto de gracia*) after the arrival of an inquisitorial tribunal into a town. After this initial grace period, public accusations were also taken in writing. This is the protocol specified by the inquisitor general fray Tomás de Torquemada in 1484 (text in Jiménez Monteserín 1980: 88, 99). The intensely cross-referenced paper and book trail is evident in inquisitorial registers such as the one generated around the Arias Dávila trials in Segovia (e.g., Carrete Parrondo 1986: 98–101, and passim).

The need for flexibility and further additions shaped the book into a dynamic entity. In cases such as the thirteenth-century *Registro de Corias*, blank spaces were left throughout the register for later additions, which were entered by different hands during the thirteenth through the fifteenth centuries.[29] The nature of account and payment books encouraged a similarly flexible model. Queen Isabel's account books, such as the so-called *Libro de asientos* (published by Torre y del Cerro in 1954), which records the payments made to various officials and servants, also followed an open-book model. The manuscript is housed at the Archivo General de Simancas (Casa y Sitios Reales, legajo 1–2) and has a large format, measuring 22.5 × 30.7 cm. Its independent units were stacked together unbound and without a cover. It was divided into individually marked sections organized by job category, with the name of each recipient starting a new folio. Several blank pages were left after each grouping of servants in order to provide space for future entries. Some group entries failed to fill all of the allotted blank folios, while others ran out of blank pages to annotate. The problem was solved by tearing the extra blank folios from one group and inserting them where needed. The manuscript was headed by an index of positions, which showed the cohesiveness of a design that was both flexible and orderly. Later legislation concerning the production and preservation of texts would continue to wrestle with a model that ranged from the flexibility of cuadernos that were custom-made out of individual sheets of paper, or ready-made cuadernos to be kept or later bound together, to the fixity of the pre-bound blank book (Rodríguez Adrados 1988: 716–28). However useful the flexible model was, a more stable and secure one would ensure the safe landing of the individual texts and records into a compact unit when necessary. The fair-copied bound volume would provide a safe repository, while also encouraging a more linear approach to the handling of the texts.

Fair Copying

The need to preserve loose documents and the important political and economic information they contained meant that documents had to be copied onto a more permanent material support, which more often than not meant a book, where the clean or *fair copy* of the document would be safely recorded. The process of the development of procedures leading to the fair copy, definitive copy, or "mundum" (also referred to as "in bello scribere") and its relationship to the first draft or original version (*minuta*) is well known (Marín Martínez et al. 1997, 2: 168–9). In the post-classical world, such need had been felt as early as the turn of the millennium, when monasteries began making a concerted effort to copy important loose documents into volumes.[30] These cartularies (also known in Spanish as *tumbos*, or *becerros*)[31] did more than save a pile of documents from destruction. In Castile, the thrust behind their configuration owed much to the impetus to gain power and property related to the economic growth of the twelfth and thirteenth centuries,[32] often in competition with ecclesiastical or state forces. This helps to explain the occasional licence taken in modifying a text to the advantage of the monastery or other interested entity.[33] But the forgeries may not have been solely the practice of ecclesiastics. The fall of the "old nobility" in the thirteenth and fourteenth centuries, and the destruction of many of their archives, prevent us from ascertaining the extent of similar practices among the nobles.[34] As in other countries, the twelfth century in Spain witnessed a rise in the production of documents that, although forgeries, were often written records of earlier donations of existing properties of the monastery. In a related development, the monarchy was looking to exercise greater control over lands and people in an effort to aid repopulation and was enlisting the bureaucratic help of religious orders in promoting a greater production of written documents (Sierra Macarrón 2002). The introduction and expansion of new orders in Spain such as Cluny and the Cistercians, was followed by that of military orders such as the Orders of Calatrava, Santiago, or Alcántara and orders of friars such as the Dominicans and the Franciscans (see, for example, Fernández Conde 1982). This coincided with a nation-building "juridical Renaissance" which entailed the writing of comprehensive law codes such as the *Usatges de Barcelona*, Alfonso X's *Siete Partidas*, Canellas's *Vidal Mayor*,[35] and to some extent the *Llibre de franqueses i privilegis del Regne de Mallorca*. Some forgeries worked alongside epic texts with the aim of nation building. This is, for example, the case of the foundational diplomas of San Pedro de Arlanza (Escalona Monge et al. 1997). An analogous interface was at play between forgeries and narratives of saints' lives, with the intention of aiding economic gain for a monastery, as studied by Dutton in relation to Gonzalo de Berceo (1984) and his *Vida de San Millán* in the introduction to his edition. The growth in documentary production at this time would result in the development of specialized terminology in Latin that would influence the vernaculars of the twelfth, thirteenth, and later centuries (Guyotjeannin 1989). The new discipline of the *ars notaria* developed hand in hand with these new documentary needs of political power (Orlandelli 1965: 349–54).

As fair copies, legal codes were presented as organized compilations of existing laws whose very material existence was needed to authenticate the book. In an overarching effort to both embody and supplant the loose document, the compilation was forced to show authoritative cohesiveness by explaining its compilatory methods. At the same time, it had to make transparent its fragmentary construction by allowing the evidence of the individual documents and their material traits to show. Canellas (1989) presented his *Vidal Mayor* as a compilation, an authoritative version of the law, and safe archive for the multitude of local laws or *fueros*, whose form and content it sought to improve and eventually supersede. The organization of *Vidal Mayor* in parts headed by titles and indexed in a table of contents for ease of use puts it on a par with similarly arranged compilations, be they legal, religious, or literary. For Ramon de Caldes, compiler of the twelfth-century *Liber feudorum maior* published by Miquel Rosell (1945), the book format provided the correct medium for organizing all the important documents ("omnia instrumenta") that were currently in a disorganized state ("in ordinatione confussa") into one volume ("uno redigerentur volumine"). The book repository was also intended to preserve the memory of the texts, "his instrumentis ad memoriam revocatis," guard them from oblivion, and serve in the settling of disputes: "tum propter eternam magnarum rerum memoriam, ne inter vos et homines vestros, forte oblivionis occasione, aliqua questio vel discordia posset oriri."[36] The gathering process that would help transpose vertical and horizontal tiras and scrolls to codex form may be seen in Caldes's depiction of the compilatory work (*Liber feudorum maior* fol. 1). Caldes described this compilatory process in the prologue to the *Liber*, a process that involved collecting and writing or registering the tenors of the instruments and other documents ("sunt hic collecti et scripti aut registrati tenores instrumentorum et aliarum cartarum"). Other twelfth-century cartularies, such as the *Libro de los testamentos* of the Cathedral of Oviedo, follow the same formula and offer a similar depiction of the individual documents.[37] Many other codices presented as compilations of legal documents, such as that of the *Consolat del mar*, or the earlier *Libro de las Estampas* or *Testamentos de los Reyes de León* published by Prieto Escanciano (1997) and discussed in chapter 5, contain illuminations portraying the single documents and/or the process of compilation as a way to transfer the authentication of the individual document to the whole codex. Further, canon law collections such as Gratian's *Concordia discordantium canonum* or *Decretum* bear the same production imprint (Winroth 2000).

This compilational and organizational process, which involved the selective copying of portions of documents and their complete integration into a radically different text, as in the *Siete Partidas*, bears much in common with other such processes. A fair-copy model was followed in grand style by the *Leitura Nova*, the sixty-two-volume systematic copy of Portuguese chancellery documents dating from the thirteenth to the fifteenth centuries, a project undertaken mainly thanks to the impetus of King Manuel I and continued by King João III (Alves 1985: esp. 95–108). The political significance of document compilation for the purposes of generating a grander narrative fed the work of authors such as López de Ayala or Fernando de Pulgar, who chronicled historical events inserting whole documents into their histories. Much like in legal texts, historiography could be shaped as a series of legal documents and, as will be seen in the next chapter, individual written accounts, threaded by a larger narrative that loomed large by encompassing authenticated independent texts and including them in a meaningful political frame.

Blank Books and the Body Politic

King Alfonso X (2004) elucidated the pressing need for the fair copying of old documents, this time with the purpose of creating a strong state bureaucracy. His *Siete Partidas* emphasized the character of the registro (register) as fair copy and as an archive for loose papers that would help prevent their destruction: "E dezimos que registro tanto quiere dezir como libro que es fecho para remembrança delas cartas & delos preuilegios que son fechos. E tienen pro por que si el preuilegio o la carta se pierde o se rompe o se desfaze la letra por vejes o por otra cosa o si viniere alguna dubda sobrella por ser rayda o de otra manera qual quier por el registro se pueden cobrar las perdidas. & renouar le las vieias" (III, 69r).³⁸ As articulated by Alfonso, the book functioned as an archive of documents, the final stage in documentary transmission, thus serving as a potential source of later copies, and thus becoming the primary intermediary in a textual tradition.

In the fifteenth century, the administration of Isabel and Fernando continued the earlier practice of the register, with added innovations required by new bureaucratic needs. The register model favoured textual fixity over flexibility, because the ability to preserve an unaltered copy of a text was of the essence to a powerful administration, as it would be to an attentive author when compiling his or her own works.³⁹ In an effort to preserve the large number of documents generated during their reign, Isabel and Fernando issued new ordinances regulating the keeping of the Court Register (*Registro de Corte*) in 1491, mandating the use of bound blank books for the (fair) copy of loose documents (letters, privileges, and others, granted to individuals), and detailing the protocol of data entry as well as the duties of the different officials involved in record keeping.⁴⁰ In his *Crónica de los Reyes Católicos*, Alonso de Santa Cruz described their use of the register understood as a *book archive*:

> Otrosí, hicieron Sus Alteças cierta hordenança de los derechos que los escribanos del reino avían de llevar por las escrituras extrajudiciales. Primeramente hordenaron que cada uno de los dichos escribanos pudiese tener un libro en que escribiese por extenso las notas de las escrituras que ante él pasasen que se ubiesen de hacer, declarando las personas que otorgasen la dicha escritura, y el día y el mes, el año y el lugar o casa donde se otorga, y lo que se otorga, especificando todas las condiciones o pactos e cláusulas o renunciaciones que las dichas partes asientan. E como fuesen escritas las tales notas, las leyesen a las partes, delante de testigos, y se las hiciesen firmar de sus nombres, e les diesen las dichas escrituras firmadas de sus nombres y signos, sin quitar ni añadir palabra de las que tienen en sus registros. (Santa Cruz 1951, 1: 300)⁴¹

It is important to note here the emphasis on entering the precise details of each document in the register – the date, place, names of interested parties and of those drawing and witnessing the document, as well as the necessity of transcribing the entire text ("por extenso") of the documents and providing exact copies to the parties involved.⁴²

The same impetus to archive fair copies of loose papers in bound books inspired the regulations for municipalities of copies of letters, albalás, and cédulas in one book, and that of privileges and sentences favourable to the city in another, with provisions that those books were to allow for continuous additions. The first book had an unspecified format ("libro qualesquier"), but the book of privileges was to be a bound parchment book ("libro de pergamino enquadernado") (Santa Cruz 1951, 1: 250). These ordinances, geared towards regulating the work of public escribanos, appear later in Montalvo's (1990) *Ordenanzas Reales de Castilla* and Hugo de Celso's (2000) *Repertorio universal* (fol. 129r). Similarly, in the 1495 specifications to the city council of Málaga, the Catholic queen and king ordered the council to have a chest with three keys, one of which would be in the hands of the council's escribano. The chest needed to hold the two types of books necessary for city governance: one containing a copy of city council privileges and authorized sentences, and a second including copies of provisions and documents given by the monarchs to the council. It was also to hold key legal codes including King Alfonso's *Siete Partidas* (text in Morales García-Goyena 1906–7, 1: 124). Similar legislation such as the 1476 *Ordenamiento* of Madrigal mandates the confirmations of privileges and *mercedes* to be copied in a separate book (*Cortes* 1861–1903, 3: 29–30).[43]

Blank books intended for fair copying were purposefully used by the administration as a way to control documentary production and halt the unlawful issuance of documents not approved by the monarchs. Isabel and Fernando assert in a 1492 letter that after being informed of the sad state of important documents in Seville, they decided to order that royal letters and ordinances be recorded in a blank bound paper book by Juan Pineda, escribano of Seville's city council. Pineda was given thirty days to assemble the book and twenty to copy all existing documents into it. Any future documents were to be copied in the same book, at the beginning of which the escribano was to insert a table of contents listing the letters contained in the book as well as a brief docket to facilitate easy reference (38v).[44] Similar directions were given to the city council of Seville for the keeping of the book of privileges and sentences, a "libro de pargamino enquadernado en que se escriuan todos los previllejos que esa dicha çibdad e su tierra tiene e todas las sentençias que en su fauor se han dado" (fol. 38v).[45] The copy of the letter bearing the instructions is headed by an explanatory rubric that briefly summarizes the nature and content of the document. It marks the inclusion of the single document – a letter in this case – into the macrotext while calling attention to the nature of the document that follows: "Carta para que se haga vn libro delas cartas en papel mayor, y otro de los preuillejos y sentencias en pargamino" (fol. 38r).[46] The book of privileges was indeed made according to specifications and is today known as the *Libro de privilegios* of the city of Seville (published by Fernández Gómez et al. 1993). Cities of lesser means sometimes had trouble fulfilling their record-keeping obligations. During a visit in 1508 the corregidor discovered that the Basque town of Lequeitio did not have a blank bound book. According to testimony of city officials, the city had no such book, only a paper cuaderno, which earned them the corregidor's reproof. Martín Pérez de Hormaegui offered a blank bound book of his own, which he had bought in Burgos, for the purpose. The book was deemed to be acceptable after it was found to be indeed blank, and parchment bound, with 259 leaves (text in Enríquez Fernández et al. 1993: 5). Similar instructions were given to other cities, such as Valladolid, where the city council's escribanos were ordered by Isabel and Fernando to keep bound books, "libros enquadernados," as recorded in the city's *Libros de Actas I* (fol. 217).[47] Earlier, the 1427 Guadalajara ordinances specified a large format for a separate registro that the escribano was responsible for bringing to concejo meetings (text in Layna Serrano 1942, 2: 527). In 1503 Queen Isabel issued a similar order requiring the use of a "libro de prothocolo enquadernado" for the escribanos of Seville (text in Rodríquez Adrados 1988: 780). In Aragon, an analogous system, albeit with its own profile, developed in cities such as Barcelona (detailed overview in Riera i Viader and Rovira i Solà 2004). The Portuguese *Ordenações Manuelinas* (1984) similarly regulate the format of blank books used by scribes. These books must be of even paper, bound in parchment or leather or a similar material, and the pages counted and signed by a superior official before any writing can begin (239–40). The blank bound book embodied the fixity and flexibility required of a medium that would best serve the needs of state and municipal bureaucracy. In addition, there is evidence of the widespread use of blank books by different types of middle-class professionals. In the inventory of Pere Posa is a half-written blank book: "Un [libre] en paper scrit la meytat de ploma laltre tot blanc e comença la primera pagina Jhs. o pater mi" (in Carreres Valls 1936: 121),[48] as well as an account of all his books ("Memorial dels llibres") written in his own hand (Carreres Valls 1936: 128; and 188–91 for a photographic reproduction of the memorial). But registers were not only fair copies of documents written on single pieces of paper. They could also be the final repository of the envoltorios discussed in chapter 5. For instance, in the explicit narrative of document production generated by the inquisitorial process against the Arias Dávila family, there are references to testimonies of witnesses preserved in envoltorios and then copied in the inquisitorial register; in n. 177: "Este testimonio se inbió de la Inquisición de Aranda y está en un enboltorio con otras testificaciones pasadas al registro" (text in Carrete Parrondo 1986: 101).[49]

However, the practice of register keeping did not always follow the neatness prescribed in all the careful regulations, for the fairness of the copy and its reliability were as good as the escribano's dedication. In 1511, royal archivist Pere Miquel Carbonell, heading the Royal Archive in Barcelona, and in charge of receiving the official registers from royal secretaries, had bitter complaints in that regard. Carbonell denounced the declining state of the registers that the royal secretaries and notaries deposited in the royal archive. The registers came bound in parchment instead of leather, many actually unbound and lacking rubrics and tables of contents, in sad shape, and poorly written (text in Martínez Ferrando 1961: 107). The range in the actual practices depended on the individual agents and on how much they were willing to put into their books.

Notarial Poems

The familiarity of various social groups with the advantages of the register model explains the easy transfer of the format to the collecting and recording of personal writings. The urban middle class, nobles, and royal escribanos and bureaucrats, such as Juan de Mena, Gómez Manrique, Álvarez Gato, or Juan Alfonso de Baena, many of whom were also involved in municipal government, were all knowledgeable about the applications of various types of material support and documentary models. Notaries transferred the official insistence on incorporating notarial signature and voice into the book of poetry by using elements such as first-person narrative, rubrics, and marginalia. This is evident, for example, in the compilation of notary Antoni Vallmanya (Cançoner J) (Beltrán 1999a: 344–8). A further blending of the different types of writing and their medium in this professional context emerges from the copies of little- or well-known poems in notarial documents and registers in Valencia (Berger 1987: 315) or in other cities, such as Toledo. The relation of poems to the notarial instrumenta in which they are inscribed merits a separate study, but their intrinsic interest is demonstrated by Ms. Egerton 482 (British Library), which bears the 1489 will, inventory, and records of the sale of the belongings of High Inquisitor Vasco Ramírez de Ribera, Bishop of Coria and President of Isabel and Fernando's *Consejo*. These documents are followed in one of the last two folios, which had been left blank, by a poem written in Judaeo-Spanish describing a Jewish bullfight (text and study in Infantes de Miguel and Conde 1997). The poem's language and its content, which deals with both highly stereotypical representations of Jews and insight into their participation in civic festivities, raises important questions regarding authorship and the meaning of textual support when poem and document are thus combined.

The presence of poems copied not only in the margins but occupying full pages of notarial cuadernos brings to mind the interweaving of archival practices of poets and notaries and suggests ways in which the single poetic paper found its way to a cuaderno. An example is that of Maestre Pedro de la Cabra, a converso doctor living in Zaragoza, who seems to have had a preference for working with converso notaries. His poem was copied in the last cuaderno (containing the recorded transactions for the year 1436) of the register of Zaragozan public notary Martín de Tarba (Marín Padilla 1998). The poem, with its six stanzas organized symmetrically on the full page, likely circulated in notarial circles. The source was probably a single sheet that originated from its author, who through his constant purchases of urban properties and other transactions kept close contact with the notarial world. Other professional works also show the wide interest in poetry and the textual mingling of different types of writing. A case in point is MN7, a medical compendium with a poem (ID4629) copied on the last leaf.

From a codicological perspective, it is common to see the use of pre-designed and often pre-bound blank books for poetic compilations (Beltrán 1995b: 254–7; Marotti 1995: 18).[50] In their similarities to the fair-copy register format, they share many traits with self-made notarial registers. Cases in point are the *Canzoniere di San Martino delle Scale* (PM1) (Bartolini 1956), as well as the so-called *Cartapacio de Morán de la Estrella* (DiFranco, Labrador, and Zorita 1989). The bound blank book model also seems to have been used in the copy of the *Cancionero de San Román* (MH1) (Beltrán 1995b: 241–6) and of *Cancionero del marqués de Barberá* (BM1). BM1 has uniform paper with a single watermark and contains groups of blank pages interspersed throughout the manuscript in order to allow for additions. It seems to have been written by only one hand at long chronological intervals.[51] As with notarial registers, in fray Íñigo López de Mendoza's cancionero (Madrid, Biblioteca Lázaro-Galdiano) (ML1), as well as in other cancioneros, such as the copy of Santillana's works in Ms. 80, Biblioteca Pública, Toledo (TP1), blank spaces were left purposefully in order to provide room for later additions while clearly marking the separation between long poems or the end of a long poem.

Individual Registers

Working nobles holding official posts and other professionals also kept rigorous records of their correspondence and of the documents they issued and received. Such was also the practice of humanists like Salutati's "quaternus privatarum epistolarum quo veluti protocollo tunc utebar" (Rizzo 1973: 339–40). Christopher Columbus is a good example of a devoted record keeper. He kept careful registers of his documents and correspondence in two fair-copy books, one dedicated to preserving copies of his letters ("libro de traslados de cartas") and another of his privileges ("otro de mis privilegios"), as he states in his letter to Nicolò Oderigo.[52] Columbus also recorded copies of his correspondence with the queen and king in a fair-copy book known as his *Libro copiador* (Rumeu de Armas 1989). Until 1500 Columbus carried these around with him, but after that date, believing that his documents were no longer safe travelling around, he deposited the chest containing these and other documents, sums of money, and other valuable objects with his long-time friend and confidante fray Gaspar Gorricio, to be kept safe in the abbey of the Cartuja of Seville.[53]

Private registers ("libro") with personal correspondence and documents bear many similarities with those maintained by Italian merchants, which noted, as Paolo da Certaldo advised, the document's content, date, and the name of the notary in charge of drawing it (text in Branca 1986a: 46n245). In Portugal, Álvaro Lopes de Chaves's (1983) *Livro de apontamentos* functioned in part as a kind of personal register of Lopes de Chaves, author and royal secretary to Kings Afonso V and João II of Portugal. Here too loose royal documents such as notes, letters, transactions, and even poetry (including Gómez Manrique's *Regimiento de príncipes*) were copied for safe keeping.

Equally fascinating is the four-volume register of Íñigo López de Mendoza y Quiñones (1440–1515), second Count of Tendilla (1996, 1). Tendilla's books, which have been discussed at length in chapter 4, contain copies of the documents – mostly letters and memoriales – generated by the Count, as the first alcaide of the Alhambra and captain general of the kingdom of Granada.[54] His register or *copiador* is similarly precise in noting the exact or nearly exact text of the letters and documents copied, displaying a keenness to provide a contextualizing paratext via rubrication. Much as in court registers, there are many self-conscious references to the material and paratextual makeup of the text. It is extremely useful to take a close look at a register like Tendilla's, because it reveals key aspects of the relation of the single document to the volume, as well as the material culture with which they are imbricated. Documents are copied in chronological order, with each page bearing a header specifying the month in which the letter was produced for easy reference while the exact date is given at the end of the letter, and even the amount paid to the carrier (e.g., 104, 144, and passim). Dockets head each document and eloquently speak of the motility of the texts by specifying the persons involved in their delivery and the different hands to which the various contents of a specific mailer (envoltorio)[55] were destined (e.g., 92, 104, 136, 144, 161, 193, 237, 286, 346, 352, and passim). Dockets also tell of the place where a specific letter or memorandum was drawn up, sometimes en route to the court, "Camino a la corte" (34). Court news learned through "the palace door" ("por nueva de la puerta de palacio," 79) is retold. Additions were stitched to the book at the appropriate page (747–8). If several copies of a letter were sent out, the text would be copied only once in the register, but any variance in text, context, or agents would be noted in the dockets. In these and other cases, dockets create a continuous whole out of the succession of documents through various techniques, not the least important being the use of the standard legal diction using deictic markers to refer to the texts that precede or follow any given document. For example, "Para el dicho Juan de Castilla, con el dicho Gamarra," or "Otra carta del tenor de la susodicha para el dicho Juan de Castylla que se junte con Pedro de la Dueña y tome el alarde de la gente de cauallo de Almuñecar. / Fecho el dicho día" (7).[56] Many memoriales were sent, often accompanying a letter and containing detailed instructions or accounts for the addressee or for another recipient to whom the memorial is to be handed (e.g., 37, 77–8, 174–5, 221, 297–8, and passim). Decisions on petitions or any additional indications were written on the verso of the relevant document (e.g., 70–1, 163, 288, and passim). When time was pressing, letters might be replied to by simply writing in the margins of the letter received (e.g., 330). Letters can be postscripted by the count and a reply requested to be written on the verso (92). Letters written to a different person accompanied others addressed to the count for his information and feedback and were copied continuously in the register only separated by an explanatory docket (e.g., 96 and passim).

Much as in official state or municipal registers, texts appear copied inside of other texts and all of them inserted in letters, leaving trace memory of their original medium within the text or in an explanatory docket, whose disappearance would lead to the erasure of the evidence of their autonomous life (e.g., 364–7). Some of these letters and reports in Tendilla's hand may have been too personal or too compromising to register, and though the fact that they were written in the count's own hand is noted, the register states that their text is omitted ("no se registró") (e.g., 173, 179, 186, 200, 205). Similarly to the *carta de creencia*, some letters were written for the sole purpose of asking the recipient to meet with Tendilla so that the business may be attended to viva voce, thereby avoiding the compromising nature of writing (e.g., 295). On the other hand, some letters written and recorded in the register were never sent out and were crossed out in the register (e.g., 301–3). Drafts may appear in the register followed by a second version, both being separated by an explanatory docket (e.g., 216). Sometimes the documents sent or received surpassed in bulk that of a single letter or set of documents and became a separate book, which was a fair copy of various documents that needed to be circulated together and presumably archived together as a separate unit. This was the case when statements were gathered from the *comendadores* with information on service, payments, and debits. On such an occasion, a fair-copy book ("en limpio") with the full account was to be sent to Tendilla (296).[57]

The making of fair copies and the keeping of registers were not the exclusive domain of state, municipal, or ecclesiastical business. They also proved useful for persons in middle and lower social strata. Texts that recount the *modus vivendi* of the middle and lower classes offer glimpses of such record keeping. Thus Celestina kept her prostitution business in order with the help of a register: "En naciendo la mochacha, la hago escribir en mi registro, y esto para que yo sepa cuántas se me salen de la red. ¿Qué pensabas, Sempronio? ¿Habíame de mantener del viento? ¿Heredé otra herencia? ¿Tengo otra casa o viña?" (Rojas 2000: 93).[58]

Personal registers proved equally useful in record keeping to higher and lower strata in their function as account books, documentary archives, and repositories of personal memories. As a crossover medium sharing in the official and personal spheres of the archive, the register model proved useful for the compilation of literary writings.

Literary Registers

The unavoidable crossover between the orchestrated method of record keeping used in bureaucratic circles and moving a loose paper of any nature into a permanent record helped establish a "dynamic book" model that operated in various spheres. This model tried to both capture the flux of textuality and create the illusion of stability while enabling the multiplicity of texts. Because of its usefulness, the register had already served as an operative model for earlier authors such as Petrarch or humanists like Coluccio Salutati, who were familiar with notary and merchant culture.[59] Notaries and merchants in turn constituted a particularly receptive public for poetry, as has been pointed out in relation to Dante (Steinberg 2007). In these and other circles, registers served the double function of exemplars and repositories of memory. The register saved short works, loose pieces of paper, and single documents from sure disintegration and oblivion not only by providing a safe repository but also by functioning as a sanctioned exemplar from which further copies could be made. Enrique de Villena (1994) clearly articulated the nature of the register as archive and exemplar or *original* in the prohemium to his translation and gloss of the *Aeneid*, which Villena still hoped could be presented as originally intended to the King of Navarre: "E púsose aquí figurado, como paresçe en este primero registro, siquier original, porque de aquí tomase enxemplo el que lo avía de poner en buena letra para lo fazer como aquí está, si viniera a caso que se pudiera presentar al dicho rey de Navarra, para quien fue començado e fecho" (7–8).[60]

Santillana analogously portrayed the poetic anthology of his collected works (cancionero) as a registro in his well-known letter-proem to the Constable of Portugal, the recipient of that cancionero. In this case, the poetic register served, if we follow Santillana's (1988) account, as a repository of works that had been dispersed mostly in separate copies owned by many different people, from whom the author had to gather them. The emphasis on a diachronic organization of its components following the poet's *curriculum/cursus vitae* is to be noted. These efforts reflect too faithfully what seems to have been common practice in the circulation of fifteenth-century poetic texts to be considered a mere literary topos:

> En estos días passados, Alvar Gonçales de Alcántara, familiar e servidor de la casa del señor Infante don Pedro, muy ínclito Duque de Coimbra, vuestro padre, de parte vuestra, señor, me rogó que los dezires e cançiones mías enbiase a la vuestra magnifiçençia. En verdad, señor, en otros fechos de mayor importançia, aunque a mí más trabajosos, quisiera yo conplazer a la vuestra nobleza; porque estas obras – o a lo menos las más dellas – no son de tales materias, ni as'y bien formadas e artizadas, que de memorable registro dignas parescan. […] Pero, muy virtuoso señor, protestando que la voluntad mía sea e fuesse no otra de la que digo, porque la vuestra sin impedimento aya lugar e

vuestro mandado se faga, de unas e otras partes, e por los libros e cançioneros agenos, fize buscar e escrevir – por orden segund que las yo fize – las que en este pequeño volumen vos enbío. (438)[61]

The original of the cancionero sent to the condestable is now lost, but Santillana later undertook his more ambitious compilation, which he forwarded to his nephew, Gómez Manrique. Related to this manuscript are SA8 and, through further copies, MN8.[62] A similar thrust would inspire compilatory efforts geared towards printed publication. Luís Vicente, son of poet and playwright Gil Vicente (1983), bore witness in his prologue to the *Copilaçam*, the printed edition of his father's magnum opus, to the compilatory labours undertaken by his father in his own hand, in regard to his own works at the end of his life: "E porque sua tenção era que se imprimissem suas obras, escreveu por sua mão e ajuntou em um livro muito grande parte delas, e ajuntara todas se a morte o não consumira" (1, 12).[63]

The features of cancioneros such as MN6, studied by Beltrán (1996), or MN15, studied by Elia (2002), which bear some of the imprint of the processes of textual production, show the marks of continuous addition of materials and thus exemplify early compilatory stages. Successive (fair) copying helped erase these marks. An illustrative example is that of the extant manuscript of the *Cancionero de Herberay* (LB2), which is a professional copy, or a copy of a copy of a lost manuscript. The original manuscript (similar to the *Cancionero de Módena*, ME1) would have actually had a loose form that has been compared to a folder, where new poems in the form of loose papers were added, either at the beginning or the end, the whole being bound together after a period of time (Whetnall 1997). Individual authorial compilations, some perhaps just a cuaderno, were copied together in a larger multi-authored cancionero, which constituted a fair copy. There is evidence that Juan Alfonso de Baena compiled his cancionero out of smaller single-author collections and in this way preserved much of the work of earlier poets such as Villasandino. A distant copy of Baena's original manuscript presented to Juan II survives in PN1, which transpires the modes of compilation as well as the process of copying a copy (probably unbound) of a fair copy more than three decades later (Blecua 1974–9; Dutton and González Cuenca 1993: xix–xxii, xxvii–xxxi).

Cancionero poets made use of various archival models, which granted them textual flexibility when compiling their works. Chronological, social, and literary criteria carried their weight, as did external circumstances. Whether a process of thoughtful arranging, as in the case of Álvarez Gato, Baena, Gómez Manrique, or Santillana, or a haphazard archival effort, as in the case of Montoro, the cancionero benefited from the register model in its use of a flexible book format as repository. Cancioneros adopted the textual fixity of the widely used blank book model combined with the register and the flexibility and convenience provided by both the booklet and the open book models. Poetic compilations were therefore open to addition but also tried to resist textual erasure or loss. The parallels between administrative book models and fifteenth-century poetic anthologies are many. The passage from loose paper or single document to book, the marks left by the touch of all those who handled the text, and the emphasis on authorship are traits that both types of book share. Both suggest the polymorphic nature of textual production, with cooperation and textual variance struggling with the desire for textual fixity, the book being both archive and exemplar. In addition, both the literary and the bureaucratic books use rubrication to delineate the borders of the text and keep the seams visible. As fair copying begins to erase the composite nature of the book, codicological markers turn into textual ones. The modes of handling loose documents and the textual marks introduced in the process of their inclusion in the book constitute an important model for fifteenth-century poets, themselves closely associated with documentary production. The principles that weighed on the creation of the personal register were similar to those of the administrative register. The need to preserve, compile, and produce a unified record, be it institutional or personal, blurred the lines between the two. Helping this overlap were the shared forces of the conservation needs of documents as a mode of self-preservation – institutional or personal – and a superimposed directional sense that threaded the individual texts. The socio-professional occupation of compilers ensured the transfer of a political outlook, as they worked in a professional and personal textual continuum.

7 Books as Memory

The nature of fair-copy codices as textual repositories was heavily influenced by the acknowledged need to preserve the compilation of authenticated legal documents for posterity.

Fair copying created a permanent record that would by design become part of both the individual and the collective institutional memory. The effort to establish and consolidate church and state institutions was fostered by the growth of literacy, which in turn promoted the written word as key to legal legitimacy.[1] This is noticeable in the language used in donations or property transfer. The necrologies that had begun to take shape in the ninth century had as one of their functions to memorialize a transfer of property or of capital. In later periods they were often referred to as *libro e memoria* (book and memory) and were used to record the benefits and privileges granted to an entity, typically a monastery, along with the names of the grantors.[2] One such book is the *Libro de memorias y aniversarios del monasterio de San Pedro de Cardeña*, noted for its important information on Rodrigo Díaz de Vivar, the famous Cid.[3] Mentions of these memory books/cartularies abound in legal texts, as in the extract of a will copied in the diplomatic collection of Santo Toribio de Liébana, where the donor asked to be inscribed in the monastery's libro e memoria (Álvarez Llopis et al. 1994: 569), or in the thirteenth-century *Tumbo menor de Castilla*, a self-named *libro de memoria* that bears the intention of creating an "everlasting memory" ("por perpetua memoria") of the royal grants to monasteries attached to military orders (Archivo Histórico Nacional, Sección de códices, sign. 1046B, fol. 452). The additional authentication provided by illuminations in cartularies and legal codes was, in the case of the necrologies or "book of memory," ratified epigraphically on the donor's tomb. This served not only to preserve memory of the donation by inscribing it in a durable medium like stone or marble, but also to ensure it entered into and remained visible in the collective memory, instantly identifying the donation with the donor. Through this public display the donor ensured that the prayers or Masses for which s/he had made provisions would become a collective responsibility by way of public memory.[4] Public inscription would thus amplify the durability of the written word.

As King Alfonso X and later Juan Alfonso de Baena made clear, the documentation of memory was unmistakably linked to power. In his paraphrase of King Alfonso X's prologue to his *Estoria de España* at the beginning of Baena's own *Prologus Baenensis*, the poet and escribano made clear the tight bond among knowledge, power, and the preservation of memory through writing.[5] Capitalizing on the importance placed on record keeping and memory, Juan Alfonso de Baena fashioned his *Cancionero* (1993) as a historical document, highlighting the value of the knowledge preserved and transmitted through writing in his introduction (4).[6] After making a long argument, Baena leaps to place poetry in a privileged sphere of knowledge: it enlightens the mind, awakens the understanding, and feeds the memory; the heart gladdens and the soul takes comfort; and all the senses benefit from its governing, sustaining, and soothing properties. In short, poetry is the ideal conveyor of historical memory because of its ideal attributes as a discourse (7). At the same time, because of its effect on the senses and the mind, poetry was the vehicle for personal transformation, thus placing the self-generated self in the midst of the political sphere of action. Garcia de Resende explained the purpose of his own compilatory efforts as a way to capture the memory of past and present. Only those more knowledgeable than him could conjure up further memories of greater deeds, upon which he did not dare lay his hands. For Resende, this provided the thrust for his efforts to put together (*ordenar*) the book, which he intended as a prompt for others to "bring to memory" ("trazer aa memoria") other great deeds (*Cancioneiro Geral* 1990–8, 1: 10–11). Resende's massive and meticulously sourced cancioneiro made it a tall order for anyone more knowledgeable to provide a greater poetic repository of memory, which, by Resende's own parameters, aggrandized the figure of the compiler.

Portable Textuality and the Remembered Self

Recording memory was a pressing need not only in the public but also in the personal sphere. The connection between memory and the self was fostered by the written word carried on paper close to the body, physically putting text and body in close and constant contact. In their breast, the young women whom fray Luis de León (1991) was trying to turn into perfect wives would carry short poems (canciones) and sonnets (278). In a similar vein, recipients showed obedience to a letter with orders from the inquisitor by taking the letter with their hands and putting it to their foreheads ("tomaron la dicha caussa en sus manos e la pusieron sobre sus cabeças").[7] The association of writing with the body (Dinshaw 1989), with manifestations such as the book shaped like a bodily organ (Jager 2000), encouraged the notion of a physical bond between body and book (or other supports), with one as an extension of the other. While the book, the written word, helped the mind to retain essential information, daily contact with the personal book (or booklet) helped promote writing as a pervasive means for the far-reaching inscription of daily life.

Personal books had an unquestionable advantage because they were indispensable for jotting down personal notes. The use by nobles and royalty of the "libro-memoria" was traced to antiquity, as Juan Fernández de Heredia (1995) noted at the end of the fourteenth century: "Et scipion oydas aquestas paraulas obseruo las bien en su libro memoria por tal que las pudiesse reportar & dezir enel senado de Roma" (fol. 255v).[8] Blank books are found in the inventory, carried out at Queen Isabel's orders, of the treasure kept in the Alcázar of Segovia: "otro libro de papel blanco pisano sin escritura con vnas tablas de papel enforradas en cuero verde" (in Ferrandis 1943: 150).[9] The use of personal blank books allowed for the recording of confidential matters at a time when conducting business required the use of writing as an extension of memory. Queen Isabel and King Fernando, as well as King João II of Portugal, used secret blank books, where they would annotate "in their own hand," matters too private for their escribanos and secretaries, such as the abilities and services performed by particularly talented people. Close collaborators were witnesses to the use of these books and could eventually be privy to their content. Garcia de Resende (1994) reported that João II used to carry a secret book "written in his own hand" where he would annotate the services rendered to him. Resende boastfully proclaimed that he had this book in his possession after the king's death. João II also had a second secret book where he noted the names, titles, and skills of members of those in his service in order to make use of them appropriately in his service:

> Antre outras muytas vertudes tinha esta singular: tanto cuydado de quem no bem servia que sem lhe pedir merce lha fazia e trazia secretamente hum livro escripto por sua mão que algum nunca ho soube senam / depois de sua morte,

no qual tinha feyto todolos homens a que mays obriguado era cada hum em sua cantidade em capitollos que dezião: "Foam me tem feitos taes serviços, lembrar-me-ha quando cousa vaguar que nelle cayba de o prover." E quando as cousas vagavam e lhas vinham pedir dizia: "Jaa a tenho dada"; e então secretamente via no libro as pessoas da calidade da tal cousa e aquella a que mais obriguam tinha a dava; e aas vezes estando as tais pessoas fora do reyno em seu serviço lhe mandava cá fazer seus despachos, de que muytos se espantavam, e foy singular vertude em que todollos boõs tinham muyta esperança de seus serviços; e este livro tenho eu em meu poder. E assi tinha outro livro em segredo em que tinha escripto todollos homens autos pera delles se servir nas cousas pera que eram, cada huns em seus titulos, huns pera capitães de cousas grandes e outros doutras somenos, outros para embayxadores, e assi pera enviadeiros, e tambem pera todollos carregos e cousas necessarias. (141–2)[10]

Lorenzo Galíndez de Carvajal (1875–8) tells of a similar use of a personal book by Isabel and Fernando: "En su hacienda pusieron gran cuidado, como en la elección de personas para cargos principales de gobierno, justicia, guerra y hacienda [...] y para estar más prevenidos en las elecciones tenían un libro, y en él memoria de los hombres de más habilidad y mérito para los cargos que vacasen" (533).[11] Other supports were used for similar purposes. Five memoranda written in Queen Isabel's hand are listed in the post-mortem inventory of her chests and provide evidence that the queen used them to make provisions for people whose names are not revealed in the inventory and that were most likely of an economic and political nature. These are far from the more substantial book that Galíndez de Carvajal chronicles and point to a light, portable and private writing support: "Cinco memoriales, todos de mano de la Reyna nuestra Señora, que aya santa gloria, el vno es en vn pliego que esta escripto vna plana y el otro esta ... plana en la vna hoja y en la otra hoja, que esta cortada, es de fasta treze renglones y otras tres fojas son escriptas todas de mano de su alteza e no son memoriales de inportancia saluo de cosas que su alteza mando proveer con vnos y con otros por entonçes" (Torre y del Cerro 1968: 6).[12]

Appealing to biblical history, Fernán Pérez de Guzmán (1965) traced the practice to Judaic practices, but noted censoriously that there was no point in writing down the services rendered to the Crown, because it yielded no benefits, as kings only reward those who flatter or please them:

La verdat e çertidunbre del origin e nasçimiento de los linajes de Castilla non se puede bien saber sino quanto quedó en la memoria de los antiguos. Ca en

Castilla ovo sienpre e ay poca diligençia de las antigüedades, lo qual es grant daño. E açerca desto falla onbre en las estorias muchas notables usanças, de las quales contaré dos. Primera, que en el tienpo que los judíos avían reyes, tenían en los armarios e caxas del tenplo libros de las cosas que cada año acaesçían e eran llamados annales, e tenían registros de los nobles linajes. E duró esto fasta el tienpo del rey Erodes el grande, el qual con terror de perder el reino e que lo avrían algunos reales, fizo quemar todos aquellos libros. Por çierto no fue alguno entre los tiranos que tanto temiese perder el reino, ca por esto fizo quemar aquellas escrituras e aun fizo matar los inoçentes, que fue una extrema e singular crueza. El segundo acto de aquel tiempo era, segunt se lee en el libro d'Ester, que el rey Asuero de Persia tenía un *libro de los serviçios* que le eran fechos *e de los galardones* que por ellos diera, sin dubda notables actos e dignos de loor. Guardar la memoria de los nobles linajes e de los serviçios fechos a los reyes e a la república, de lo qual poca cura se faze en Castilla, e a dizir verdad, es poco nesçesario, ca en este tienpo aquél es más noble que es más rico. Pues ¿para qué cataremos el *libro de los linajes?*[13] ca en la riqueza fallaremos la nobleza dellos. Otrosí *los serviçios* no es nesçesario de se *escrivir para memoria*, ca los reyes non dan galardón a quien mejor sirve nin a quien más virtuosamente obra, sino a quien más les sigue la voluntad e los conplaze; pues superfluo e demasiado fuera poner en letras tales dos actos, riqueza e lisonjas. (18)[14]

Pérez de Guzmán may have been protesting the same kind of unjust practices that drove him to withdraw from active participation in politics. With regard to genres, Pérez de Guzmán seems to be contrasting two types of books: a book of lineages or *nobiliario,* and a book of merits and rewards, a private libro de memoria. His disparaging remarks about the second type of record criticize the equation of nobility with wealth (see chapter 1) and thus point to the uselessness of a meritocracy that was subjective and unstable, inscribed in a like support, a personal memory book. The memory book of queens and kings would help conceptualize nobility as an unstable quality too dependent upon royal individual perceptions of aspiring members of their inner or even outer circles. His *Generaciones y semblanzas* constituted, in his estimation, a first attempt at the creation of a contemporary book of lineages and a more viable means of documenting worth. Pérez de Guzmán shows the importance of controlling the creation of an institutional memory and exemplifies the tensions among the vying forces competing for such control.

Remembering was crucial for Queen Isabel, who seems to have surrounded herself with papers, booklets, and books for such a purpose.

Isabel used loose papers that she would put in her sleeve to similarly act as little reminders: "Y de la Reina Católica Doña Isabel se dice que cuando gobernaba con el rey don Fernando su marido, se le cayó acaso un papel de la manga en que tenía escrito de su propia mano: 'La pregonería de la ciudad se ha de dar a fulano, porque tiene mayor voz'" (as told by Galíndez de Carvajal 1875–8: 534n).[15] This practice is strikingly similar to advice given to Italian merchants. Paolo da Certaldo, in his *Libro di buoni costumi*, suggests having a piece of paper or a booklet on which to write little reminder notes. His recommendation to always carry the paper in one's money purse is intended to make it easier to find the paper and have constant access to its contents (text in Branca 1986a: 29n136). Further evidence for Isabel comes from the enumeration of the books and papers found in the queen's chests and caskets after her death. Queen Isabel's inventory shows several blank books or booklets that she used to jot down notes ("cinco libritos para escribir memorias" [five little memory books]).[16] These small memory books could have been regular paper books, but perhaps, in light of the description found in the inventory, they were more likely libros (or libritos) de memoria made out of wood, ivory or similar material, and/or paper, coated with ink or burnish and then written upon with a stylus. These are amply documented in sixteenth-century inventories and literary texts.[17] In these "books" writing could be erased by smoothing the surface, which could be reused like a wax tablet. They provided the same portability and versatility as a small blank book or piece of paper, though the intrinsic impermanence of the text copied in them necessitated, if the text was to be preserved, its eventual transfer to a regular book archive. Similar books used for drawing appear in the queen's inventories: "Mas dos libros de debuxar, vno negro de nueve tablas con su çerradura e cabo e otro blanco de otras nueve tablas con su çerradura e cabo de latón."[18] The use of "memory" wax tablets by fifteenth-century authors is documented in Juan de Lucena's (1892b) *Libro de vida beata*. In it, the gloss to the term "tablillas ceradas" provides additional information on this medium – which was intended to safeguard the privacy of writing – and its relation to memory: "Tablillas de cera otro tiempo vsauan los cónsules de Roma, y oy usan muchos, donde comendauan a la memoria las cosas que querían que no viniesen en público, porque luego las rematauan" (141).[19] In his *Libro de vida beata*, Lucena has Alfonso de Cartagena portray Juan de Mena as writing on wax tablets whatever Cartagena says or does; this discourages Cartagena from improvising a poem for fear that it would be entered into the private tablets and thus stolen from him: "Jamás laureado poeta vi tan puntoso quanto tú. Con tablillas ceradas razonas conmigo; arráyasme luego sy fablo, sy digo, sy río, si juego, si burlo contigo. Por pocas te faría una copla; mas temo que no me la notes. Arrebatar la palabra, y roerla, más es de can que de ombre" (141).[20] Mena's ability to take full advantage of the flexibility and portability of the tablet medium was inconveniencing Cartagena, for whom such impromptu writing amounted, as he said (half) in jest, to stealing and gnawing, an activity more apt for a dog than a human. Casually jotting down an authored text on a transient medium for the purposes of memorial inscription was convenient for the scribe but degraded him in the eyes of the author when the scribe acted beyond authorial control. Further, the textual appropriation of another's work in a private medium was problematic because it could eventually find its way into (unauthorized) publication. Acting as an independent agent and using portable writing supports, the scribe no longer ruminated and wrote, but rather mangled and stole the author's texts. However, Cartagena's satirical representation of Mena's writing modes failed to provide a compelling argument against scribal authorship. The "laureate poet" did not build his reputation by plagiarizing Cartagena. The use of flexible writing supports, though threatening to other authors, allowed constant poetic activity and authorial independence.

Self Support

In Castile, the use of the "libro e memoria" by a middle class of merchants and professionals, including converted Jews or conversos, is attested to by the anti-Semitic mock document in which King Juan II allegedly gave a nobleman what were perceived as converso privileges, the infamous "Traslado de una carta de privilegio que el rey don Juan II dio a un hijo dalgo" published by López Martínez (1954): "E asímismo que fingidamente podades entrar en la iglesia y lugares sagrados sin ninguna devoción, llevando en lugar de oras o psalterio el libro e memoria de las rentas e alcavalas que tenés arrendadas a vuestro cargo, fingiendo que resáis los psalmos penitenciales, como lo facen y acostumbran muchos de la dicha generación de los marranos" (386).[21] This citation points to the small and flexible format of these memory books, closer in size to the ubiquitous books of hours than to their larger monastic counterparts.[22] The size and portability of both prayer and personal account books caused obvious anxieties, if we are to judge from this text, as participants in religious services were rendered unable to judge the contents of the book in hand and were thus deprived of a form of control over beliefs and thoughts. This anxiety worked concomitantly with a similar one occasioned by the rise in silent prayer ably studied by Saenger (1989). The small format of "pocket" books (Petrucci 1995: 169–235), epitomized by Luis Milán's diminutive *Libro de motes*, facilitated the portability of the written word and the further inscription of identity, with the possibilities it offered for social and/or ethnic profiling, as in the case of the text cited above. Large-format account books, however, belonged to institutions and conveyed a grander sense of the importance of its business. Juan del Encina, in the prologue to his 1496 large folio *Cancionero* (1989), capitalized on this perception when he called it a book of new accounts ("libro de nuevas cuentas"), defending his decision to publish his works, some of them circulating in badly disfigured versions, in order to silence those who thought him only able to produce rude and comical poetry. His cancionero was thus a way to explain and bring closure to the past and to start anew in a bibliographically grand manner.[23]

The continuities among book models enable the conceptualization of a personal accounting book as an embodiment of the self that relies on various textual supports. For poet Juan Agraz, the sum of his life took the shape of an account book ("el libro de mi cuenta"), where he kept a tally of the debits and credits of his life. Additional records came from the devil in the form of a booklet ("cuaderno") relating Agraz's shortcomings, which proved lengthier than the defending angel's much shorter documentation in the form of a sheet ("su plana") filled with Agraz's Christian deeds (ID0380 vv. 29–32, 53–8). An anonymous caballero sent his collected works, copied in his own hand ("de mi mano trasladadas"), to a lady whom he had just started "serving." In spite of the stated fact that part of the love complaints in the book had been inspired by other ladies, the grand total of the blame is assigned to the cancionero recipient alone: "y avnque a otras se presenta / parte aqui de mis querellas / al rematar de la quenta / la suma de todas ellas / a vuestra merced se asienta" (ID0228 vv. 6–10).[24] In this case, the autograph cancionero functions as an account book of documented love debts. When handed to the lady, who is named the wholesale debtor, she becomes solely responsible for payment. In content and form, the cancionero becomes a constant reminder of the debt, which is heightened by the skill and effort involved in its poetic encoding. However, the ultimate embodied document may be the person fashioned after the most powerful exemplar, as found in ID4378, penned by converso poet Pedro de León and studied by Perea Rodríguez (2005). In this openly messianic poem, Christ consults his registro and orders a faithful copy of himself ("traslado"): "El glorïoso Mexxías / requiriendo su registo / con todas su gerarchías, / ordenó que, en nuestros días, / un traslado fuese visto / elegido para Cristo, / ungido por ser propheta / para cavallero e Rey" (vv. 13–20).[25] The resulting *traslado*, the poem goes on to explain at length with a profusion of detail, is none other than King Enrique IV, the identification having already been provided by the initial rubric.

Documenting Memory

Because the preservation of documents and the documentation of experience were important for both institutions and individuals, many types of texts blended the personal and the political spheres. A memorial (memorandum) or *relación* was a freestanding text to which any of the interested parties could contribute when an escribano or notary was in the process of preparing a legal document.[26] As such, it became inserted into the final document, which through such means became a co-authored text (Bono Huertas and Ungueti-Bono 1986: 42). In this manner, individually produced accounts constituted a contribution of personal to institutional history. Personal accounts in the form of memorial or other such documents relied on a flexible medium and were usually not longer than a few pages. The format of a memorial was flexible enough to fit the needs dictated by the content as well as its intended use. In their administrative use, memoriales could show strong affinity to account records, as shown by the five-page document regarding payments to the *alfaquíes* of Granada published by Albarracín Navarro (2002), which is written by two hands and visually similar to Gonzalo de Baeza's books of expense accounts. Fernández de Oviedo (1870), describing the complex system of record keeping used in the establishment of Prince Juan, offers a glimpse of the uses of registers, account books, and memoriales, as well as the scrupulous copying and cross-referencing to which they were subject. According to Oviedo, memoriales were used to itemize lists of clothes and other objects in the prince's chamber for accounting purposes and as a way to receive and grant petitions, often of monetary value (e.g., 30, 35–7, 42, 61–3). Their use in providing narratives or directions too lengthy or specific to be trusted to anything other than paper suggests further affinities to a larger archive such as a register or other repositories, where, depending on their importance and content, memoriales could ultimately be copied. As first-person accounts of experiences or thoughts, they fit the self-referential writing of the "autobiographical subjects" studied by Smith and Watson (2001: 2–3, 16–21), with a clear political or economic bent. This is the point of reference for Leonor López de Córdoba's famed self-account, which is laced with notarial language and references to "escriptura" and "instrumento" (Lacarra 2007).

A memorial by Hernando de Zafra, converso and secretary to King Fernando, exemplifies the elasticity of these texts and their development into an introspective genre that would be captured by literary authors, themselves memorial artisans. Zafra's memorial, ostensibly addressed to King Fernando, whom he refers to as "su altesa," is a first-person narrative that in chronological order gives an account of his key role in the conquest of Granada, his loyal and uninterrupted services to the Crown thereafter, and the impact on his personal finances, ultimately arguing for a reward in the form of a lucrative position.[27] Memoriales as short eyewitness accounts served as sources for chroniclers and historians, as evidenced by the narrative quoted on chapter 1 where Fernando de Pulgar received the first-person account of the battle from Enrique Enríquez, King Fernando's uncle and his chief major-domo. These history-making memoriales conveyed the personal experience and perspective on a given event, a fact recorded by chroniclers trying to tie them all into a larger and politically meaningful narrative. Diego de Valera (1927) in his *Crónica de los Reyes Católicos* refers to varying accounts of given events depending on the authorship of a specific memorial. Thus, referring to the capture of the king of Granada, he states: "Y en la prisión del rey ay esta diferencia del memorial del conde al del Alcayde de los Donzeles; porque el conde dize que quando los moros huyeron que el rey de Granada fué el postrimero" (167–8).[28] A similar system is mentioned in one of the letters (Letra XXXIII) of another royal chronicler, Fernando de Pulgar (2007), who thanked the Count of Cabra for his memorial relating the count's victory in battle, which Pulgar explicitly said he would incorporate into his own chronicle. Interestingly, Pulgar also refers to a previous narrative of the same battle that the Countess of Cabra had sent to Queen Isabel, who in turn had forwarded it to Pulgar to use as a source for his chronicle (108–9). The first-person or eyewitness nature of these texts justified their use as historical sources of a larger narrative, the official account penned by Pulgar and sanctioned by Queen Isabel. In so doing, they make evident that even grand narratives like chronicles are multi-authored endeavours and the text itself an edited (fair) copy of sundry single documents threaded together by an imposed narrative line. The fragmentary draft that has survived in the Archivo General de Simancas studied by Fernández Gallardo (2004) is witness to this methodology. Chronicles are the result of textual cooperation that was by design fraught with multiple subjectivities. They share the constitutive importance of this layering of texts with other types of writing. The memorial nature of the narrative could be played to its best advantage in individual or noble chronicles, as is the case for example of the *Victorial*, which author Gutierre Díaz de Games fashioned as a reminder of services rendered or "memorial de servicios prestados," as noted by Gómez Redondo (2001: esp. 200–2).

When it involved a compilatory process, the memorial often merged with the registro to create institutional memory. This model informed the new genre of history proposed by Fernán Pérez de Guzmán (1965), who by his own account sought to improve previous modes of historiographical writing in Castile. Taking an idea sanctioned by the authority of Guido delle Colonne, he chose to fashion his famous *Generaciones y semblanzas* as a *registro o memorial*, providing insightful profiles of notable aristocrats and monarchs (3), a model that overlapped with that of the books of lineages discussed above.[29] As a practice, it also had antecedents in the earlier *anales*.[30] Andrés Bernáldez (1962), who allegedly used his grandfather's register model as self-appointed memory keeper and political commentator of the Catholic monarchs' rule, described his work as a *memoria* produced with privileged first-hand information (23–4). For Bernáldez, a memoria was a collective affair, subject to modifications by more trustworthy witnesses (24), thus combining the personal account with that of others in order to create a sanctioned narrative. Also a combination of personal and institutional memory, Melcior Miralles's (1992) *Dietari* constitutes, according to its editor, a cross between a diary and a chronicle and as such was saturated with an ideology that steered its meaning. Miralles, chaplain to King Alfons V of Aragon (el Magnànim), included a fair copy of a collection of his own holograph notes,[31] ending his account of witnessed events by promising further entries: "Lo que en apres se seguira e saber porem, metrem en libre de memoria" [Whatever happens after this and comes to my attention, I will put it in the memory book] (407). Royal archivist Pere Miquel Carbonell was charged with keeping a similar dietari during his service at the Royal Archive in Barcelona, among other writings and documentation. His writings are a rich mix of official and personal business, including poetry, historical notes, free-flowing personal and literary lucubrations, and colourful commentaries on the mandatory reduction of his scribal fees, all inserted in the official documentation under his charge (texts in Martínez Ferrando 1961: esp. 105–8). Of a similar nature is a book by a member of the middle class, Garci Sánchez, *jurado* of Seville, most likely a converso.[32] Garci Sánchez's ambitious work starts as a larger world history narrative to become minutely and densely filled with eye-witness detail of political events in the central decades of the fifteenth century.

The relevant role played by bureaucratic personnel, particularly notaries, in the development of a personal written memory weaved through official narratives or historical records, protocols, and registers, has been pointed out for Catalan and Valencian texts (Escartí 1998). Similar practices are evident in other Iberian cities such as Seville, which already in the thirteenth and fourteenth centuries kept a book of remembrances ("libro de las remembranças") (Ostos Salcedo 2005: 96). The escribanos had the charge of documenting the memory of a community in detail. In the *Ordenamiento* of the *cortes* of Valladolid in 1312 the escribanos were ordered to keep a log of town events ("todos los ffechos que acaesçieren en ssos logares") in their registers so as to keep the king informed upon demand ("porque me den rrecabdo ende cada que gelo yo demandare") (in *Cortes* 1861–1903, 2: 209). The 1503 ordinances that Queen Isabel issued for the escribanos of Seville similarly commanded the escribanos to accept additional documentation to be included in their books as "registro o memorial" (text in Rodríguez Adrados 1988: 781). Escribanos were accountable as the holders of textual memory, being responsible for recording public and institutional events, enabling the monarchy to use the information for its own political purposes. The same system of register keeping as a political tool was in the hands of those specifically appointed to create an official narrative. In his *Crónica de Enrique IV*, Enríquez del Castillo (1994) begged forgiveness for any shortcomings regarding the earlier and more prosperous part of the king's reign. As a justification for any omissions, the chronicler complained of the theft of his possessions and of all the registros that were to provide the materials for the chronicle when he was imprisoned in Segovia: "donde me rrovaron, no solamente lo mío, más los rregistros con lo proçesado que tenía escripto de ella, visto que la memoria, segund la flaqueza humana tiene mayor parte de olvidança que sobra de rrecordaçión" (132).[33] The theft of the registers meant the vanishing of institutional memory, which could not be exclusively buttressed by individual powers of remembering.

The Inquisition as a large, ever watchful institution displayed similar textual strategies. The Inquisition used memory books (libros de memoria) as the repository of inventories, drawn up by the judges (*Jueces de Bienes*), of the goods confiscated from the accused.[34] However, immediate textual control was for all practical purposes, in its basic layers, in the hands of the escribanos, who saw the paper in front of them and their registers as a medium of their own, to be used for recording not only official but also personal business and observations. The fifteenth-century notarial cuadernos and assorted loose records in Seville's archives extracted by Bono Huertas and Ungueti-Bono (1986) hold a rich mix of information in the form of marginal notes, including: records of the public ceremony of the baptism of forty-eight converts to Christianity (plus "children and older women") (61); punishment of two women for lesbian practices (68); the burning of forty-seven people and "reconciliation" of twenty-three punished by the Inquisition (121); a mix of legal, personal, and family notes (182–3); a notice of thirty-five people burned by the Inquisition (188); a notice of deaths from "pestilence"; a disgruntled note on obeying new royal ordinances concerning the escribanos complaining that these must obey the Queen in spite of the Devil ("a pesar del diablo") (196–7); and a lengthy note on processions of people condemned by the Inquisition (241). Personal and socio-political history was thus written by escribanos framing the official records that were under their charge. However, access to writing allowed individuals to keep personal records of the events surrounding inquisitorial activity, such as the one dating from 1485 to 1501 published in Horozco's (1981) *Relaciones* and cited in chapter 2. The pervasively vigilant Inquisition was in turn being watched, the eyewitness accounts producing their own narratives.

Poetic compilations illustrate authorial skill by intersecting administrative memoranda with other types of discourse. In a transposition of the socio-political power structure, authors of bureaucratic or politically minded memoriales penned poetic renditions, creating verse accounts of services rendered to a lady who would presumably be compelled to produce a reward upon receipt. Modelled on the practice of writing memoranda to the Crown, poets such as Diego de Valera (ID2121), Juan del Encina (ID4459), or Jorge Manrique (ID6151) wrote verse memoriales that emphasized the nature of their poems as repositories of information and a reminder directed to a potentially forgetful lady. These verse memoriales show the effects of authorial control over a text written for the improvement of the recipient's memory. Through the process of (self-)publication, personal reminders entered the public domain and in the process authors became aggrandized by their mastery of the mixed discourse of bureaucracy and poetry. Juan del Encina remarked on this relationship between poetic writing and memory. The forgetful lady will always have the poem as a memorandum of the poet's suffering in his love for her – "contareys por este cuento / de aqueste memorial" [you can count on the account / of this *memorial*] (ID4459 vv. 39–40) – and so will, we may observe, a great number of readers. Fray Íñigo de Mendoza's poem ID1949, addressed to his sister, doña María de Herrera, abbess of the important

Huelgas Monastery, is punctuated with references to "memoria," calling to "Acordáos" or "Acuérdeseos." The surviving memoriales give an idea of what these poems may have looked like when they arrived in their recipient's hands.[35] In these poems, the emphasis ultimately rests on the author's worth and what is due to him, their publication amplifying the extent of individual memory.

Accountable Memories

The heavy use of blank books and blank registers in public and private record keeping translated into a booming business for booksellers, who, along with blank books, registers, and ledgers, also sold writing materials such as paper, parchment, and pens.[36] The 1504 post-mortem inventory of Joan Santalínia, a *pelaire* (wool comber) and shopkeeper in Castellón, shows books, booklets, and single sheets of paper, both written and blank, kept in different pieces of furniture (table and armoire drawers) in the home's entryway (Navarro Espinach 2006: 165). Other inventories show a similar display of different types of reading and writing materials appearing in various rooms (from the entryway to the kitchen, to the study, bedchamber, and various other rooms) and the use of writing in the private homes of small merchants and artisans in Valencia (Aparici Martí 2008). A similar range of supports, including papers, booklets, small booklets, and blank books may be found among Italian merchants, and the extended use of account books is evident in Italian, Catalan, and other European merchants.[37] Italian *ricordanze* like those of Giovanni di Pagolo Morelli or Bonaccorso Pitti (published by Branca 1986a) and domestic memories inscribed in notarial chronicles (Zabbia 1998) share important characteristics with similar recording practices among the middle classes in other parts of Europe.[38] In addition, there are striking parallels between the work of García de Salazar, a noble who had fulfilled escribano and notary public duties (Avenoza Vera 2007), and that of the ricordanze published by Branca (1986a).[39] Because of the need for record keeping, the middle class of artisans and merchants was heavily invested in writing (e.g., Aparici Martí 2008, Caunedo del Potro 2006, Navarro Espinach 2006), their books constituting a mix of business and personal records. The weaver ("texidor") Johan Fretero's "libre memorial, capbreu et, o, repertori" (Aparici Martí 2008: 31), or the account book (*libro e razo*) of one Valencian merchant, Gaspar Trinxer, which starts out in 1497 (Mandingorra Llavata 2002), exemplify the books' range. Pezzarossa (1979) has rightly pointed out the interconnectedness among chronicle, diary, memorial, memory book, and *ricordanze*, and thus the difficulty in studying these textual practices as independent genres (96), as may be further gleaned from the Valencian texts studied by Escartí (1998). As part of a textual continuum, any professional or frequently used book could harbour the self's private memory. Of particular interest are the manuscript booklets of notaries, where various texts relating to the exercise of the notarial profession, such as legal texts and formulas, were joined by others relating to the individual's family life or having practical interest – i.e., recipes for ink – mixed together in a sort of personal miscellany or commonplace book (as studied by Durán Cañameras 1955). An example may be found in Ms. Ripoll 140, of the Arxiu de la Corona de Aragó (Durán Cañameras 1955: 149–51). The two short booklets (in tira form, measuring 315/320 × 115/120 mm) of merchant Pere Seriol, dated in 1371, mix business accounting with blank spaces and a short chronicle of his (unrequited) love life, complete with sightings of his beloved and a copy of the different love messages that he wrote for her on the walls of streets that she frequented (Gimeno Blay and Palasí Fas 1986). The textual continuum constituted by personal and institutional memories is likewise evident in cancioneros such as BU2, CO1, EM9, MN19, MN56, MR2, MR3, PN2, and ZZ3 (table of contents), which contain miscellaneous prose texts, including letters, a public oration, historical accounts, political commentary, legal documents, *memoriales*, treatises, and literary texts, among the poems.

In their mixed content, personal memory books share some traits of the medieval *florilegia* and poetic and prose miscellanies[40] such as Boccaccio's *zibaldoni*.[41] This type of textual association also recalls the sixteenth- and seventeenth-century memory book that can contain prose alongside verse and letters, and obviously the commonplace books or memorial books of the same period.[42] Sixteenth- and seventeenth-century "memory books" show the pervasiveness of textuality through the different social strata and the coexistence of letters, verse, expense accounts, and other sundry texts.[43] The book of memory, in Cervantes's (1994) *Rinconete y Cortadillo* (598), for example, is the place to list the petty crimes to be committed for the week and their payment. Cervantes (1998) has his don Quixote write the letter to his beloved Dulcinea in his memory book, which Sancho will take to her, after having the letter transcribed onto a separate piece of paper in a more suitable format (286, 295). In Lope de Vega's (1988) *La Dorotea*, verse gets written in a memory book as well (259, 330). Modern letter writing, the development of autobiography, and other forms of narrative owe much to these material developments as they contributed to the progression of the relationship between the self and the written page.[44] This also points to continuities in the relation between poetry and memory book already present in Dante's (1960) famous reference to the "libro della mia memoria" in his *Vita nuova* (153).[45] This well-known reference makes evident the early connection between the personal notebook, similar to the memoriales or the memory books, and the poetic compilation.

These texts and the register practice described in the previous chapter show the close connections between register and memorial, with their similar function as a written record of personal and institutional or collective memory. Both foster the creation of a narrative that emphasizes individual agency in record keeping and retelling. The connecting thread from hand to pen and paper, and thence to the interests they served, rested on the combination of technical ability and political reliability. The traits that developed as a necessity for the construction of a strong bureaucracy, combined with those that served the needs of commercial and individual household record keeping, facilitated individual applications that kept pen, paper, and texts not only close at hand but also close to the heart.

The practices described in this chapter give witness to the use of writing, pen, and paper as an extension of memory. Poetic anthology, account book, and personal book all intersect and facilitate the practice of authorship, obeying similar forces that require the use of a flexible medium, one that could incorporate increased uses of textuality. These book models also address specific needs of textual preservation, as well as support the cultural and cognitive developments that foster the identification of text and self. The needs of new communities of readers and writers, alongside those of the bureaucrats, intellectuals, middle class, and an increasingly strong monarchy thus converged in the formation of flexible and personalized book models that would fit new textual needs. In this sense, while a growing use of documents encouraged literacy, the opportunities for authorship encouraged the production of documents,[46] creating a spiralling effect of paper proliferation. When collected for a specific purpose, these papers and scattered notes became powerful constituents of a personal narrative.

8 Arranging the Compilation

Moving the text from roll, booklet, or loose paper to a book and creating an organic whole relied on careful threading to open up the text, hold it together, and secure its meaning in the hands of an attentive compiler. The individual support of the copied text would endure an erasure that it resisted by creating a textual imprint of the vanished medium, effectively narrating itself and its (past) situation. Turning a pile of sundry papers into an organic whole created the need for a meaningful paratext. The paratext, as much as the text, bears the mark of the material components of the poetic anthology. One key aspect of the paratext are the rubrics or short introductory titles that convey important information about the poem, including the circumstances of the poem's composition, its author, genre, or any element that will help connect the abstract or obscure content of the poem to a concrete material object, body, or situation necessary for the production of meaning. While highly dependent on its contextual situation, cancionero poetry uses a conceptual and abstract language, rhetorically colourful, that can be almost impossible to comprehend with precision unless the extratext is understood. The function of the rubric, specific to the processes of poetic compilation, can be better demonstrated by close examination of an example like the following poem:

> si la carne que rresçiuo
> encarna mis huesos por vos
> mas areis que hiço dios
> de encarnarme siendo viuo
> o que misterio hareis
> señora tan conocido y señalado
> que a todos dezir podreis

que primero fui nasçido
que encarnado. (ID0211)¹

The highly abstract poem plays with the concepts of flesh and incarnation (*carne-encarnar*), a skilfully developed polyptoton (*derivatio*), common in fifteenth-century Spanish poetry and suggestive of a religious (and possibly blasphemous) reading in which the lady and the poet take turns at performing the divine mystery of incarnating or becoming incarnated. The humorous deception of the reader and thus the one-upmanship and wittiness of a seemingly compromising text is not properly understood without the rubric: "Copla de vn caballero para vna dama porque le envio carne de membrillo estando flaco."²

Similarly, poem ID0248 (MN6e) suggests a warfare motif, perhaps a metaphor for the love conquest:

Sostener la gentileza
con tan flaca fortaleza
tengolo por cosa vana
porque por la barbacana
La dara naturaleza
Asi quedara la gala
con deffensa harto mala
y de ser hondo el fossado
tanbien seruira descala
por parte de lo arrugado
si el conbate es apretado.

However, the rubric surprises the reader again and attempts to fix the meaning by transcending not only the obvious literal interpretation, but also the decipherment of the metaphors: "Del Almirante de castilla a don Joan de mendoça viendole vn dia rapada la barba porque se pareçian muchas arrugas en el rostro."³ The reader is thus ushered into a reading of the poem and into an appreciation of the ingenious double entendre of the Spanish "barbacana" as "barbican" and as "barba cana" or "grey beard," both serving a similar purpose of supporting potentially unstable structures.

As can be inferred from the examples, the poem is based upon an intended ambiguity made possible by the abstraction of its elements. Such a characteristic is one of the main tenets of cancionero poetry. The rubrics function as a reading aid and at the same time help situate poems within social and literary transactions, framing them with crucial information not necessarily provided by the poems' text. Besides historically situating political, religious, or moralizing poetry, many rubrics lay open the grounding in daily life of poetic composition: the poet's lady friend asking him for stanzas to read (ID0705); the dog that ate the piece of meat that belonged to a lady and her sister in Murcia (ID0995); the return from hunting followed by a game (ID3127); an invitation to dinner (ID1896); an ill choice of attire (ID5982) – all provide grounds for poetic invention. As the poem textualized and strove to turn daily life into an abstraction, rubrication developed as an attempt to recover the concrete materiality of the written record and of the situation it textualizes.

The following pages will examine rubrication as a key element of the poetic compilation. Some of the more important functions of the rubrics are to articulate and structure the poetic book, provide and control the interpretation of the poem, and inherently mark distance from the author, thus bearing the mark of publication.

Modes of Rubrication

Organizational principles used in the twelfth and thirteenth centuries for academic and other types of books continued to be adapted to the needs of book production and reproduction in later centuries. Titles, tables, and various forms of commentary were used, but the old terminology was adopted to reflect new practices. The term "rubric" as part of the lexicon used by humanists such as Salutati no longer necessarily meant "title or mark in red" (though this colour was sometimes used), but simply "title."[4] Fernández de Santaella (1992) noted the original meaning in his *Vocabulario eclesiástico*, while also assigning the rubric the role of summary header: "Rubrica. ce. femenino genero me. pro. el color bermejo o linea o regla o filo teñido colorado para hazer differencia o la tal señal o tintura. Sapientie .xiij. tambien se dize rubrica la summa del libro o del capitulo. la qual se escriue de bermejon o bermellon" (fol. 155v).[5] The rubric as title was consequently defined as a paratextual marker and a reading guide that summarized the content of the text.

Both "rubric" (*rúbrica*) and "title" (*título*) are terms that can adequately be used to refer to the headings of a poem, while the Provençal terms "vida" or "razo," with which fifteenth-century rubrics have been compared, are not found in common poetic terminology in the Spanish texts of the period. The term "rubric" was used in fifteenth-century manuscripts and early printings, such as Enrique de Villena's (1994) translation with glosses of the *Aeneid* (7–9), or in treatises such as the Spanish translation of Gordonio's *Lilio de medicina*, or fourteenth-century authors such as don Juan Manuel, in his *Crónica abreviada*.[6] The term "rubric" as a synonym for "title" appears explicitly in works such as Eiximenis's *Libro de las doñas*, where the appropriate "titles or rubrics" are listed at the beginning of each part, and may be seen, for example, heading the fourth treatise of the second part of the book that deals with the seven mortal sins. In this sense, rubrics and titles not only introduced the content of the text that followed but also marked the division of the macrotext into parts, providing structure and aiding comprehension. Juan de Mena (1989) explained the role of title rubrics in poetic books in his *Omero romançado*: "Pues agora, muy esclaresçido Rey y señor, fize algunos títulos sobre çiertos capítulos en que departí estas sumas, aunque todos los poetas, segunt la sobervia y alteza de su estilo proçedan sin títulos; pero añader los he yo por fazer más clara la obra a los que en romançe la leyeren" (338).[7] Explanatory rubrics indeed delineate and articulate the different summaries of Mena's work, a compendium of the *Iliad* in the vernacular.

The use of titles as division markers is frequently specified in the prologues of different kinds of works, and chapter designators (*congrua capitulorum designatione*) are advised for translations of classical texts, as eminent converso author Alfonso de Cartagena suggested to Pier Candido Decembrio (text in González Rolán et al. 2000: 388). He gives a similar recommendation for the compilations of declamatory letters, which should include a brief summary before each letter in order to help the reader to better understand each text. In this rendition, the rubric goes beyond being a summary to being an explanation of the text, the key to a text of otherwise obscure meaning and a way to enlighten the reader's understanding. Cartagena justifies this effort by stating that a text will not be well understood once it is far removed from the place and time of composition: "Quo fiet ut disputationes ipsas clarius intelligere ualeant, cum plerumque contingat quae in una prouincia uel tempore notissima sunt, in aliis prouinciis uel saeculis ignotissima esse" (text in González Rolán et al. 2000: 430).[8] Thus, the "little preface" or *argumentum* heading each letter is needed in order to make meaning available to those who may be distant in time and place from the text's original production locus. From the division into parts, titles, and laws common in legislative texts such as the *Siete Partidas*, to the graduated structure of chapters, titles, and parts in other works, the markers also function as a finding aid and, at the same time, as a widely recognized mode of referencing authoritative texts. Alfonso de Cartagena (1969) introduced a similar structure into his translation of

Cicero's *De inventione*, dividing the book into main parts called *títulos* and those into sections or *capítulos*, as he explains in his introduction to the work (31). Likewise, Juan de Mena (1989) quoted extensively from Isidore of Seville's *Etymologies* in his prose commentary to his long poem *Coronación del Marqués de Santillana* by referencing the titles in Isidore's work (155, 160, 161, 174, and passim). The structuring of a work out of a collection of smaller parts brings to light the role of the titles in the compilatory process as part of the intended use of the work. This was of particular importance during a period of intense textual production that obeyed both political and administrative forces, growing on a par with cultural activity.

The Accessible Text

Several textual and critical frameworks have been identified as the generating principles for paratextual strategies that aim to make the text accessible and easy to use. The medieval *accessus ad auctores*, the scholastic prose introductions to Latin texts,[9] as well as the collections of *exempla* and hagiographic literature, may have provided the thrust for early explanatory titles such as the Provençal *vidas* and *razos* (Holmes 2000: 188). The practice of ordinatio and compilatio,[10] the *sermo brevis* (Branca 1986b), and the new model of prologue that appears in the thirteenth century (Minnis 1992) also affected rubric practices of the period. An important concomitant development in the academic book in the twelfth century was a loss of pre-eminence of the margins of the text as "biblical authority" (*auctoritas biblica*), yielding to a greater importance and autonomy of rubrics that mark sections and note their content (Nascimento 1995). Thomas Aquinas's *Summa* epitomized academic ordinatio (Marichal 1990). The needs of the mendicant orders helped bolster this book structure, while developments in university book reproduction had an impact on liturgical books and other types of texts.[11] Clearly, these different practices could be combined and reworked. For instance, Dante took one further step by combining the vida and razo tradition with numerical disposition and the scholastic *quaestio* (D'Andrea 1980). The system underlies the structured organization of the works of Latin authors, such as Vincent of Beauvais, and of vernacular ones, such as Boccaccio's *Decameron*, Gower's *Confessio amantis*, and Chaucer's *Canterbury Tales* (Nascimento 1995: 244–5; Parkes and Doyle 1991).

Early compilers often specify in the prologue the organizational protocols followed. Ramon de Caldes, the twelfth-century compiler, made transparent the rationale behind the paratextual organization of his cartulary in his prologue to the *Liber feudorum maior*, explaining that his choice was brought on by sheer necessity because of the large volume of texts. He related how he conceived the work, organized the documents in two volumes instead of one, and then divided each into sections marked by titles, stringing the terms *incepi-inceptum-ordinavi-ordinatum-divisi-divisum-distinxi-distinctum*, parts of an ordered process that made for a happy ending ("fine felici consumavi") (text in Miquel Rosell 1945, 1: 1–2). Similarly, the preface to another cartulary, the *Tumbo A* of the Cathedral of Santiago de Compostela, explained its division into five books (*libri*), with each book divided into *tituli* for easy reference ("per intitulacionem facilius inuenire") (text in Lucas Álvarez 1998: 47–8). From the thirteenth century onward rubrication became frequent in copies of didactic treatises (including religious, moral, and scientific). The same attempt to enable access to meaning appears in Latin devotional texts such as the ubiquitous books of hours, in which the rubrics may be written in the vernacular in order to help the user navigate the different sections of the book. One such book of hours is Newberry Library Ms. 39, which has Latin text and Catalan rubrics.

Important intersections between the production of poetic and legal compilations have been identified, particularly in the case of individual authors such as Guiraut Riquier and others (Holmes 2000: 118). Rubrics helped in the archival process of single documents and worked as finding aids. Once a document was drawn, part of its processing usually involved writing a brief summary of its contents in the form of a rubric in the margins or as a *brevete* at the bottom or top.[12] The registers or other escribano books usually bear a short summary of the content of the document they introduce, briefly stating the nature of the business, a name, date, or some other indicator of its contents, as is the case in the cuadernos of the proceedings of city council meetings in Guadalajara (see López Villalba 1997: 52 and passim). These rubrics could indicate, in a manner that closely resembles those found in cancioneros, the brief form of the name of any of the persons concerned (e.g., "Baraxa," "Herrera," "Luzón"), be it the giver or the receiver of the document, or the type of document (e.g., "carta," "carta de pago").[13] Further, the inscription of a document or its draft in the escribano's book involved a rubric system indicating the nature of the document as, for example, a sale, debt, lease, or will, along with annotations indicating the issuing, distribution, or reception of the text (Bono Huertas and Ungueti-Bono 1986: 42). Some of these tags are identical to those found in cancionero rubrics, where poems may be inscribed as a *testament* (will), *almoneda* (public sale), or carta (document or letter). It was often the same hands that produced the notarial document that also wrote the rubrics. This practice may also be seen in Spanish cancioneros, which can bypass a specialized rubricator in favour of having the task fulfilled by the various hands that copied the poems, as is, for example, the case in the *Cancionero de Palacio* (SA7) (Tato 2003: 508).

Guiraut Riquier's patron, Alfonso X (2001), explained in his *General estoria* the need for an apparatus that would facilitate access to the text: "Todo omne que alguna razón quiere contar de guisa que ayan ende sabor e aprendan y los que lo oyeren deve fazer en el comienço sobr'ella todas aquellas maneras de departimientos que sopiere por que los omnes la puedan entender mejor" (2: 424); and "E estos departimientos de las razones d'esta estoria por libros son porque los qui los leyeren que non tomen ende enojo de luengas razones. Por esta razón misma son los títulos e los capítulos en los libros, e por departir por y razón de razón e por los títulos ir más cierto a la razón que omne quiere en el libro" (1: 525).[14] Building on the many different reading aids added to the academic book during the twelfth century (Nascimento 1995; and Parkes 1991a), the thirteenth century saw access to texts facilitated via a detailed ordinatio, an apparatus of titles, chapter divisions, rubrics, and tables of contents (*tabulae*).[15] With regard to rubric practices, Vincent of Beauvais explained the relation of tabulae and rubrics: the former served as a compilation of the book's tituli, a handy reference to its different divisions and topics (Parkes 1991a: 63–4). Rubrics and other paratextual elements proved equally useful to vernacular authors such as Martín Pérez (2002). The preface to his fourteenth-century confessional treatise establishes a division of the work into three parts, where rubrics allow easy navigation of the text. A table of contents (*tabla*) heading each section contains the rubrics found in the text along with a short summary of the part pertaining to each. The preface delineates the cross-referencing role of the rubrics inside the text vis-à-vis those of the table of contents, emphasizing that the former should be corrected to match the latter when necessary. The author also encourages the individual owner of a particular codex to introduce foliation (11–12).[16] In fifteenth-century legislative works such as Díaz de Montalvo's (1990) *Ordenanzas reales de Castilla*, titles were listed in the table of contents, which provided a guide to the parts and titles of the compilation of laws ("tabla de los libros & titulos desta copilaçion de leyes") (fol. 1r) [table of contents of the books and titles of this compilation of laws]. The letter sent by Isabel and Fernando ordering the escribano mayor of the city council of Seville to keep a register with the copy of the letters, ordinances, albalaes, and cédulas presented to the council offered a comparable specification regarding a table of contents (text in Fernández Gómez et al. 1993: 77).

A similar, though not identical, relationship between rubrics and *tabulae* is evident in fifteenth-century poetic compilations such as those by Juan Alfonso de Baena or Juan del Encina. In Encina's 96JE, the table of contents refers to the rubrics and summarizes or in some cases paraphrases them. For instance, poem ID4445 is headed in the compilation by the rubric "Juan del enzina a una dama que le pidio una cartilla para aprender a leer" [Juan del Encina to a lady who asked him for a cartilla in order to learn how to read], but is referenced in the tabla as "Un a.b.c. de amores a vna dama" [A love alphabet for a lady]. There are some instances of the opposite practice, by which the tabla provides a title lacking in the cancionero, as in the case of ID4447, which is introduced in the tabla as "Otras de amores," but appears headed only by the author's name in the compilation. Only the *villancicos* are referenced by the first verse. The tabla breaks the continuous reference to the body of the compilation in the list of villancicos, which are listed by topic in the tabla, though arranged differently in the cancionero. The folio numbers are given correctly in the tabla and easy reference is uncompromised. While enabling discontinuous reading of the poems, this encourages a linear reading of the tabla and is intended to better stir the initial interest of the reader. The table of contents heading a cancionero, be it manuscript or print, not only served the primary function of useful finding guide but also served as an advertising tool for the book's contents.

Making Poetry Accessible

The long tradition of compiling, assembling, and organizing diverse individual texts, in both Latin and the vernaculars, is essential to understanding the rationale behind the compilation of poetic compositions.

Enabling access to the poetic text was greatly assisted by the solutions devised to facilitate reference in scholarly, legal, and other types of works. Latin methods of textual access provided models for vernacular genres. However, poetic anthologies were subject to specific forces and required a custom-designed apparatus. The time and place of the gathering of a compilation also determined the specific modes of paratextual arrangement. Several key factors need to be considered: first, the material and autonomous nature of the poetic paper, as well as its life before it became part of a compilation; second, the uses of that paper vis-à-vis the specific modes of dissemination and reception, and, in that vein, the dynamic of a written versus an oral medium; third, the cultural specificity of a particular corpus and the effect that space and time coordinates have on the other two factors.

Perhaps the best-known vernacular tradition of textual access is that of the Provençal vidas and razos, which provide short introductions to the life of a given author (vida) and the circumstances of composition of a particular poem (razo). Of particular relevance to the study of the vidas and razos are their ties to the transition from an oral to a written transmission in the Provençal lyric,[17] though some critics see evidence of a written tradition from the outset (Poe 2000). Both the material support and the modes of diffusion encouraged a negotiation of meaning that rubrication was meant to help navigate. Poems would have been composed in writing and then copied in rotuli, with the tradition of reading vidas and razos before an oral performance coming at a later stage.[18] An oral contextualization of troubadour author and poem before a public performance was supported by a specific cultural milieu that relied on the distribution facilitated by jongleurs and the use of the written rotulus as a memory aid.[19] Appearing as an already constituted genre in the thirteenth century, the vidas and razos may postdate the poems by several decades. The thirteenth century has been postulated as the period in which authorial compilations were formed, after an earlier period of oral distribution but written composition (Meneghetti 1984; Tavani 1969). Latin poetic compilations had seen a rapid growth from the middle of the twelfth century, with a few witnesses containing the work of a single author or that of a poetic school, often surviving in the form of a loose sheet or a booklet (Bourgain 1991: esp. 63–80). Some well-known compilations of the period include the *Carmina burana* and Petrus Riga's *Floridus aspectus*. Early on, rubrics had an important macrotextual function in the creation of a coherent structure from the collection or compilation of "micro" gatherings of poems by an author (the *Liederbücher* and *Gelegenheitssammlungen*) in the thirteenth century.[20] There is clear evidence of dissemination through booklets of the works of authors such as Peire Cardenal (Brunetti 1993; Holmes 2000: 9). In the early Italian lyric, emphasis on a written transmission was complemented by the crafting of the poetic manuscript for visual reception in what H.W. Storey (1993) has cogently described as a visual poetics.

Locating rubric practices geographically and chronologically helps interpret their function. The presence of vidas and razos in manuscripts produced in places such as thirteenth-century Italy, distant in time and place from the initial circles of poetic production, makes perfect sense (Brea 1999). A vanishing but prestigious culture would warrant such archaeological interest stemming from the periphery (Brea 1999: esp. 37–8). An early oral "official interpretation" would be aimed at the heterogeneous public of the feudal courts and would look to streamline potentially divergent interpretations and provide a univocal reception (Meneghetti 1984: 279). This was all the more critical once the poem left its immediate time and place of composition (Riquer 1995: xiii; and Holmes 2000 for Italian reception). The vidas and razos were intended to fulfil thirteenth-century Italian audiences' need to imbue formal poetry with an external meaning,[21] thus effectively resisting a de-spatialized reception. The need of increasingly literate audiences for referentiality would account for this "realistic" tendency of thirteenth-century art and literature, and would help explicate the need for a contextualized reception of vernacular poetry (Bäuml 1980: 262). In many cases, the vidas and razos provide summaries based on information easily extricated from the poem's text, as seems to have been the composition strategy for Uc de Saint Circ (Holmes 2000: 44; and Riquer 1995: xvi), although in a good number of notorious cases they present crucial new information that cannot be attributed to a textual or rhetorical origin. This spatial and chronological distance would give the razos a textual tradition of their own, as they would be the product of a unilateral effort on the part of one author or a limited number of authors (Avalle 1993: 107–11).

Quite different from the Provençal *cançoners*, thirteenth-century Galician-Portuguese poetic compilations or cancioneiros also utilize paratextual strategies to address the needs brought on by written dissemination. Rubrics would have been introduced at least in part by the cancioneiro's compiler in order to provide the reader with a key to the understanding of the text (Gonçalves 1991: 455). Two geographical centres and periods stand out in a discussion of the textual transmission of thirteenth-century Galician-Portuguese rubrics. The first period would have extended until the end of the thirteenth century, with Galicia and Castile as its centre. The second period, beginning in the fourteenth century, would have displaced the centre to Portugal (Lorenzo Gradín 2000). Cancioneiro rubrics have been categorized into "identificatory rubrics," which identify a poem's author or genre; "explanatory rubrics," which provide a relevant explanation for the correct understanding of the poem; and "codicological rubrics," which furnish details regarding the material production of the compilation.[22] Cancioneiro rubrics often provide information that cannot be extrapolated from the poem and that is key to the correct understanding of the poem or its social or political significance. They show a textual relationship with the poem, alongside extratextual contextualizing efforts, as they bridge the macro/textual divide by characteristically including part of the beginning of the poem in the body of the text (Brea 1999: 46; Lagares Díez 2000, 1998; Livermore 2000). Rubrics in thirteenth-century Galician-Portuguese cancioneiros play a central role in the study of verifiable authorial compilations and in cancioneiros produced at later dates when oral transmission of contextualizing information had already dwindled away. The addition of a paratext here too marks a rupture with oral performance. These rubrics were composed in close relationship to the development of a written tradition and, in some cases, were subject to the control of authors who recognized the value of asserting control over the meaning of their poetry.

Although King Alfonso X's theory of access as stated in the *General Estoria* referred to a large prose text (as cited above), very similar organizing principles may be seen in his poetic works. Rubrics in the *Cantigas de Santa María* reflect the specific uses of a devotional collection of Marian miracles and praises. They offer a quick summary of the miracle while also functioning as an efficient finding aid for the user searching for a specific miracle or theme (Brea 1999: 48). Recent studies have shown that the copyists of the *Cantigas* created their own rubrics when preparing the text. The single booklets, leaves, or *rolos* they used had no rubrics, and the first copyists worked without an exemplar. *Cantiga* rubrics present a lack of textual uniformity that may be explained by the early dissemination of copies of poems without rubrics.[23] The lack of textual uniformity in the rubrics preceding individual miracles in Berceo's *Milagros de Nuestra Señora* has also led Bayo (2000) to hypothesize an early stage of dissemination of the individual miracles through single leaves or booklets (*cartiellas*) before they were compiled into a miracle collection (76). Rubrication here would then mark the (long) paths of inclusion of the single poem or cluster of poems into a macrotext.

Modes of rubrication in fifteenth-century poetic texts respond at least as much to the material needs of textual production of the time as to any conscious imitation of earlier contextualizing practices. Titles can be found to serve similar purposes as the Provençal vidas and razos, but their use was "institutionalized" with the wider practice of poem collecting both in manuscript and in print anthologies (Marotti 1995: 218–19). The mix of oral and written dissemination of the lyric in the vidas and razos, as well as the literal references to the voice in the poetic text, gave way in the fifteenth century to a prevalent visual reception of the written poem, as well as written transmission. A literate public would be more likely to receive a text visually and thus require a visual reading guide, i.e., rubrics, much as a changing readership had shaped the techniques of textual access of the thirteenth century.[24] These changes inspired efforts to organize coherently fifteenth-century cancioneros in order to help the readerly process (Beltrán 1998: 68). The more overt the written nature of the text, as opposed to a pseudo-orality, the more active involvement in the text is required of the reader (De Looze 1991: 168). Rubrics, tables of contents, and continuous foliation appear in response to these changing needs of visual readers. The parallels among different lyric corpora may thus be explained more by the presence of similar axes of coordinates to which writing practice acts as a catalyst than by an attempt to replicate the textual elements of an earlier tradition, although the influence of those elements should not be denied. The progressive shift from oral poem to written poetic compilation was supported by a wider access to writing and the ubiquity of a paper culture propelled by the literate communities described in the previous chapters.[25] These and other aspects of fifteenth-century cancionero lyric point to similarities with sixteenth-century poetry while displaying some continuities with secular poetic written traditions, breaking a Zumthorian conceptualization of a circularly homogeneous pan-oral Middle Ages.[26]

Social Rubrication

The relationship of text to context and to medium was marked through a careful rubric system. The hermeneutic need for access to each poem would arise with each reading. Rubrication recorded many of the multiple acts of explanation that took place as often as a poetic paper was distributed and read, while at the same time exposing the need to fixate meaning for the purpose of dissemination. It tried to remedy a hermeneutic profusion, by which a reader far removed from the place where the text originated was free to read it however s/he wanted or, conversely, was left unable to appreciate the poem and its meaning. Rubrication shows a negotiation of literary authority among author, reader, and the other hands that may handle the text in ways that point to similarities with the glosses that surround a central text, particularly as developed in the early modern period. It appeared at exactly that point as an attempt to bridle hermeneutic profusion and retain control of the text while making it accessible. The cryptic nature of poetry and the test that it exerted on the reader's ingenuity elicited a continuous need for contextualization in each act of reading and copying. The abstraction involved in conceptual poetry challenged the ingenuity not only of those readers who enjoyed a greater immediacy to the poem and the textual community in which it was written and circulated, but especially of those who were chronologically, physically, and socially displaced from the text, its textual community, and its extratextual setting. The purposeful obstruction of meaning was postulated during the fifteenth century as a mark of inclusion in a socioliterary and political elite, as authors such as Enrique de Villena and Alfonso de la Torre attest. Villena (1994) theorized poetic obscurity in gloss number 50 in his translation of the *Aeneid*, stating: "los *dezires poéthicos* fablan por tales encubiertas que a los non entendidos paresçe escuro e velado e a los entendidos claro e manifiesto" (43).[27] That this was done purposefully is apparent in Alfonso de la Torre's (1991) *Visión deleytable*: "ca non sería bueno que el çiente e el ydiota oviesen manera común en la fabla [...] E aún por esto no sola mente fue neçesario el fablar secrestado e apartado del vulgo, mas aún fue neçesario paliar e encobrir aquéllos con fiçión e diversos géneros de fablas e figuras" (1, 126–7).[28] This philosophy of literary language, similar to that of the later Baroque period, called for the kind of apparatus established by the rubrics to make the text accessible and to ground the textual within the extratextual. The subtle irony is that, in so doing, the rubric defeated the purpose of the obscure language, facilitating its access for Torre's *ydiota*.

Forming part of the circles in which contextualizing information was transmitted brought instant credibility to the compiler and his work. Escribano, poet, and compiler Juan Alfonso de Baena showed himself a legitimate rubricator through his privileged position as both witness and participant in poetic production. As the efforts to provide useful information in his cancionero clearly show, in many cases, and at least initially, the contextualization of the poem could take place outside but alongside the paper medium. As studies from other European traditions have shown, the copyist sometimes took the rubrics from an exemplar, while at other times the information provided would come from oral or written firsthand information that originated in circles in close proximity to the author.[29] Encouraged by the intense oral and written transactions of the time, extratextual information was certainly discussed orally before becoming part of the anthology's paratext. Traces of this practice can be found in the rubrics of the *Cancionero de Baena* (PN1), notably in those preceding Villasandino's poems, one of the earlier cancionero corpora included in PN1, which received varied and at times conflicting interpretations.[30] By his own account, Baena gathered these interpretations from different sources, which he marked with a distancing "dizen que" [they say] or "es opinión de otros" [others are of the opinion], at times subordinating them to his own judgment: "pero non sse puede creer que lo el feziesse" [but it is not possible to believe that he wrote it], thus: "Este dezir dizen que fizo el dicho Alfonso Álvarez de Villasandino al Rey don Enrique, padre del Rey nuestro señor, quando estava en tutorías; pero non se puede creer que lo él feziesse por quanto va errado en algunos consonantes non embargante qu'el dezir es muy bueno e pica en lo bivo" (ID1199); or also "Esta cantiga muy sotil e famosa fizo el dicho Alfonso Álvarez de Villasandino por amor e loores dela dicha doña Juana de Sosa e por que gela mandó fazer el dicho Rey don Enrique el viejo es opiñón de otros que la fizo a la Reina de Navarra" (ID1168).[31] Baena's rubrication of Villasandino's poems shows the treatment of a much-admired earlier poet who may not have left a self-rubricated compilation of his works. Baena's rubrics not only play an organizational role, but also point to a readerly demand for an authoritative interpretation, which was the more elusive the further the poem was removed from the immediacy of the author and his or her circle. Being able to rubricate, then, would prove an intimacy with authorial processes of poetic invention. The contextualizing information heading Villasandino's poems offered by Baena in PN1 differs from that found in other cancioneros, such as MN33 (see for example ID1204) or in any of the other copies, all of which are witnesses to active poetic circulation. Baena's efforts in his *Cancionero* (1993) to provide detailed rubrics were warranted by the intended broad readership, well beyond the king and queen, revealed by his introductory remarks: the queen's "dueñas e donzellas de su casa" [ladies, married and single], the prince, and all important members of the nobility and clergy, which he then lists in detail by title, along with all the members of their households (2). Baena's target public was ostensibly the nobility but was consciously projected to include the larger and heterogeneous group of those who were in contact with noble households.

Instances found in other poetic compilations similarly point to the addition of rubrics after the text had circulated. This is the case, for example, in ID7612 (ZZ10), where the rubric narrates the interesting occasion at length and adds yet another point: the particular comments by Queen Isabel upon reading it. The queen, using an expression used by conversos, is said to have called the author "thief in his own house" because of the humorously disparaging tone of his verse against another converso: "Diegarias contador mayor de los Reyes Catolicos caso un hijo o sobrino suyo con una parienta del cardenal don Pedro Gonçalez de Mendoza. Conbido para Segobia todos sus deudos: olvidose o hiçose olvidado de Rodrigo Cota el Viejo natural desta ciudad de Toledo. Sentido della celebro la boda con ese epitalamio. Leyendole la reyna Ysabel, dijo que bien paresçia 'ladron de casa.'"[32] In this case the obvious advantage of presenting the poem contextualized within its royal reception was to garner close attention for the poem, while flattering the queen's wit.

Tensions between authorial contextualization and social appropriation are present in both manuscript and print cancioneros in relation to a variety of situations, including social conventions and urban transactions. A relatively common occurrence meriting a poem is the visit of the poet to a lady in her sickbed. The nature of such interaction helps to illustrate the relationship between cultural practice and the textual variation present in the rubrics heading different copies of a poem and those of different but similarly themed compositions. What may be seen as a writing exercise around a rhetorical topos in fact responded to the social conventions associated with the rights of visitation to the sick in conduct (and religious) handbooks and to the healing benefits of poetry. The *Flors del gay saber* (Anglade 1926: esp. 37–9), as well as Deschamps's (1994) advice to "lire aucun livre de ces choses plaisans devant un malade" (66), had already theorized the curative applications of poetry. As mentioned in chapter 4, Alfonso X was cured by the physical powers of his book of Marian cantigas, which was brought to his sickbed (cantiga 209). The therapeutic power of poetry was often summoned when illness afflicted, and poems or books of poetry were sent to the sick, as in the rubric to a poem by Juan Fernández de Heredia (1955): "Juan Fernández estando enfermo envió a pedir a Andrés Martín Pineda le enviasse el libro de sus coplas y enviándole el libro le envió estas coplas" (170).[33] That illness provided an acceptable excuse for heterosocial written and oral communication can be seen through many cancionero compositions (such as ID0715, 0108, 0222, 0344, 3105, 4738, 5256, 5777, 6888). The didactic treatise *Castigos y Dotrinas que vn sabio daua a sus hijas* advises against such visits, cautioning about the potential for the erotic overtones of male access to a lady's bedchamber when the lady was in bed:[34] "Otrosí aueys de guardar para ser onestas, que, mientra estouiedes en la cámara dormi(da)s y mucho más enla cama, no consintays que entre ninguno a vosotras avnque sea de vuestra casa saluo vuestras mugeres y moças, ca sería cosa muy desonesta estar vosotras en la cama y hablar con ninguno" (280).[35] The advice was not always heeded, as is evident in poem ID3105 by Juan Álvarez Gato, which bears rubrics that attest to the fact: "Juan aluarez porque la vido mal en la cama y denamorado y de turbado no la oso hablar ni pudo" (MH2); and "Otras suyas a una señora que vido en la cama mala" (11CG, 14CG).[36] On the slightly differing accounts offered by the two sets of cancioneros, MH2's longer and more detailed version may be trusted to be a more accurate contextualizing depiction since this manuscript is Álvarez Gato's personal compilation. The print 11CG emphasizes the generic theme of the poem while doing away with the more personalized information, which may not have been available to the compiler or not understood to be of interest to a wider audience. Another compilation, Pedro Manuel Jiménez de Urrea's personal print cancionero (13UC), which was published under close authorial supervision, contains a similar composition. Poem ID4738 offers minute details on the type of illness afflicting the lady (measles), which also strikes the poet following contact with the lady, as asserted by a second poem (ID4739). The rubric preceding a poem on the same subject in 14CG (ID6888) does not provide any detailed information and turns the poem into a generic theme. A similar loss of the contextualizing details that closely link the poem to the immediate moment of poetic reception is found in the long rubrics heading the print publication of Juan del Encina's plays. The later printed versions, far from the author's control, lose all the details that can anchor the work to a specific date and place (Pérez Priego 1992: 348). This variation makes evident that the many readings to which a poem is subject are contingent on the degree of authorial control of the means of dissemination, be they manuscript or print, and the level of interest that specific details may arise when the poem moves further away through the hands of compilers, editors, and readers. When removed from the immediacy of author, detailed rubrication becomes dependent on the concerted effort of socioliterary networks that have a vested interest in hermeneutic access to a particular corpus. This made the need for control by authors and compilers all the more pressing.

Ordering and Compiling

Access to poetry followed similar principles as prose ordinatio and compilatio, due to the fact that many of the *ordenadores* such as Enrique de Villena involved themselves in different types of discourse, both prose and verse. Theories of textual accessibility call attention to the role of the learned facilitator or ordenador. Villena (1994) explained early in the fifteenth century the role of titles and other elements of ordinatio used in his translation of the *Aeneid*:

> Llama en este lugar a las rúbricas de los libros e capítulos e distinçiones argumentos, por cuanto quiere dezir argumento aquello mediante lo cual la verdat de la cosa saber se puede. E por eso en la lógica los silogismos, que son formados de sus premisas con la consecuençia para saber la verdat de alguna

cuistión ho dubda, se llaman argumentos [...] E Ovidio en los títulos que fizo a los libros de la Eneida, epilogando lo en cada libro contenido, siquiere coligendo, llámalos argumentos. E d'esta postrimera manera se entiende aquí. E los argumentos qu'el dicho don Enrique aquí puso en los comienços de los libros non son tales como los de Ovidio, porque le paresçió que eran mucho obscuros; por ende, púsolos por más llana manera. E los de los capítulos e distinçiones eso mesmo. (58–9)[37]

Rubrics were used as a summarizing device that enabled access to the meaning of an obscure text. Accommodating an increasingly literate public necessitated the wider introduction of textual aids such as rubrics, which explain "in a simple manner" the more difficult text they precede. The rubricator claimed a privileged exegetical position because of his command of specialized knowledge. Visual signs as hermeneutic tools of the ordenador elevated the latter to the role of learned mediator between the text and the uninitiated in art (*artifiçio*) or science (*regla sçientífica*), as Villena (1994) himself stated: "E, demás d'estos párrafos, sean los vocablos señalados por letras mayores significadas e de amarillo insignidos, ansí que todos sin confusión vengan a notiçia del leedor, que alguna dubda notable non quede por solver e paresca claramente lo que la textual texedura contiene e se vea el artifiçio del ordenador reluzir con çierta regla sçientífica e plazible" (58).[38] The craft of the ordenador shines in the weaving of the text, the "textual texedura" referred to by Enrique de Villena that produces a well-ordered and pleasurable result.[39]

Contextualization and attribution of the fifteenth-century poem could take place through different means before being archived. The *compilador* (compiler) often supplied contextualizing reading aids in the form of rubrics or titles. The work of the rubricator as a separate agent in the manufacture of the codex could instead be undertaken by the copyist[40] or, as we have seen in the cases of Baena and others studied here, by the compiler him/herself and the author as self-compiler. In the case of the Italian tradition of Boccaccio's "summary rubrics," it was also scribes, albeit non-professional ones, working in geographic areas removed from the author's circle that showed the strongest effort to pay tribute to a great author through lengthy synopsizing rubrication (Branca 1986b: 28). Awareness of the importance of rubrication compelled authors to take control over access to their texts. Baena, author not only of the poems that appear under his name, but also of the elaborate rubrics that head the poems included in his Cancionero (PN1), saw his role as one of both author and compiler: "Johan Alfonso de Baena, escrivano del Rey, actor, componedor e copilador d'este presente libro" (637). The expression "componedor" also conveys the meaning of "author" either in a dual expression – as used by Baena "actor, conponedor," and in other fifteenth-century authors such as Teresa de Cartagena (1967), who refers to herself as "abtora o conponedora" in her *Admiraçión Operum Dei* (113) – or by itself, as used by Gómez Manrique, himself a "componedor" (ID3398, v. 25). In the brief introduction preceding the prologue heading his *Cancionero* (1993), Baena presents the volume to King Juan II and clearly identifies himself, Johan Alfonso de Baena, the king's escribano and servant, as author, compiler, and overall creator of the book ("fizo e ordenó e compuso e acopiló") [made, ordered, composed, and compiled] (1), which he tags as "compiled writing" ("acopilada escriptura") (2). This task proved tiresome enough to inspire the two verses that close the introductory section after the table of contents ("tabla"), where Baena complains of his pains ("Juan Alfonso de Baena / lo compuso con grand pena" [Juan Alfonso de Baena / composed it with great pains] 10). The verb "ordenar" seems to suggest writing according to certain rhetorical, poetic, or otherwise specialized or technical rules, as evidenced by its often being qualified by an adverb, as in a rubric heading another cantiga by Villasandino, which is announced as being "muy sotilmente ordenada" [very subtly ordered] (ID1160). Beside the obvious influence of the concepts of ordinatio and compilatio, there is also a clear overlap with escribano language. This may be observed, for example, in the language of the 1455 will of Cordoban veinticuatro Pedro González de Hoces, which states that the best remedy upon death is to have one's will written ("escripto e ordenado" doc. 43 app. in Ostos Salcedo 2005: 352, similarly in doc. 46 app., 367). In an application for a royal escribanía in 1496, the letters of support repeatedly highlight the ability of the candidate, Alfonso de la Barrera, to work in the escribanía and produce (*ordenar*) all types of writings ("escripturas") (document in Bono Huertas and Ungueti-Bono 1986: 395–6).

In the introduction to his *Cancionero General* (11CG), Hernando del Castillo explains the genesis of the multi-authored cancionero as a process of compilation ("conpile un cancionero" [I compiled a cancionero]) that entailed an ordinatio, an editorial work by way of corrections ("y assi ordenado y corregido" [and thus ordered and corrected]), finally producing a fair copy ("acorde sacar en limpio el cancionero" [I decided to produce a fair copy of the cancionero]). The publication contract for the Cancionero General also presents Castillo as a compiler: "vos, Fernando del Castillo, avés copilado este original" [you, Fernando del Castillo, have compiled this original].[41] His prologue further provides a telling illustration of the production of a poetic book within the framework of the at-times-conflicting forces driving multiple authorship and editorial credibility. Following a standard authorial topos, Hernando del Castillo urges that the reader, the Count of Oliva to whom he dedicates his book, or anyone else,

> passe los ojos por esta lectura y mande corregir y enmendar en ella lo que yo por ventura en prejuyzio de alguno o no pude o no supe corregir ni mejor ordenar. E si alguna cosa el más claro ingenio de vuestra señoría o de los otros lectores hallaren mal puesta o mudada [...] o variación en los títulos de aquellos suplico a vuestra señoría y ruego a todos me perdonen y enmienden lo que bien no les parescera. E el que hallare agena marca en sus obras que la raya y ponga la propia y haga lo mismo el que la suya sin ninguna hallare [...] E escuseme tambien la manera que tuue en la recolection destas obras que con toda la diligencia que puse aunque no pequeña no fue en mi mano auer todas las obras que aqui van de los verdaderos originales o de cierta relacion de los auctores que las hizieron por ser cosa casi impossible segun la variacion de los tiempos y distancia de los lugares en que las dichas obras se conpusieron. E porque todos los ingenios de los ombres naturalmente mucho aman la orden y ni a todos aplazen vnas materias ni a todos desagradan ordene y distingui la presente obra por partes y distinciones de materias. (fol. 1v)[42]

Castillo goes on to detail the sections into which he has divided his cancionero and the key role that titles and authors' names played in its organization, with the table of contents ("tabla") functioning as a finding aid at the beginning of the volume. He makes clear that rubrication, by "marking" and giving titles, is part of the role of the compiler, with help from the author and from what has been referred to as "social rubrication" above. Collecting poems from various sources and times, correcting them, and making them accessible through rubrication were primarily the responsibility of the compiler, though ownership of the volume by gift or purchase extended that duty to the owners. The figures of authors, compilers, readers, and owners were thus intermingled, and their different roles overlapped.[43] The credibility of the compiler relied greatly on his or her ability to claim proximity to authorial circles. Detailed contextualization via rubrication was for Castillo a compiler's task to fulfil the expectation of readers who were distant in time and place from authorial circles.

The ability to create a good compilation of texts from various authors, particularly if one of those authors was the compiler himself, was grounds for claiming authorship of the whole volume. Alfonso de Cartagena states as much when writing to Pier Candido Decembrio about the latter's now lost *Declamationes*: "ac licet diuersorum scriptorum et tuae commixtae epistolae sint, tuum tamen integre uolumen potest uocari" [and although the letters are a mix of different authors and your own, however, the volume can be called entirely yours] (text in González Rolán et al. 2000: 416). A similar editorial logic probably lies behind the fact that Hernando del Castillo's publication contract names him alone as the beneficiary of the sale of his *Cancionero General* or the publication mode of Baena's cancionero, which we still to this day call his *Cancionero de Baena*.

Many individual works may form an organic whole thanks to the textual stitching provided by the rubrics, which make a book out of bits and pieces. Nevertheless, as suggested by modes of textual handling explored in the previous chapters, this apparent eventual containment of the poetic text in an archive was but one step, by no means linear or permanent, in rubric practices. The intervention of the ordenador, componedor, and compiler, as separate agents or combined into one, created a rich net of possibilities that could be superimposed upon one another and that maximized textual dissemination but could be detrimental to authorial control.

Rubrication and the Self-Made Author

Manipulating self-authorizing strategies such as rubrics also meant that authors could seize power over the organization of their book of poetry through the creation of a narrative in what constituted a selective self-anthologizing process. This was all the more desirable because of the inevitability (desired or not) of publication, in manuscript or in print form. The use of the rubric or title as a marker of exclusive authorship was explicitly recognized in Fernández de Santaella's (1992) *Vocabulario eclesiástico*: "Titulus. li [...] Jtem el nombre o titulo puesto al libro quasi tutulus de tueor. por que deffiende el auctor dela obra que no se atribuya a otro."[44] Rubrics mark authorial ownership while attempting to prevent misattribution.

Poets left self-referential markers when creating a textual apparatus for their anthologies. For example, some rubrics were written in first (authorial) person, as is the case for the poets Carvajal and Santa Fe. Tato correctly infers that in Santa Fe's poems, explanatory rubrics containing information not forthcoming from the poem's text came from an authorial compilation or a gathering assembled in close proximity to the author.[45] In rubrics, the authorial marker may appear in the choice of possessive morphological forms, as in ID2837 ("A mi señor Juan de Haro" [To my lord

Juan de Haro]), although some first-person markers may reflect the competing voice of the compiler, as in ID2831 ("Ciertos consonantes creo entresacado de vnos antiguos que conpuso vn hebreo").[46] Authorial self-referential morphology can appear in the third person as well, as attested by Álvarez Gato, or Gómez Manrique's compilations. The presence of several rubrics in the author's voice may indicate an original authorial anthology even if it is integrated within a multi-authored cancionero. Such is the case of Carvajal, of whom we find authorial rubrication in MN24 and RC1, and of whole compilations such as the Catalan *Cançoner des Masdovelles*, where the voices of the compiler-author read as a poetic diary rich in extratextual and chronological references.[47]

A lack of rubrics carrying authorial attribution has been pointed to as a sign of the codex's proximity to circles where that information would be unnecessary, since it would have been conveyed in non-textual or unwritten ways and be known by those involved in the paratext (Beltrán 2004; Whetnall 1997). The absence of rubrication has been linked to proximity to the author in cases such as the *Cancionero de Herberay* (LB2), where the wife of author Hugo de Urriés functioned as protocompiler, as may be inferred from the intimate nature of some of the poems. A female compiler lies also at the origin of Cerverí de Girona's poetic book, which was slowly formed thanks to the compilation work of his lover.[48] The absence of rubrics would presumably suggest an expressed desire to limit the dissemination of the work. In works destined for publication, unwanted rubrication plagued authors, who looked disparagingly on the actions of the many intervening hands. Fernando de Rojas (2000) denounced unwanted non-authorial intervention in the titles of books given to the printing press in the prologue to the *Tragicomedia* version of *Celestina*: "Que aún los impressores han dado sus punturas, poniendo rúbricas o sumarios al principio de cada auto, narrando en breve lo que dentro contenía: una cosa bien escusada según lo que los antiguos escritores usaron" (52–3).[49] The clear danger of losing control over one's work clouded what could otherwise be bright authorial prospects of recognition and rewards. For these reasons, control over rubrication was as important as control over texts.

Rubrics, Publication, and Property

The precarious nature of loose papers particularly worried authors of short pieces, such as poems, letters, or other writings. These anxieties and the desire to establish a poetic identity that would garner obvious social, political, and economic advantages help explain certain material features in the design of single-author cancioneros and some authorial clusters present in mixed-author poetic compilations. One such feature is the constant and seemingly over-repetitive rubrics stating the author's name before every single poem and at the head of every single folio in Encina's print cancionero or in Álvarez Gato's manuscript compilation. This practice seems to serve various purposes, all geared towards the establishment of a strong authorial persona and the demarcation of literary property before copyright. The very material and real concern of pages being ripped from a book, copied, or lost from a worn or unbound volume would prompt poets to take extra precautions in marking their literary property. The almost immediate loss of authorial ownership of unmarked poems presented a sizeable problem in the fifteenth century, which worsened in the sixteenth and seventeenth centuries and remains to a large extent unresolved today. In the sixteenth and seventeenth centuries, with few compilations that are autographs or focused on a single author, the bulk is the work of compilers that follow their personal taste, with little regard for textual matters or correct author attributions (Sánchez Mariana 1994: 122–3).

Absence of information about the author's name could easily lead to the poem's attribution to a different and perhaps better-known author.[50] The theft of poetic authorship is in some instances chronicled in the rubrics. Poem ID1899 is headed by a rubric, differently worded in the various manuscripts, that states: "Anton de montoro a juan poeta por vna cançion que le furto & la dio a la Reyna" (LB3, SV2).[51] In the versions found in MP2 and MN19 the poem is addressed to Queen Isabel, but the terms of the accusation remain: "Montoro a la Reyna sobre que Juan de Valls fijo del pregonero dixo que habia fecho unas coplas que Montoro ficiera y le embiava" (MN19) (see also MP2).[52] Juan de Valladolid's reply, poem ID2722, tried to deny the charges, and in ID2723 Montoro wrote a counter-reply. The immediate loss of literary property when the poem was made public was at times used to the poet's advantage in cases of slander, as in ID3016, where Antón de Montoro advised Román to publish some slanderous verses under someone else's authorship in order to avoid due punishment (MN19, vv. 10–11): "publicadlas por agenas / y guardaos de las setenas" [publish them under someone else's name and avoid the punishment]. Conversely, a rubric could reveal the author's name while informing of its intended anonymous circulation, as in ID7812: "Carta de Amores sin firma ni sobrescripto por que no se sepa para quien es ni cuya es conpusola villalobos para su señora violante artal telas de su coraçon."[53] Rubrics thus expose the tension between the desirability of authorial control and the subversive force exercised by the dissemination process over any claim on literary property.

Rubrication is thus a key structural element in the process of including the single text or the cluster of texts into a paratextual whole. Rubrics help organize and create an organic whole out of single papers and booklets. When the poem was taken out of the immediate situation from which it arose, the rubrics were a reminder of the extratextual and extracodicological world where the poem lived as they attempted to make the text accessible to readers detached from the poem's immediate circles of production and circulation and also to lock the text's meaning in an attempt to resist divergent readings. The potential for a wider reception compelled the poet to secure control of his or her work through the introduction of self-exegetical elements such as the rubrics, which would also enable control over the meaning of his or her poetry. Readers and compilers joined the at-times-random process of poetic dissemination, leaving their marks on the poetic paper. Rubrics show that all these attempts were simultaneously encouraged and resisted. The proprietary impetus leading to the copyright process and the purported social projection of the printing press were long anticipated by the rubrication and compilation processes, and the need for textual control did not wait for technology.

9 The Book of Fragments

Accounting for the (Social) Self in the Fifteenth-Century Poetic Compilation

The central role of poetry in social interaction helped propagate the poetic paper in ways that were unforeseeable and that escaped the author's control, as well as that of other hands. The sheer pervasiveness and the number of papers encouraged the search for fitting repositories, particularly as such papers were actively sought and their content valued. Securing a poem conveyed the need to know its reference points in order to better understand it. When copying the poem in a textual archive, contextualization took the form of a rubric. However, when the texts are arranged mindfully, contextualization by way of rubrication takes a macrotextual role that can go beyond the marking of the individual text. Rubrication can help create a narrative thread that marks cohesiveness and turns the compilation into something larger than the sum of its parts. The narrative thread responds to an assigned logic that the author superimposes over the individual texts as they are organized into a coherent body. The organizational criteria and the storyline along which the compositions are ordered respond to personal, social, and political considerations. In the absence of a strong storyline, be it in authorial or other compilations, the elements reinforced by the rubrics still give a penetrating insight into the compilation's meaning.

It has been noted that the thirteenth and fourteenth centuries witnessed a steady increase in the author's "autobiographical intervention," related to the practices of scribes and to a notarial and pre-humanistic culture (Huot 1987a, Petrucci 1992). The self-referential revelation that emerges in many cancionero poems is in line with a similar one found in other administrative and courtly environments in later centuries, with Hoccleve and Gringore as cases in point.[1] The model followed was not new. Similar authorial control over the fashioning of the poetic "book" and thus of the poet as "author" can be seen in some troubadour razos, Dante's *Vita Nuova*, Guittone d'Arezzo, Francesco da Barberino, and, paradigmatically, in Petrarch.[2] Dante's *Vita Nuova* and Petrarch's *Canzoniere* follow a carefully designed organization imposed over a set of previously written poems (Sturm-Maddox 1980). At the end of the thirteenth century, Guiraut Riquier organized his *Libre* in a similar way, though relying not so much on a weighty prose base like Dante, but rather on rubrication. Riquier's book, though explicitly intended as an exemplary art of poetry directed to Alfonso X, was fashioned towards the end of his life as the personal itinerary of a fine artist who prevailed in the middle of social and personal adversity (Bertolucci Pizzorusso 1978: 254–8; 1991). Poetic compilation and self-conscious autobiography could become meaningfully combined with the help of rubrication in a self-canonizing strategy. In this way, rubrics could function as an exegetical apparatus in line with glosses and other types of commentary as a means to monumentalize the text and (self-)canonize the author.[3] This overarching model could have been operative in the fourteenth-century *Libro de buen amor*, a work in which rubrics play a key role, or in some manuscripts of *Troilus and Criseyde* (Boffey 1994). Later, Christine de Pizan, Deschamps, or Gower, pursued a similar self-construction (Kendrick 1992), as did Gringore and Lemaire, who sought to shape themselves as authors securing their continuous presence throughout their work (Brown 1995: 56, 239, and passim). The narrativization of identity, as Ricoeur (1992) has pointed out, includes attentive care of grammatical elements such as pronouns (esp. 137–8, 168), of particular interest in the case of first-person narratives studied here such as notarial documents, memoranda, and poems. Rubrication thus stands as a key macrotextual structuring element in the formation of the book of poetry by creating a narrative that helps piece together a self-account and the generation of an authorial self out of individual texts.

In the fifteenth century, these trends join those that drive the renewed popularity of biographical narratives represented not only by collections of biographical profiles and genealogies such as Fernán Pérez de Guzmán's *Generaciones y semblanzas* or Fernando de Pulgar's *Claros varones de Castilla*,[4] but also by chronicles portraying the character and deeds of a single noble or monarch, collections of female biographies such as those by Álvaro de Luna or Diego de Valera, or even autobiographical narratives of a notarial nature such as Leonor López de Córdoba's or religious ones such as that of *conversa* Teresa de Cartagena. Nader (1979) has pointed out that a new tendency to provide personal details in biographies championed by authors such as Valera, together with its emphasis on the exemplary function placed on biography, is tightly related to ideological leanings of the period that deny the divine origin of the state and emphasize the actions of the individual (26–7). In fifteenth-century biography as a genre, the intended monarchic control of institutional memory clashed with individual biographies of nobles who emphasized their own lineage, as well as other group traits (Fernández Gallardo 2006), as is evident in other individual chronicles, such as Álvaro de Luna's or the Condestable Miguel Lucas de Iranzo's. These books took a further step that bridged the practice of the personal register with a more narrative form. A clear example is García de Salazar's *Istoria de las bienandanzas e fortunas*, written in 1471–6, where a family register was melded with other types of historical and literary writing, such as nobiliaries, chronicles, and fictionalized historiography. Poetic compilations worked in a similar manner by functioning as stylized accounts of the self and as legacies with a socio-political meaning. The social configuration of the self, shaped through interaction, brings about various modes of self-presentation and transformation. Three fifteenth-century cancionero poets help explore the different uses of these rubrics. The following pages will examine the book-building strategies of Gómez Manrique, member of an important noble house, and nephew of the Marqués de Santillana. In contrast, the use of a storyline in Álvarez Gato's cancionero and the detailed rubrics that provide the introductory apparatus to many of Montoro's poems reveal the strategies of authorial self-constitution, particularly in the case of Álvarez Gato, as well as the power of social networks in the canonization of middle-class poets.

A Noble Story: Gómez Manrique

Gómez Manrique's poetry is preserved in two major compilations: manuscripts MP3 (Biblioteca de Palacio, Ms. 1250) and MN24 (Biblioteca Nacional, Ms. 7817).[5] These are the product of a careful compilatory process supervised by Manrique himself spurred by a request from King Afonso V of Portugal around 1465 and by the secretary whom the Portuguese king sent to Ávila in order to put pressure on the poet (though the compilation did not materialize then). As Manrique (2003) narrates in the introduction addressed to Rodrigo Alonso de Pimentel, Count of Benavente, who had also requested a "copilaçion de mis obras trobadas" (97), the poet began gathering his works, which lay forgotten in his chests, and thus proceeded with the compilation process: "E delibrando de conplir su mandamiento, fize buscar por los suelos de mis arcas algunas obras mías que ally estauan como ellas mereçíam, e procuré de aver otras de otros, mal conoçedores de aquéllas, que las tenían en mejor lugar. E asy començe a fazer vna copilaçión dellas" (105–6).[6] In the same introduction, Manrique presents himself as organizer and "componedor" of his own compilation: "yo he deliberado de amenguar a mí por conplazer a vos y conplir vuestro mandamiento; cunpliendo el qual, le enbío con este mi criado esta copilaçión de mis obras que con tantos afincos me ha pedido, que estouiera mejor ronpida que copilada; la qual, por mal que vaya escrita y ornada, commo lo va, yrá mejor que ordenada ni conpuesta, porque la escritura y ornamento, tal qual lo verá, avrán fecho más sotiles ministrales que lo es el conponedor" (108).[7] Manrique here differentiates between the material execution of the manuscript, copying and ornamentation ("escrita y ornada"), at the hands of skilled artisans ("sotiles ministrales"), and the process of literary composition, compilation, and organization of materials ("copilada," "ordenada," "conpuesta"), which he clearly attributes to himself as "conponedor" who set out to make a compilation ("fazer vna copilaçión"), resulting in a "copilaçion de mis obras." Santillana (1988) had similarly compiled his works, which had been scattered through many different hands and copied in different cancioneros of various ownerships ("libros e cançioneros agenos, fize buscar e escrevir") and which he procured in order to organize them chronologically ("por orden segund que las yo fize") in response to a similar request from the Condestable de Portugal, as he narrates in his *Prohemio* (438). This seems to have been the customary method to compile scattered letters, papers, and documents, and was used by other eminent authors such as Pier Candido Decembrio, who mentioned going through a similar process in his correspondence to Alfonso de Cartagena (text in González Rolán et al. 2000: 422, 426). Because of the time lapsed between the composition of the piece and the time of compilation, the chronological organization might lack precision, allowing other criteria to overlap in the organizational frame. The alleged reason for initiating a compilation was socioliterary and political. In Manrique's case, the noble status of the intended recipient(s) of the compilation is corroborated by the luxurious execution of MP3 and, to a lesser degree, that of MN24. Like Baena, Manrique (2003) foresees publication to a noble audience when he states in the introduction: "Mas commo estas mis obras, viniendo a poder de vn señor tan grande commo vuestra merçed es, en cuya casa tantos parientes y nobles concurren, a los quales de neçesydad han de ser publicadas" (107).[8] Manrique also sees his cancionero becoming part of the private book collection kept in the count's private quarters when he humbly beseeches him to have his book locked in his chamber ("quiera tener este libro asy çerrado en su cámara" 109). Manrique adds a defence of the importance of knowledge for the caballero, while arguing a lack of formal instruction in the art of poetry, making him an "inspired" poet. Therefore, Manrique fits the ideal caballero and poet in possession of God-given knowledge and familial *habitus*, having studied and exercised both his intellect and his body in military service. His argument, carried through the length of the letter, of the benefits of learning for the (military) caballero is subtly supported by the noble web, almost a genealogy traced by way of noteworthy examples, beginning with his brother Rodrigo Manrique, Maestre de Santiago, his uncle, the Marqués de Santillana, and the count himself, to which he carefully adds his own name. This genealogy will be highlighted in the organization of the cancionero.

The question of lineage and legacy was important for Manrique as he had had to face the bitter experience of burying all his children, sadly lamented in his letters. In 1480 his only son as well as one of his daughters died; in the following year his wife became very sick, for which reason Queen Isabel called him urgently to court, writing a note to him "de mi mano" to urge him to visit his ailing wife, doña Juana de Mendoza, the queen's *camarera mayor* (letter in Paz y Meliá's edition of Manrique's [1991] works 2, 316–17). As the letters published by López Nieto show, Gómez Manrique was in poor health during the last years of his life and spent some time in the town of Mazarambroz, possibly during the end of 1488 and the beginning of 1489, because of the bout of pestilence in Toledo during that period. In his will, written in 1490, Manrique named Ana Manrique, daughter of his dead son, his heir, leaving her to the queen's care (text in Rivera Garretas 2007: 147–8, 167).

Gómez Manrique's cancionero MP3, a beautifully illuminated luxury codex that was to contain his magnum opus, was compiled some time after 1481 (Beltrán 1998: 58). It is the work of a poet and statesman at the height of his career undertaken towards the end of his life (he was born around 1412 and died some time between 1490 and 1491). Nearing the end of his life and with literary and political success on his side, Manrique set about to publish his compilation and dedicate it to a powerful noble. He had a clear model in his uncle the Marqués de Santillana's own compilation, which Manrique himself had requested and received from him.[9] The monumentalizing and self-authorizing effects of such a compilation, particularly when beautifully executed, were clear to Manrique. The compilation also allowed him to present himself as a learned and lucid politician by including his long poems containing advice to rulers (*Regimiento de príncipes*, dedicated to Princess Isabel and Prince Fernando, and the "Esclamaçión e querella de la gouernaçión" for Archbishop Carrillo). The expectation that his cancionero would dwell in noble households may have been matched by someone who hoped for a noble title, a not uncommon corollary to a life dedicated to military and political service furthermore augmented by cultural worth. Queen Isabel had shown her early appreciation for Manrique's poetry when, still a princess, she had asked him to write some *momos* (mumming) to celebrate her brother Prince Alfonso's birthday on 15 November 1467. Perhaps more importantly, Manrique had defended Isabel's rights to the throne from the beginning, and the new queen and king had recognized his services by making him corregidor of Toledo and therefore giving him one of the more sensitive and powerful positions in that key city. His admonition in *Regimiento de príncipes* to put able men in public posts as alcaldes, jueces, or corregidores or suffer God's punishment obviously benefited him once Isabel and Fernando became monarchs (vv. 343–51). Manrique (2003) bragged about his illustrious lineage in his address to Princess Isabel and Prince Fernando in the dedication of his *Regimiento de príncipes*, and although he lacked the grand status of his ancestors, he proudly declared that he was not deprived of many immaterial gifts, such as the love for his ancestral nation (625).

Gómez Manrique's cancionero follows the well-known chronological linear design and, as the worthy nephew to a poetic master who had used a similar organization in his own compilations, begins with an introductory work on poetic theory dotted with autobiographical references. A quick survey of the rubrics in Manrique's cancionero shows an attempt towards a global organization based on an overall chronological progression. This design was clearly aligned with institutional and monarchic milestones reminiscent of the dietaris, anales, and registers of escribanos and officials in the service of the Crown. It served to form a narrative that was sanctioned by the state. However, in addition to the fact that many compositions cannot be precisely dated, the chronology is not a steadfast structuring principle in Manrique's cancionero. This practice is similar to what can be observed in home registers, where documents that are close in date can sometimes lack strict chronological order. As an example, ID3378 (number 95 within MP3), whose text dates it on 1468, appears before ID3379 (n. 96) and ID3380 (n. 97), dated in the prose introduction to the poems on 1467.

It is clear, however, that Manrique's labour as an "ordenador" brings him to group certain types of poems according to their genre, as in the case of the canciones, or their addressee and/or interlocutor, as in the case of the exchanges with Juan de Mazuela at the beginning of the cancionero, or similarly with others further along the compilation. In spite of the courtly amorous poetry addressed to his wife, doña Juana de Mendoza (ID1874) (n. 39), the cancionero is in fact strewn with various such compositions addressed to ladies whose name the poet does not reveal and which contain professions of love or love laments. The chronological order is underlined in the poems addressed to the successive monarchs and great political figures for whom Manrique writes. The introduction mentions King Afonso V of Portugal and his keen interest in Manrique's poetry, as well as Manrique's important uncle, the Marqués de Santillana, with whom he exchanges resounding *arte mayor* verse (ID3350, ID3351) (numbers 45, 46). There follows a composition addressed to King Juan II on the occasion of the birth of Prince Alfonso (ID0410) (n. 47 in MP3), dated on 1453, as well as others addressed to Queen Juana (ID0412) (n. 65), King Enrique IV's contador mayor, Diego Arias (ID0093, ID0094) (n. 80, 81), Princess Isabel (ID3378, ID3379, ID3380) (n. 95, 96, 97), Prince Fernando's teacher, Francisco de Noya (ID3383) (n. 100), Prince Fernando (ID2946) (n. 117), Princess Isabel and Prince Fernando (ID2909, ID1872) (n. 118, 119), and finally King Fernando (ID1883) (n. 133). This last composition was written around 1477 when he was already corregidor of Toledo, as he states in the last verses addressing the king (vv. 54–6). Following this composition is poem ID3398, directed to Juan Poeta and protesting in biting verses Poeta's requests for gifts in kind. Given the overall chronological arrangement of the cancionero, it is not unlikely that the poem was written around 1481 or 1482 since it refers menacingly to the work (with fire) of the Inquisition in Seville and the excessive rains of that year (Vidal González thinks it too harsh of Manrique [Manrique 2003: 348n129]).

The cancionero strategically closes with Manrique's famous political poem on government, his *Querella de la gobernación* (ID0096) and Pero Díaz de Toledo's prose introduction to the work (ID3399), which underline the poet's political weight. Poem and prose commentary are marked as an independent unit by the beautifully illuminated page that heads it (fol. 491). There is also a loose cyclical chronology that follows the main festivities of the liturgical calendar, mainly Christmas, with the famous *representation* written for the monastery in Calabazanos at the request of his sister, *vicaria* in the monastery, in addition to several seasonal pieces like *estrenas* and, to a lesser degree, *aguilandos*, dedicated to various figures, as in the short series ID3354, ID3355, ID3356 (n. 58, 59, 60). There are fewer compositions dedicated to the Passion, including one dedicated to his wife on the topic of the sorrows of the Virgin (*cuchillos de dolor*) (ID3368) (n. 82). This "institutional and monarchic chronology" is underpinned by a persistent interest in indicating the full names and noble titles of his addressees, referring to his close family members by their degree of relation to himself and their noble title, rather than simply their name. This is a clear attempt to display his *linaje* for the benefit of his readers. Other poets, such as fray Ambrosio Montesino (08AM), attempt to do the same but highlight their socio-political connections with royalty and high nobility instead of less lofty family ones. In MP3, there is a succession of compositions addressed to the Marqués de Santillana, his uncle ("su tío); don Diego de Rojas, Marqués de Denia, his nephew ("su sobrino"); the Condesa de Castañeda, his aunt ("su tía"); the Conde de Paredes, his brother ("su hermano"); the Conde de Treviño, his brother ("su hermano"). These rubrics suggest that the poems were sent as independent papers to each addressee and marked as such in the cancionero in its function as register, similar in many regards to noble registers such as Tendilla's. The tone and intention of such poems and others addressed to powerful figures such as Diego Arias, various monarchs, or Archbishop Carrillo, contrast with those addressed to poets with whom he is in open sociocultural and economic competition, such as Juan Poeta, over whom he is trying to establish his superiority by virtue of his lineage. In this vein, it should be noted that there are missing poems by authors such as Juan Poeta to whom Gómez Manrique responds or addresses but whose work he often fails to include (e.g., the series ID3386, ID3387, and ID3376, ID2966, ID2967, ID3377).

Manrique's rubrics do not generally include much contextualizing information, their content being reduced to the information of the names of his recipients, genre designators, and very little else. There is a strong concern, as witnessed in other authors, about marking authorship, albeit in a less repetitive manner than that present in Juan del Encina, Juan Álvarez Gato, or even Antón de Montoro. Much like other cancioneros, his rubrics also present the cataphoric and anaphoric deixis, as well as adjectives and intertextual connectors such as "otra," "el mismo," "del mesmo," or "por ella" (in reference to a lady mentioned in the previous poem), which suggest that the rubrics were written linearly and that the cancionero was also enabled for a linear reading, or at least a reading of whole clusters together, as one rubric would refer to the next or the previous one. Through this means, rubrics enable the creation of a narrative thread that joins individual compositions. The nature and emphasis of this narrative frame, in conjunction with the poems' content, are controlled by the author if he is also the compiler. This narrative is, in Manrique's case, more similar to Santillana's than to Álvarez Gato's, Baena's, or Montoro's, and is marked by its insertion in a power structure. As will be discussed below, the interest in constructing a self-revelatory narrative thread will be of more interest for other authors with a sociocultural program of their own.

Following a line similar to that of his compilation, Gómez Manrique also constituted himself as a poet and statesman through inscriptional strategies outside of the book. Marking his urban government jobs and poetic skill, a poem by Gómez Manrique was inscribed on a wall by the staircase of the town hall (Casas Consistoriales) in Toledo, the city where the poet held the public post of corregidor. Manrique's coat of arms, which also fronts his cancionero (MP3) accompanied by his rhymed motto, forms a situation-dependent macro-sign with the building, to which the poem points as its visual device, the steps ("aquestos escalones"). In the poem, Toledo's ruling elite, the "noble men" addressed in the first two verses, are asked to leave behind fear, ambition, greed, and other negative forces as they ascend the steps. The individual interests must give way to the common good. The men are compared to the building's pillars supporting "the richest ceilings," a clear political allegory of their function as supporters of the Crown.[10] Engraved or painted, Gómez Manrique's device ("divisa") not only appears on the Toledan concejo's staircase, or his cancionero, but also on a number of objects listed in the 1490 post-mortem inventory of his belongings, including silver serving dishes, religious ornaments in his chapel, as well as various armour pieces (Caunedo del Potro 1991: 98, 103, 107–8). These arms, as symbols of himself, were to carry his memory onto that of future generations well after his death. In his will, Gómez Manrique stipulated that his heir should adopt his name and last name as well as his arms (text in Rivera Garretas 2007: 177). In the same will he left careful directions in relation to his burial site in Calabazanos and the position of his arms and device, as well as his wife's (text in Rivera Garretas 2007: 148, 155). Like the noble names that dot and thread his cancionero, the presence of the poet's arms in civic and religious architecture, poetic compilation, armour, and material everyday objects bolster Manrique's noble and proud lineage. By emphasizing the symbols of lineage and power in all objects that conveyed his legacy, Manrique fashions himself as a statesman and poet with important political and family connections, as well as a notable member of an important lineage. His poetic and government activities serve to reinforce his political abilities.

Manrique's poetic production reveals a sustained attempt to display the cultural worth that a noble with key positions at court and city would deftly transform into socio-economic advantage and vice versa. The insistence in highlighting lineage connections and ties with high nobility and monarchy appear accentuated in the connective thread that weaves the compilation of loose poetic papers and gives body to an ideology that marks the passage from paper archived at the bottom of a chest to the luxury codex. Along with this process, his cancionero also ties personal to institutional history in a fashion akin to the official books of royal chroniclers, archivists, and secretaries.

Juan Álvarez Gato: The Compilation as Legacy

The labour involved in the constitution of the self through a memory of single episodes shows clearly throughout Álvarez Gato's cancionero, but the whole work is an attempt to present a unified narrative that is in fact a bio-fictitious construct rather than a strict life chronology.[11] Sidonie Smith's (1987: esp. 45–8) conclusion that all autobiography is in fact fictitious invites a close inquiry into the selective process of self-account.[12] In addition, the construction of a retroactive memory conveys the selection of past events in light of present principles and beliefs (Ross and Buehler 2004: esp. 32–3). The organized selection of independent experiences and unconnected moments is threaded by way of a narrative line that organizes the fragmented remembrances of the autobiographic memory and confers teleological causality retrospectively.[13] Derrida (1988) has pointed out that the other involved in this explanation is transformed into a cosigner, thereby exposing the testamentary nature of the first-person legacy (50–1). Contemporary critics since at least Lejeune (1982) have noted the juridical character of autobiography, both in a testamentary sense as well as a memorial, confession, or witness account, Foucault (1986) having already included the legal responsibility of the text among the characteristics of the "author-function" (145 and passim). In its textual format, Álvarez Gato's cancionero bears strong ties with the personal or individual household register of a bureaucrat involved in city government and royal service.

Although several of his poems are included in a number of multi-authored compilations, Juan Álvarez Gato's cancionero has been preserved in a single manuscript (Ms. 9/5535, Real Academia de la Historia) (MH2). This manuscript, including the rubrics, was most likely produced under direct authorial control and likely with his intervention (Márquez Villanueva 1960: 202; Beltrán 1998: 62). As is common in many single-authored cancioneros, rubrics refer to the poet in the third person, even when they contain information that only the author could provide, therefore originating directly from him. The manuscript is a modest endeavour, vastly different from presentation copies or the luxurious codices of Gómez Manrique or Santillana and closer to a household register.

Judging from the poems that can be dated in MH2, there seems to be an attempt towards an overall linear chronological organization, though the individual poems in some particular cases may not follow it strictly.[14] This is in keeping with personal-official register practices such as those exercised by Tendilla (reviewed in Moreno Trujillo et al. 2007: 75). Álvarez Gato's compilation is imbued with the optic of a bureaucrat who is a close collaborator of the monarchy and also a central figure in civic government, living at a time of momentous changes in the socio-political and religious climate. At the same time, his converso roots encourage taking a problematic stance on these events and a reflexive outlook on matters of self-representation. His compilation, presented as a life chronology, is in fact a retrospective construct imbued with the political and socio-professional positioning of the poet towards the end of his life. The manuscript portrays key events in the poet's life from a retrospective outlook articulated through a careful organization and contextualization of the materials. The compilation of a life narrative out of individual texts or textual fragments results in a literary construct presented as a memory exercise in an attempt at explaining and legitimizing the self.

The retrospective nature of Álvarez Gato's book is evident throughout the manuscript. There are telling instances about the time of compilation. The rubrics, having been inserted at the time of compiling, usually narrate the context in the past, which differs from the present tense commonly used in the texts they head. Fernán Álvarez Gato's letter praising the "arçobispo de granada" (fol. 87v) and defending him against the accusations of heresy is headed by the rubric "esta carta se hizo en vida del arçobispo" [this letter was made during the life of the archbishop]. The rubric thus refers to Talavera as already dead and should therefore be dated after 1507, while the text of the letter predates it. Taking into account the known date of some of the texts in MH2, such as ID3164 (written between 1497 and 1507), letters 7 (dated 1492), and 9 (dated 1499), or the three last letters, which state they were written during the poet's old age, along with the codicological evidence, it is possible to ascertain that the compilation was undertaken during the last years of the poet's life (Beltrán 1998: 61–2, Márquez Villanueva 1960: 203). This labour is directly alluded to in the rubric introducing poem ID3121, which laments that only two stanzas of a longer work (*tratado*) were to be found at the time of compilation: "de vn tratado que hizo sobre esto no se hallan agora mas destas dos coplas."[15] This points to a retrospective effort that exposes the distance between the events and the compilation of the written documentation. As this example shows, rubrics expose the gathering efforts of independent pieces while bearing witness to the loss that the passage of time can have on the material safekeeping of poems. Thus, the compilation appears as an archiving effort that can postdate poem composition by many years. This implies a design that is motivated by a thoughtful retrospection and by an attempt to organize scattered poems into a coherent life narrative and to give it closure.

The political and ethical dimensions of the presentation of the self are negotiated through the strategic design that sustains the autobiographical compilation. The first part of the cancionero is said to coincide with the first period in the life of the author and introduces an autobiography that moves from youthful beginnings to the later moralistic prose through a middle stage of political, moralistic, and religious poetry. The manuscript is thus designed to mirror a biographical trajectory from light-hearted youth to the moral gravity of old age. The first few folios missing, the cancionero is explicitly divided in fol. 60v into two main parts by the central rubric heading the 68th poem (ID3130). This lengthy rubric states that the author has repented for his misdeeds and is leaving the world in order to lead a new spiritual life. The cancionero presents a youth who progresses to political gravity and becomes spiritualized, a rejection of worldliness shaped as a personal transformation. The rubric explains that the poet, "having known the world and experienced all estates and having reached and enjoyed much of what can be gotten from it," saw that it was all working for the damnation of his soul and wanted to "disrobe" from all vanities and sins of his youth both in his deeds and in his poetry ("en las moçedades asy en el trobar como en los efetos de sus obras liuianas"). The rubric introduces the short poem on the subject that follows, declaring Álvarez Gato's resolution to bid worldliness goodbye ("hizo esta copla al mundo despidiendose") and stating his resolve to lead a new spiritual life ("tomando nueva vida espirytual"). The shift is ratified at the end of the poem by a second rubric that points to the codicological division mirroring the fundamental change in the poet's life ("Daqui adelante no hay cosa trobada ni escrita syno de deuoçion y buena dotrina")[16] and which is presented as a textual unfolding and reading key. This division of the cancionero has traditionally been interpreted literally as a reflection of a supposed crisis suffered by the poet in 1471 as a consequence of the intense persecutions of conversos.[17] This surmise is based on the attribution of a version of the first lines of the central poem several centuries later to the Conde de Belalcázar when he entered religion in 1471. This poem was supposed to have accompanied a letter attributed to Álvarez Gato and addressed to an anonymous friend who was entering religion found in an eighteenth-century manuscript, which presumably would point to the Conde de Belalcázar in 1471. Beyond the as-yet-unproven links, upon a closer reading, the letter reveals itself as a reflection on a critical change in the friend's (not Álvarez Gato's) life. This helps us question the notion of a purported crisis in 1471 and invites further inquiry into the meaning of the structural division of the manuscript, which was devised at the time of compilation and towards the end of the poet's life a few decades later. Further signs that the central rubric/s and poem are a literary division shaped as a life fiction are to be found in the poems themselves. One example is poem ID3110, which appears in the first and thus youthful part of the cancionero but was written, according to its rubric, during his old age: "Juan aluarez siendo viejo." Álvarez Gato's book is the result of a careful restrospective compilation of the texts written during his life. As a personal register, it creates the impression of a true chronology, but the clear presence of a teleological narrative thread exposes its nature as a personal history with common points to *dietarios*.

Far from the physical and spiritual retirement from court in 1471 that the cancionero is thought to suggest, the poet was actually busily performing bureaucratic tasks at court, working in the 1480s as royal contino ("contino de la Casa de su Alteza") and escribano de cámara, and, according to a 1495 document, as the queen's major-domo. Further, the last decades of the fifteenth century did not bring a total separation from courtly poetry for Álvarez Gato, given that it is after that date that the poet most likely wrote the amorous and courtly poems for Queen Isabel. As Weissberger (2004) has shown, authors surrounding Isabel's court played a key role in the construction of her queenly power. In a long, revelatory, and compromising poem (ID1951), dated between 1474 and 1480 (Jones 1962), actually left out of MH2, Álvarez Gato declares himself hopelessly in love with Queen Isabel and portrays her in messianic terms. A poem in the first half of MH2 may also have been written for a queen ("la mayor"), whose identity is concealed (ID3096). The danger of daring to write love verses even when they helped the poet's self-promotion was that it could expose the writer as a bad poet and, perhaps worse, as a heretic in the eyes of God: "por vos so mal trobador [...] y por vos señoras vos / me hize erege con dios" (ID3126, vv. 103, 115–16). Later censors would convey the

Inquisition's confirmation of Álvarez Gato's self-perception by modifying two blasphemous, hyperbolic lines in one of his compositions featured in the grand *Cancionero General* (11CG).[18]

In the 1480s Álvarez Gato was also busy defending his urban properties and was involved in a dispute with the concejo in Madrid regarding his right to open a door onto the central plaza de San Salvador and other rights relating to his house and the square.[19] The *Libros de acuerdos* record the 1497 concejo's decision to give the poet an important role in the solemn ceremonies prepared after the death of Prince Juan, charging him with ordering the coffin and overseeing the procession. Towards the end of his life, Álvarez Gato had become relatively prosperous, reaping the benefits of a lifetime of service and the different sources of income from his various positions. This time of relative financial prosperity, with his time freed from urban and courtly duties (a document dating from March 1510 states that he had retreated to his home due to old age) enabled Álvarez Gato to put his affairs in order and think of his legacy. To this effect, in the Monastery of El Parral in Segovia in 1509, he instituted the patronato of the chapel of San Salvador, destined to be the family's burial site. At around the same time, he also appears to have prepared his will, to which he added a codicil in Madrid on 13 January 1510, his death occurring some time between 3 April 1510 and 1 January 1512 (Márquez Villanueva 1960: 40–2, 384–7). On 20 April 1514 the *Libros de acuerdos* mention Álvarez Gato's will in relation to the moneys for Masses he left to the church of San Salvador in Madrid.

While the last years of Álvarez Gato's life were ones of relative economic prosperity, they were also filled with unsettling social and political events. These were advancing precipitously during the last decade of the fifteenth century and the first of the sixteenth, with the arrival of the Inquisition in Madrid in 1490; the first *autos de fe* in neighbouring Toledo; Lucero's activity in Córdoba, particularly horrifying in the years 1504–5; and the inquisitorial persecution and subsequent death of the poet's great friend fray Hernando de Talavera in 1507.[20] Álvarez Gato's compilation of his own works must inevitably have responded at least in part to the persecutory environment against conversos and, more immediately, to the establishment of the Inquisition in Madrid in 1490, which he witnessed first-hand in the special meeting of the concejo.

It is at this point in his life when, retired in a relatively comfortable socio-economic situation and in a privileged political position, but with the bitter notes of the frightening socio-political climate, that the poet formed – compiled – his personal cancionero. His book is a personal memory and testamentary legacy for his family and the institutions that gave him socio-professional legitimacy but also threatened the already precarious social order. This outlook tints the retrospective compilation of the cancionero and helps explain the meaning of its structure and central division, which is justified again in the book's concluding poem.

Álvarez Gato's cancionero is in fact only the first part of the compilation, which is referred to as *libro* by the author and whose ending is marked by a closing poem and what may be the author's authenticating signature. The rubric, poem, and signature that close Álvarez Gato's libro in the last folio of the compilation (fol. 138v) are of crucial importance because of what they reveal about the compilation model. The end rubric reflects the common notarial practice of "closure" of a libro by means of an *explicit* that marks the official conclusion of data entry and formally puts an end to a codicological unit (Ostos Salcedo 1994: 205). The reference to the libro and to the one who wrote it ("el que le escriuio") (ID3171) (fol. 138v) is followed after the poem by what seems to be Álvarez Gato's signature, written in the same hand and ink as the rubrics that bear the author's name throughout the manuscript.[21] The duality of Álvarez Gato's libro is underlined in the poem that concludes it: "este libro va a meytades / hecho de lodo y de oro / la meytad es de verdades / la otra de vanidades / porque yo mezquino lloro / que quando era moço potro / syn tener seso ninguno / el cuerpo quiso lo vno / agora ell alma lo otro" (ID3171) (fol. 138v).[22] It is a similar association as that conjured up by Santillana when justifying with Paulinist doctrine the nature of the love poetry written during his youth in the introduction to his cancionero, an explanation similarly claimed by Fernando de Pulgar (1982) in one of his letters (Letra XXIX, 105).[23] In Álvarez Gato's poem, the dichotomies of mud-gold, truth-vanity, youth-old age, and body-soul serve to underscore a dualistic codicological ordinatio that makes the beginning and end of the libro two extremes balanced by a dramatic centre as a focal point. The rubric that heads the poem suggests that the structure is rather more complicated, dividing the book into three parts: "porque comiença este libro en coplas viçiosas damores pecadoras y llenas de moçedades y prosiguiendo habla en cosas de Razon y al cabo espirituales prouechosas y contenplatiuas hizo el que le escriuio esta copla."[24] The end-of-life assessment specularly reflects the structure of the cancionero and divides the life and poetic trajectory of the author into two main parts that were held together through a progression towards "reason": a vain youth thoroughly occupied in the needs of the flesh and a maturity that holds the hermeneutic key that exposes the error of past vacuities and shows the phenomenological and epistemological superiority of the spirit.[25] The central and last rubrics show the nature of the explanation of the self. This explanation is granted in view of another or others who, either explicitly or implicitly, elicit such account and tint its mode of execution.

Throughout the manuscript, the autobiographical note is underscored by a concerted effort to make pervasive the figure of the author. A key strategy is the addition of authorship marks that link all the poems and the appropriate prose pieces. In addition to contextualizing rubrics that head virtually every single text, there appears a second rubric stage intent on marking authorship undertaken linearly in a single operation. Álvarez Gato's name appears mostly centred at the head of the page almost in the form of a running title. In the prose section of the manuscript containing the letters, his name appears written on the margin next to the introductory rubric to each letter, wherever it may begin, frequently in the middle of the page. Álvarez Gato's presentation of himself as dexterous in the mores and discourse of the (court and urban) elite, requisite of those in a situation of upward mobility, rests in part on the progression of the stories linked by rubrics that weave one poem with the next. The picture that emerges from the accumulation of the various story threads is as complex as the many strokes that bring it to life, but the compiler's skill shepherds the isolated poems or clusters into a larger narrative.[26] This is accomplished through a chronological organization marked in the rubrics through connective, deictic markers that highlight the creation of the rubric apparatus as part of the process that weaves individual units into a larger macrotext. Clusters such as ID3076, ID3077, ID2933 or ID3085, ID3086, ID3087; and ID3097, ID3098, ID3099, ID3100, ID3101 refer to an "esa señora" or "aquella señora" mentioned earlier, creating a series of linked scenes that centre around an episode of the poet's love life. They are also supported by the identity of the lady, whose name remains undisclosed. The rubrics provide the short story in which the poet pursues the evasive lady, who responds with a negative, causing the poet to renew his efforts to contact her through his writing. This elicits a further reaction from the lady as well as a male friend, who was also able to read the poem, the narrative pointing to well-established modes of publishing. A glaring difference with Gómez Manrique's cancionero is the amount of contextualizing detail provided by the rubrics, an obvious attempt at creating a powerful narrative with attention to each constitutive segment and an effort similar to that present in Montoro's corpus.

Ending the first half of the cancionero, other sets of rubrics showcase the poet's personal, literary, and political friendships with Gómez Manrique

(ID1016, ID1017, ID2944, ID2945), Jorge Manrique (ID3122, ID3123), and Hernán Mejía de Jaén, who was regidor in Madrid and a close friend of Álvarez Gato, a dedicated *madrileño* (ID3112, ID3113; ID3125, ID3126, ID3127, ID3128). In these poems, as in many others throughout the libro, Álvarez Gato shows a clear awareness of his poetic worth. Wedged in this last section is a long poem with a long initial rubric and a complex internal rubric system. While the overall section obviously highlights Álvarez Gato's illustrious socio-professional contacts, the composition shows poetic solidarity in the form of defence of poetic worth in skilled authors of lower social status, such as Antón de Montoro or Mondragón, with an argument in favour of poetic worth over socio-economic standing (ID3117), a topic clearly reminiscent of similar arguments regarding the nature of nobility by fellow converso Diego de Valera, and of poetics as God-given knowledge by Juan Alfonso de Baena.

Once the carefully threaded storylines establish the author as a seasoned poet in the first part of the cancionero, the manuscript moves on to poetic texts that present a weightier politician, royal adviser, and overall man of confidence to the Crown. Álvarez Gato's commentary on the political unrest under Enrique IV is showcased by the series ID3112, ID3113, ID3114, ID3115, and in the second part of the cancionero by the long poem ID3142. The benefit of such criticism coming out of the pen of one of Isabel's men could only elevate her subsequent rule and fashion him as her long-time supporter. The inclusion of poems on King Enrique's reign in both parts of the cancionero points to a desire to create a balance and a logical transition. In spite of the mirage of the central break, the poems provide continuity between the two parts of the poet's life (courtliness and moral gravity) and monarchic power (Enrique to Isabel). In so doing, Álvarez Gato is actually highlighting his political role and suggesting that he is the strong force that keeps the two extremes together. Poems ID3110, ID3111, which precede the poems on King Enrique and have a religious content, support this structural function by foreshadowing the religious poems in the second part of the cancionero. A similar function for providing structural equilibrium appears in the short poems or *letras* copied on both parts of the cancionero. The letras inscribed in various objects, such as a cape, a helmet, or a necklace in the first part, have their counterpart in the banner, the mirror, shield, or tomb in the second half. Although appropriately placed according to the overall scope of each part of the cancionero, the letras underscore the culturally meaningful inscription of material life, which subtly undermines the intended spiritual sublimation of the second part of the manuscript. Further, the inscription of individual texts in material culture helps legitimize the poet as an agent of graphic acculturation and as a player embedded in the forces that shape urban and court politics and society.

The presence of political criticism directed towards Enrique IV in both parts of the cancionero points to the glaring absence of any political commentary regarding Isabel and Fernando's rule. However, it is possible to see the overall criticism of Enrique IV's rule as a subtle warning against the perils of poor government by subsequent monarchs. The cries against bad government that carry forward to the second half of the cancionero suggest a larger warning against a monarchy that nullifies the individual, a point made eloquently and forcefully in poem ID3142. The poems in which Álvarez Gato defends the need for the monarch and his subject to observe mutual faithfulness (ID3114, ID3116) have their counterpart in the long prayer on the pacification of the kingdom ("sosiego del reino" ID3142) in the second half of the cancionero. Similarly, letter 7, written on the occasion of the assassination attempt on King Fernando in 1492, put like ideas in the words of a "simple man." According to this letter, the support and protection that royal subjects owe their king needs to be reciprocated by the monarch's similar protection of the loyal subject. The monarch also needs to examine his conscience and "correct any negligence so that these are purified and cleared." In other words, the loyal and virtuous subject will not betray his monarch, but knows how to recognize bad government. These underlying ideas create a continuum that joins the political gravity of the poet with his later ascetic life.

The second part of Álvarez Gato's libro has a marked religious character, with many compositions dedicated to Marian and Christological themes, a poem to Hernando de Talavera, himself a poet of repute (ID3164), as well as others, including poems on the theme of *de contemptu mundi*. The asceticism that ends the cancionero is continued and intensified through the collection of Álvarez Gato's grave and moral prose that follows the cancionero in the manuscript. First appear the doctrinal letters with a moral and religious content, which include the letter in defence of Talavera written by Fernán Álvarez Gato, and which in turn are followed by the graphic ladder of spiritual ascent to perfection (presided at the top by a heart and the apt inscription "Accedet homo ad cor altum, et exaltabitur Deus," from Psalm 63:7–8). After these, we find the copies of letters and writings by religious figures, Álvarez Gato's book ending with the closing rubric and poem in fol. 138v referred to above. In the manuscript, this is followed by the anonymous *Breve suma de la vida de fray Hernando de Talavera*, a biography of Talavera written by a different hand. The

Breve suma is part of an attempt to make a case for the saintliness of Talavera's life, thus constituting the logical end to Álvarez Gato's autobiographical compilation. In the codex, Álvarez Gato's asceticism gives way to Talavera's saintliness.

The final *ordinatio* that may be seen at the codicological level is in line with the carefully crafted design that structures Álvarez Gato's cancionero and converts it into a legal memory. The inclusive structure creates a larger storyline of which Álvarez Gato's poetic autobiography is the first stage. An autobiographical account is followed by a biography with which it is intimately tied by virtue of their points of contact, both vital and ideological. Both constitute a memory of testamentary value in which they coexist, calling attention to the value of the manuscript as notarial *instrumentum*. This was obviously intended by the poet when he was preparing his political and literary legacy. It also must have been particularly important for his descendants in the long centuries marked by religious persecution in order to be able to abate all suspicion of heterodoxy. Politically and socially, it also served as a legitimizing tool for Álvarez Gato and his heirs. The manuscript as document rests heavily on the poet's professional, ethical, and political stature. This calls attention to the important role that compilation processes play in the creation of meaning. The power to project that meaning into the future is rooted in the nature of the compilation as legacy and its use as notarial *instrumentum*. The model of the *escribano* as an accountable and personal author mingling individual notes with institutional memory was evident in the author's creation of a testamentary fair copy of the textual life of the self.

Antón de Montoro and the Manifold Corpus

Converso mediano Antón de Montoro, tailor, used-clothes tradesman (*ropero* and *aljabibe*), and urban dweller well acquainted with power circles, wrote a substantial body of poetry that survives in several cancioneros.[27] None of these is the single-authored manuscript combining personal, family, and institutional memory that we have seen in the work of Álvarez Gato, Gómez Manrique, or Santillana. They are, rather, testaments to the popularity of Montoro's work and its wide circulation in copies that bear the mark of the many hands that participated in their transmission. The detailed rubric system implies that the corpus in its multiple incarnations underwent intense circulation. Even in the same family of cancioneros, such as that represented by SV2 and LB3, rubrics are not identical and indicate that they may have originated in different, though similar or related, exemplars (Severin and Maguire 2000: 21–2). The manifold nature of Montoro's corpus invites us to consider the forces behind the various compilation efforts that formed each collection and the mediating role played by the rubrics. The flexibility of a variegated corpus is controlled by means of a system of rubrics that provide ample contextualization, but also help construct specific accounts of poet and texts, highlighting the social and material environment. The conspicuous effort to contextualize virtually every one of Montoro's poems is common to all manuscripts, which situate the poetry in an Andalusian and, more concretely, in a Cordoban environment, the poems being an active element in urban networks and transactions.

The cancionero housed in the Biblioteca Nacional, Ms. 4114 (MN19), contains the largest corpus, followed by MP2 (Biblioteca del Palacio Real, Ms. 617), SV2 (Biblioteca Colombina, Ms. 83-5), LB3 (British Library, Egerton, Ms. 939), and, with a smaller number of poems, RC1 (Biblioteca Casanatense di Roma, Ms. 1098), PN10 (Bibliothèque Nationale de Paris, Ms. 233), and SA10 (Biblioteca Universitaria de Salamanca, Ms. 2763). Montoro's poems also appear, though more sparingly, in manuscript cancioneros such as HH1, and in print cancioneros such as Hernando del Castillo's *Cancionero general* (11CG). As scholars have indicated,[28] there is a clear relation among some of these manuscripts that can be established on the basis of the order of texts within the compilation and through particular readings in some of the poems and their rubrics. Thus a connection can be made between MN19 and MP2; SV2 and LB3; and RC1 and PN10; with SA10, in its current state of disorganization, bearing no resemblance to any of the other manuscripts. In spite of their similarities, the variant readings in the text of both poems and rubrics make it impossible to construct a coherent *stemma codicum* (introduction to Montoro 1990: 35–9), which points to the active and varied copying activity and dissemination of Montoro's poetry.

Even when we allow for codicological disarray, the organization of Montoro's poems does not seem to emphasize a timeline or biopoetic chronology. In MN19, for example, poem ID2729 (n. 193) dated 1457 appears quite later in the manuscript than ID3035 (n. 179), dated in 1461; ID1930 (n. 107), which decries the Cordoban pogrom of 1474 appears before both poems and others with an earlier date, such as ID1918 (n. 139), which deals with events occurring around 1448 (though this composition precedes ID1911 [n. 153], which comments on events from 1453). Further, ID1912 (n. 165), which praises Queen Isabel, appears considerably later in the manuscript.

The overall content of the manuscripts is of significant value as it provides insights into the circulation and reception of Montoro's poetry. MN19, for example, appears to reflect the intellectual circle that surrounded Archbishop Carrillo. The codicological contextualization of Montoro's corpus in cancioneros such as LB3, a humble quarto characterized as having strong pro-Isabel leanings (Severin and Maguire 2000: 13) compiled around 1480, is noteworthy in that a large number of Montoro's poems are copied as a compact corpus and bracketed by some of the more canonical works of noble and prominent poets. Both SV2, compiled in the second half of the fifteenth century, and LB3 point to a keen appreciation of Montoro's poetry, on a par with that of some of the famous compositions of the time such as Pérez de Guzmán's *Vicios y virtudes*, Santillana's *Proverbios*, Mena's *Laberinto*, or Gómez Manrique's *Coplas a Diego Arias*, with which he shares an important portion of the manuscript pages. This "manuscript contiguity" puts Montoro on equal footing with well-established authors, many of them noble, and points to the equalizing power of the compilation, a literary democracy of sorts, overcoming the social forces that attempt to push him towards or portray him as belonging to the middle or lower strata, a tension never voluntarily resolved by Montoro himself.

In each of the main cancioneros, Montoro's corpus forms a tightly knit unit delineated by a strong emphasis on the figure of the author and by markers that guide a linear reading. After clearly indicating authorship in the rubrics heading one of the first poems in the corpus, the rubrics alternatively use macrotextual referential markers sending the reader to the previous poem(s) for the identification of headers such as "otra suya." Further, references among poems occur not only in regards to author name or genre, but also to other names. For example, in LB3, ID2744's rubric "a don pedro" is followed by the rubric heading the poem that follows, ID1789: "otra a el." The different deictic references in, for example, "Otra suya al dicho corregidor porque le mando que jugase a las cañas"[29] (ID1791, LB3), can only be understood after reading the rubric of the previous composition, which states the name of the "said corregidor" ("gomez dauila"), and after searching for (or having read) even further up in the manuscript the name of the author. For the understanding of "otra," however, we must refer to the expression commonly used in notarial books such as inventories to mark the additions of "items" and similarly in cancioneros as a marker for an entry of "another" poem. SV2 provides one further sign that Montoro's corpus would be read linearly, almost as a multiform long poem much like the other ones that grace the manuscript. The table of contents that heads the manuscript lists Montoro's corpus as a unit, devoid of specific folio information for each individual composition. SV2 similarly constructs a self-deictic rubric system that assumes not only a linear reading but, much like the other manuscripts, also a linear rubrical composition. After stating the poet's full name in the first few individual rubrics, SV2 uniquely favours a familiar use of the first name ("Antón") in the rest, as well as deictic connectors such as "otra suya" or, for the author's name, "el dicho Antón." The use of the poet's first name as the only authorial marker in many of the rubrics suggests a familiarity with the poet that may signal proximity and also act as a social marker. The use of single names, mainly last names, as a short and convenient mode of referring to an easily identifiable poet in close proximity (other examples include Guevara, Puertocarrero, and Tapia) is normal currency in cancioneros, but the exclusive use of the first name is fairly uncharacteristic of cancionero usage, being generally relegated to familiarly designating persons in a particular (lower) social station. In MP2, compiled around 1560–70, the author's name is characteristically revealed by a general rubric that marks the corpus at the beginning – "Obras de anton de montoro y Por otro nombre llamado el Ropero"[30] – while virtually all of the compositions are headed by an explanatory rubric but lack any individual authorial marks aside from the name of the poet that appears at the top of each page as a running title. In this manuscript, the compositions are intended to be read linearly and as a single unit, almost like a long poem composed of smaller parts. This cancionero places an emphasis on (powerful) dedicatees, whose names are given at the beginning of many of the rubrics, creating a repetitive pattern, as is evident in the first compositions: "al dicho Don pedro," "al corregidor," "al cauildo de Córdoba," "a Dom pedro de Aguilar," "A la Reyna Dona ysauel." In MN19 authorship is fastidiously marked, virtually all rubrics bearing the name "Montoro." As an exception, the acrid exchange with Román refers to him alternatively as Montoro or el Ropero, providing a clear marker for an independently circulated cluster. Other manuscripts, such as SA10, however, do not show concern for naming powerful dedicatees, favouring the use of connectors such as the repetitive "otra" and placing the poet's name "Montoro" at the beginning of most rubrics; on the other hand, rubrics in RC1 and PN10 repeat "otra suya" after the initial introduction of the poet's name. These connectors indicate that the rubrics were written or rewritten at the time of compilation and impose a linear (though not exclusively) reading that follows also a linear copy of mostly independent compositions as well as some that would have circulated clustered together in thematic groupings, in one or more leaves. They also lay bare the expectation that these poems will be archived and transmitted together. The transition from independent poem to compilation establishes the dependency on the macrotext, where reading is assumed to take place in a more linear fashion with each composition building upon the last. The main thread in these manuscripts is clearly the name of the author, who is to provide an assumed coherence to the range of topics. By creating a unifying thread, even if not a single storyline, these rubrics are producing the expectation that the poems will have shared traits, an expectation strongly supported by the emphasis on the authorial figure. The different types of organization imply an attempt to mark out a set of poems under a unifying authorship and point to a poetic unit's dependence on a macrotext. They imply careful compilational efforts and the independent circulation of the different incarnations of the corpus. In spite of the (self-)deprecatory tone that Montoro's poems and those addressed to him can at times convey, the analysis of the compilation efforts surrounding his corpus reveals the establishment of a strong authorial figure aggrandized by a detailed apparatus of rubrics. In addition, the multiplicity of compiling and contextualizing efforts provide clear evidence of the lively interest in his poetry and the wide circulation of his poems as those of a canonized author.

The amount of detailed contextualization provided in the rubrics of Montoro's works is evidence of the interest in the poems and the author. The rubrics highlight the professional activities of the author by noting transactions regarding fabrics (i.e., crimson), garments (doublets, cloaks), sewing and payments (albalaes, libramientos), as well as political commentary. They acknowledge invitations to dinner or deal with gifts of food (a quarter of a mutton, sardines, dried meat, pork) or the potential gift of a horse. Many of his poems were directed to those in his urban network, as well as members of the political elite.

The imposed construct of the linear reading of Montoro's corpus is exposed by the existence of several poetic clusters that evince the independent and varied life of micro-units before being incorporated into a larger archive. This is exemplified by the three compositions (ID1925, ID3012, ID8801) that appear as a series in MN19 revolving around garments ("jubón" and "manto"). This association is not present in all manuscripts, poem ID1925 appearing independently in 11CG, 14CG, 19OB, or the cluster in a shortened version in MP2 (ID1925, 8801). Some clusters contain exchanges with other poets, such as one with Gonzalo de Monzón that consists of three poems (ID1779, ID1780, ID1781, in SA10). Montoro's exchanges with Santillana and Mena appear in different manuscripts, undoubtedly as a way of showing Montoro's poetic worth. This emphasis is clear in MP2, which opens the section of Montoro's works with two poems addressed to Santillana and one to his escuderos, or SA10, where Montoro replies to a question posed by Juan de Mena to Santillana. The two poems directed to Santillana and his escuderos (ID1896, ID0174, ID0175) may have circulated as a cluster, which appears, albeit with the poems in different order, in MN19 and MP2; and the two poems to Santillana (ID0174, ID0175) appear together in LB3, PN10, RC1, and SV2. However, Santillana's poetic exchanges with Montoro have been excluded from Santillana's own grand cancioneros. An important cluster in MN19 contains the exchange between the comendador Román and Montoro. The series has a general title that marks it as a defined cluster of ten poems, ending with the finale to the series ("la question," as ID3016 calls it) in which Montoro acidly acknowledges Puertocarrero's notice of Román's departure from Córdoba (ID3016, ID3017, ID3018, ID3019, ID3020, ID3021, ID3022, ID3023, ID3024, ID1931). A few other clusters stand out in MN19, such as Montoro's exchanges with Juan Poeta, Alfonso de Velasco, or Juan de Córdoba, and some matching compositions with a similar topic (e.g., ID1901, ID1902: on going or coming from "la vega de Granada") (also in MP2) or a common addressee (ID1784, ID2735, written to a caballero who had given him a libramiento for some wheat, which the caballero's wife tore). In addition to clusters, there are a number of thoughtfully arranged series, as is the case in LB3, where the last few compositions are directed to "don pedro," the powerful don Pedro de Aguilar, (d. 1455), fifth lord of Aguilar and member of the prominent Cordoban oligarchy with which Montoro had a close relationship. LB3 and SV2 have two unusually long series (for Montoro's textual tradition), one dedicated to the corregidor of Córdoba Gómez de Ávila (ID0180, ID1791, ID1900, ID2729), and another also to don Pedro de Aguilar (ID0170, ID2736, ID2737, ID2738, ID2739, ID2740, ID2741). These point to concerted efforts to agglutinate the poems according to a worthy dedicatee, replicating a recurrent emphasis on social connections within an urban setting.

Few of the poems can be exactly dated, but there are important compositions denouncing pogroms against conversos in 1473 (ID1930, in MN19, MP2, denouncing the "destruction of the Cordoban conversos") and 1474 (ID1924, in MN19, MP2), where Montoro decries the atrocities committed against the conversos, "esta gente convertida," in the 1474 sack of Carmona ("el robo de Carmona"). These events are vividly described in historiographical writing of the time such as the *Crónica anónima de Enrique IV* (1991, 2: 398–408, 447–8). In addition, one poem (ID1933) that only appears in MP2 addresses Queen Isabel as it speaks against the rough prosecution and punishment of conversos by the Inquisition. This suggests a date of around 1478 or 1479 or perhaps 1482, if the reference is to be understood to be about the introduction of the Inquisition in Córdoba. The queen spent about two months in Córdoba in 1478 (22 October–15 December) and returned for a longer stay in 1482 (22 April–1 October), giving birth to her daughter María on 28 June of that year (Rumeu de Armas 1974: 73–4, 103–6). One of these stays in Córdoba would have given Montoro the occasion for such an address. In contrast, less polemic poems, such as ID1899, where Montoro denounces to Queen Isabel Juan Poeta's poetic theft, appear in MN19, MP2, SV2, and LB3. In addition, ID1918, an earlier poem dealing with another political crisis dated around 1448, the death of the two comendadores, only appears in MN19 and MP2 (and MN33). Of the two compilations including the political poems, MN19 comes out of what has been argued to have been a converso circle surrounding Archbishop Carrillo, who had spent some years in opposition to Isabel and Fernando. MP2 is a late compilation that shows an overall interest in collecting both canonical and little-copied pieces. The dates of compositions and manuscripts suggest that inclusion or exclusion of polemical pieces into the cancioneros was the result of editorial or compilational politics. The dates cited above indicate that Montoro's corpus accrued through his old age until his death some time between 1483 and 1484. He had made his will in Córdoba in 1477. It is clear that poems written towards the end of Montoro's life augmented his circulating corpus, adding new layers to an already complex poet, and providing a different angle from which to read his poetry and understand his persona, particularly at a time when the poet would have been seeking some closure. These additions seem to have rested largely in the hands of those copying, reading, and circulating the poems with the intention of emphasizing select meaning. In this way the ostensibly polymorphous corpus could use its fluidity to its advantage, re-imbuing the compilation with meaning and allowing it to be constantly renegotiated.

Overall, the rubrics in Montoro's corpus present many scenes in which poetry is very closely attached to a material situation or object, a celebrated characteristic of Montoro's poetry. The detailed rubrics tell of the concrete context surrounding poetic composition, providing vivid details of the poet's domestic and public life, his professional and literary contacts, or his reaction to the political climate. In so doing they weave a biographical profile that is not based on a life chronology, but rather on independent scenes and insights into the socio-political and literary climate that congeal around a powerful authorial figure. Montoro's rubrics show a wide web of connections with different people, some of the more important ones being the constant and abundant poems written to two members of the Cordoban oligarchy of the Aguilar, don Pedro and don Alonso, as well as sources of urban power, such as the cabildo de los abades de Córdoba, or Córdoba's corregidor Gómez de Ávila. Other people who are the objects or addressees of Montoro's compositions include the comendador Román, Juan Avís, the poet's own servants, as well as many others of various social levels, from the heavy drinker Langosta to King Enrique IV, Queen Isabel and King Fernando, and many others. The rubrics heading Montoro's poems in the various manuscripts present texts that are destined to foster vertical ties of solidarity with important members of local and central government. His poems also present him in a relation of both solidarity and competition in his relations with people with whom he competes for material and economic resources. The rubrics heading Montoro's poems show intense social connections fostered by means of constant literary transactions conducted within a wide socio-political spectrum where conflict and solidarity are shown to emerge out of social proximity. The emphasis on the specific placement of the poem, not only within a social network, but also within its material context (dinners, clothing, yards of fabric, dried meat, wheat, wine, horse, mule, sardines, or pork) expose the tight relation that poetry has with the everyday life and the social, political, and economic networks that sustain it. It is important to note that references to these commodities are written in regard to mostly urban exchanges in which Montoro interacts with a wide variety of people, some of whom are business relations. As such, they are documents that mark such economic exchanges and show Montoro's political connections. The various compositions referring to his professional toil or to his (unwilling) participation in war clearly position him as a mediano. References to material objects and petitions come linked to documents: *carta de pago* (ID3009), libramiento (ID1784), *cobro* (ID1907), albalá (ID2729, ID3035), bill or price list (ID1926). Montoro's negotiation of his social status, a position in between artisan and merchant as well as caballero put him at a closer distance to urban oligarchies and, to a certain degree, state government. Montoro situates his poetry professionally and responds to repeated attempts to reduce him to a subaltern status – the pressure for him to go back to his sewing discussed in chapter 2 – that would deny him the ability to write or that would render any uttering as reprehensible, both in a rhetorical and in an ethical sense. He shows his ability to be conversant with administrative language, political commentary, and stylized poetic discourse while displaying his status as mediano. Manuscripts SV2 ad LB3, however, present a poet stripped of his political weight but on an equal footing with canonical authors and their grander compositions. Montoro's poeticized documents (seen in chapter 6) show the advantage of being "administratively literate," as well as the documentary grounding of poetry, which Montoro used deftly. The obvious previous life of the poem as document served the dual purpose of giving the poetic paper an added value and also of placing it within larger textual practices, thereby helping to establish the author's cultural authority. Montoro's poetic corpus makes explicit like few others the tangible ground upon which negotiation of socio-economic status takes place, as well as the fundamental role of poetry in the establishment of such status. This relation of poetry with its concrete environment is not solely characteristic of Montoro's poetry. What stands out in his poetry is his persistence in making them visible within his poems and by means of a minutely contextualizing rubric system, a characteristic that he shares to a considerable extent with Álvarez Gato. Rubrics play a fundamental role in the construction of a poetic personality clearly designed by the author, although not completely controlled by him, and further reinforced in the dissemination networks in which his poetry circulated and which mediated the modes of textual access to the poems. They also show an emphasis on pragmatic elements and on the relation of poetry to that which is tangible – in other words, to the materiality of poetics. The variety of compilation efforts that the cancioneros display presents a corpus that is mediated by the many hands that participated in its dissemination. This creates an immediacy to the author, his material context, and the web of socio-political relations that supported him, granting him an importance that other authors were perhaps slower to achieve. The careful use of rubrics in all manuscripts containing Montoro's poems belie the apparent haphazardness of the archival efforts in the constitution of a personal register. The rubrics point to an involved archival process that monumentalizes Montoro even as it announces his mediano status.

The emphasis, along the lines of Santillana's or Gómez Manrique's approach, on the sublimation of poetry implied weeding out the tangible elements of poetry that effectively removed generalized access. Conversely, the use of detailed contextualizing rubrics would ensure that difficult conceptual poems would be understood in wider circles and that the author would thus gain larger appreciation. By placing themselves in the midst of the social and political conflicts of the time and the tangible realities of everyday life in which writing is situated, poets such as Montoro or Álvarez Gato conceived of their poetic activity as record, repository, and at the same time instrument of such conflicts.

Conclusion

The previous chapters have followed the journey of the text from pieces of paper to book as it was propelled by the social, political, and cultural forces of the time. The resulting picture has shown the rapid movement of the text and the constitution of textual agency. While the focus of this book has remained on the fifteenth century, it also paints a wider landscape in which the continuities throughout centuries form strong links that belie claims of ruptures and boundaries. The manuscript practices described here bear strong parallels with those already established for the Golden Age and, it could be argued, later centuries. Print practices continue those developed in manuscript culture without creating an alienating breach either in periodicity or in modes of re/production. By paying attention to the imbrication of textuality and socio-professional networks, the focus has shifted towards the concrete uses of the text both in interaction and in the constitution of the self. Further, this focus suggests enduring forces that bear upon current developments in the use of texts and communication technologies.

Because of its acknowledged liminal position between the Middle Ages and the Renaissance, the fifteenth century exposes the pitfalls of periodization. As pre- or proto-Renaissance, this border century exposes the problems and value judgments associated with rigid periodization. This is framed by the very relational nature of the "Middle Ages," regarded by definition as what it is not, neither classical nor modern, in fact a perpetual "other" whose very value lies in its alterity, only able to engage our attention through its unknowability. We look at the Middle Ages, but only with averted eyes. It has been for centuries the negating backdrop against which other periods affirmed themselves. Modernity is dynamic, historicist, self-reflexive, and in a discontinuous position with traditional "agrarian societies" (Giddens 1990). Burckhardt argued that the modern age was embodied by the Renaissance, particularly that of Italy. In the link between modernity and mechanized technologies the printing press has been identified as the first communication technology to bring about the severance between traditional thought and modern self-reflection (Giddens 1990: 77). Key changes in reading and writing have been attributed to the printing press and opposed to a fundamentally oral Middle Ages (see, for example, Kennedy 1984).

The writing practices studied in these pages show that the textual developments associated with the humanistic Renaissance, with later governments such as that of Felipe II (the *Rey Papelero*), and with the printing press, cogently studied by Bouza Álvarez (2001) and Rodríguez-Moñino (1968), were in fact present in the fifteenth century. The forces involved in the textual profusion of the fifteenth century were not a prescient move towards or a gateway to a full bloom or a "rebirth" of civilization, but formed a complex web of social, political, and cultural developments in continuity with both earlier and later practices – a complex but continuous trajectory, rather than a "swerve." Abbas (2002) has stated that "Otherness lost its innocence as a result of the colonial experience" (226). If this is the case, modernism may need to reconsider the ways in which it appropriates or disowns other periods. Knowledge of the past must become part of modern self-reflexivity. Rather than obeying period-specific value judgments, it is worthwhile to take the gaze away from the muddling darkness with which periodicity has typecast the Middle Ages and look closely at textual agents and their practices along a wider chronological expanse. The study of their hands, their networks, situations, and materials reveals much about resilient continuities. Reshifting the focus towards networks of textual exchange brings into relief the autonomous workings of writing. The physical toil, the materials of writing, the involvement of human agents, as well as the specific solidarities and conflicts they generate have a logic of their own that, though unquestionably situated in a larger historical context, creates strong threads and continuities through expanses of time that widen when the focus shifts. The object of such shift is not to disregard change along a chronological continuum, but rather to identify the inner dynamics of specific groups strongly vested in writing activities and the situations that sustain them and constitute their particular culture. When we do this, some of the tenets of strict periodicity begin to crumble. In this regard, such practices as rubrication in the process of book formation may claim continuities with the "modern" process of re-embedding the contexts of action that the act of dissemination had rendered disembedded and faceless (Giddens 1990: 80–8).

Writing as a physical and intellectual activity incorporates not only literary but also legal, economic, and historical texts. This approach is as much a departure from Weberian perspectives that regard bureaucracy as static as it is from modernist theories of pre-modern organizations as non-dynamic. It is grounded in the identity of the human agents involved in textual production and the similar nature of their materials, and focuses on writing networks and materials rather than on the imprint that their belonging to a particular period may carry. This approach does not deny the effects of such imprint. Rather, it claims that when the focus shifts, some period differentiations may begin to erode or simply may no longer appear pertinent. Without denying the differences in manuscript and print modes of production, we need to acknowledge the ramifications of the continuities between modes of reproduction as well as the endurance of the manuscript medium for centuries after the advent of print, going beyond the substitution of one for the other and the construction of their break as a sign of modernity. Common issues such as the wide circulation of poetry, patterns of manuscript circulation and publication, the concomitant issue of anonymity and problematic attributions of authorship, and the preoccupation with the graphic aspects of writing and its physical nature are present in startlingly similar ways in what we call the medieval, Renaissance, Baroque, and later periods. The social projection of writing and the networks of authors in contiguous social levels and continuous contact lend a distinctive set of characteristics that appear throughout the periods but that have been interpreted in fragmented and differential ways when approached from the standpoint of period specificity. Labels create boundaries.

Writing has its own rules and does not depend exclusively on external power structures. A disembodied authorship creates its own logic and its own hierarchies. The possibility of occasional writers and the multiplicity of textual agents help aggrandize authorship but at the same time undermine it by decentring the author. These are conflicting forces. Authors try to control the production, reproduction, and dissemination of their texts, while other agents effectively diffuse such efforts and remove the text from the author. Agency is communicable and contested. It is also grounded in the material aspects of production. Through these means, the text becomes its own agent; it has iterative powers that use but are not wholly dependent on human agents. The processes of (re)embedding the text in circumstance, as it moves through networks of exchange and the efforts for the preservation of individual and institutional memory, help the text's pervasive endurance, of which the book is a common repository. But in addition to its function as textual archive, the book can be the instrument of textual submission or control, an attempt to bridle the text, to bring it into confinement.

Textual agency rests heavily on other agents – material, human, or environmental – and thrives in social networks. In the fifteenth century, these networks functioned not only at court but also in cities, placing the "civic" in Elias et al.'s (2000) "civilizing process." However, the importance of mastery over highly regulated discourse may not be exclusively identified with the move towards self-restraint of a curial nobility, but may be more the product of interaction within and among heterogeneous networks of constituents in constant negotiation of their social position and of their self-identity. This means that the process of self-generation is not exclusively dependent on changes in the court, but rather on larger opportunities for social mobility and on a destabilization of the notion of social, political, and personal categories such as nobility, middle-class and urban commoners, and caballería. The intricacy of the elements that weigh on human interaction can push and pull the lines that divide social groups until they become less straight, inviting what Wright (2009) has called an "integrated analytical approach." Although socio-political stratification works in the symbolic realm, it rests on regular interaction within networks of human agents in concrete settings. Transactions reinforce existing relations but also push them in one or more directions. In addition, the different groups are not in a hierarchical relation regarding aesthetic or cultural values. Non-elite middle-class groups look to be on an economic par with the urban elites, but they do not seek incorporation nor do they necessarily appropriate its values. Counter to Bourdieu (1993), most agents in fifteenth-century poetry cannot be characterized as "possess[ing] all the properties of the dominant class minus one: money" (165). But neither can, it may be argued, members of the Generación del 27 or authors such as Martín Gaite, who were all members of an affluent bourgeoisie. Because of the ability of the poem to become its own agent and potentially destabilize the social structure, the poetic elite does not exactly correspond to the political elite, though there is a clear pressure to create that overlap. There is a theoretical attempt to control aesthetic discourse that is thwarted by actual capabilities of agents. Further, in their overarching attempt to regulate discourse, poetics and rhetoric also observe the need to produce remonstration, to write of reprehensible social behaviour, and thus open the possibility for negative discourse, which in practice is extended to slander. This creates the paradox of dejected discourse being used to accrue personal worth. The identification of levels of discourse with social class thus encounters the opposing path of entering into offensive and slanderous discourse as a part of a poet's range. Slander and inappropriate discourse belie homogeneous definitions of distinction or prestigious discourse because, although socially and politically destabilizing, they can be viable means for accruing poetic worth.

Finally, the physical elements of the writing process and the materials of writing make for a richly textured text layered with meaning. Much like earlier poetic networks of exchange, current social electronic networks rely on communication through visual images and short and witty texts that are shared and spread to an expanding number of contacts. The progression through the centuries from the prevalence of large-format books housed by institutions, such as monastic libraries and universities, to the growth of personal libraries has its corollary in the body becoming the repository of the small-format book or text, instead of the institution (though institutions such as court libraries and museums as separate architectural units continued to grow from the fifteenth century onward). These developments are mirrored by current technologies of the word such as computers, which have gone through a similar development as their demand, format (from mainframe to personal computer to tablet and smartphone), availability, and price have made them more widely available. Tactile contact with the writing page (or screen) is guaranteed by touch screens and keypads that are either incorporated into the screen or mounted underneath. From the large mainframe computer, to the home personal computer, and to the personal electronic memory book small enough to be held in the hand, the word continues to inscribe daily life and to inspire flexible, portable, and ubiquitous modes of authorship and textuality.

Notes

I would like to thank the National Endowment for the Humanities for a Summer Stipend that allowed time for research on this book, and to Purdue University for providing me with a Center for Humanistic Studies Fellowship that enabled me to spend time writing. Charles Faulhaber read an earlier version of chapters 3 through 8 and patiently answered my questions when I was working on the translations of the Spanish texts. Sílvia Oliveira answered my questions on the translations of the medieval Portuguese texts. Óscar Perea Rodríguez read a draft of chapters 1 and 2. Alicia Raftery helped prepare the index. To them goes my sincere gratitude for their generous help. To my family I am indebted for their unconditional love and support.

Introduction

1 [You have made me happy beyond measure. To see your collected poems like this, almost on the verge of a book is what I wanted. Because in this way your stature as a poet is fully in view, complete, varied, most faithful, and rich all at the same time.] This and all other translations in this book are my own.
2 Salinas (1992) makes this clear: "Para mí lo único claro es que necesito publicar los versos nuevos, buenos o malos, porque como me suele suceder cuando tengo acabado lo que yo creo un libro, si no los publico se me interponen, como un estorbo, y no escribo más. Siempre me ha pasado así. Es decir, necesito quitármelos de enmedio" (194). [To me, the only thing that is clear is that I need to publish the new verses, good or bad. As it usually happens when I have finished what I consider to be a book, if I don't publish the poems, they become obstructive, like a hindrance, and I don't write any more. This has always been the case with me. That is to say, I need to get them out of the way.]

3   In these pages, the term "extratextual" refers to the material context of the text outside its medium, while "paratextual" refers to the situation of the text within its medium.
4   For recent studies emphasizing paratextual analysis, see Arredondo 2009.
5   [I moved the chest of drawers on Monday, when Soledad came to visit, after she left, because the conversation that we had stirred a lot of things in me. It could be turned into a chapter of the notebook [...] Afterward, I jotted down some notes from the conversation and dated them. I think that was the last time I wrote something. Not in the notebook, but on loose papers. Where did I put them? Dealing with loose papers is awful. "I should make a fresh copy," I remember thinking while moving the chest of drawers. "It is all a matter of continuing to write, stitch by stitch."]
6   Insightful overviews of medieval Spain, including the events discussed in this brief introduction, may be found in Edwards (1982, 2000, 2004), Ladero Quesada (1986, 1992b, 1999a), O'Callaghan (1990), and Teófilo Ruiz (1994). Further information on specific subjects may be found in the following chapters.
7   I would like to thank Charles Faulhaber for his help with these data.

1. Poetry, Bureaucracy, and the Social Order

1   [as Your Worship fulfils your job, and I mine, it is no wonder that my hand is covered with ink and your foot in blood.] Pulgar (1943) used this personal narrative in his *Crónica de los Reyes Católicos* (2: 77).
2   [The very subtle escribano / who works night and day / with his nice writing set / on smooth Tuscan paper /... / in his rich writing / of unequivocal and pure letters / well written in his own hand / certainly not that of a commoner.] This and other citations of the *Cancionero de Juan Alfonso de Baena* refer to the 1993 edition. I use the poem ID numbers and manuscript abbreviations found in Dutton 1990–1. In this book, quotes from cancionero poems refer to Dutton's edition unless noted otherwise. Most poems and manuscript descriptions can also be easily accessed through http://cancionerovirtual.liv.ac.uk.
3   [Since my honour increases / with riches and advancement / for praising the great worth / of he who is gracious and courtly / and is close to the king.]
4   [so that I, my lord, may write, / I, the lowest among your people, / things that may amuse you / my hand, your captive.]
5   [and I, though the lesser of these members, know how to exert myself to serve my Prince, not only with physical strength, but also with my mental and intellectual powers.]
6   [the warrior neither eats nor sleeps, and never has a moment's repose, keeping watch on others, or afraid of being ensnared; they serve in war but they never become rich; they never prosper and always work; they continuously make

their dwelling in campaign tents or portable tents or pavilions, or sometimes under tree branches or under their shields; at times they are freezing with cold and at times burned by the sun; when they are lightly wounded they die in the camp for lack of remedy.]
7 [You show great spirit, my Juan de Mena, by so extolling arms. Your flesh has become thin from long vigils behind a book, but not hard and callous from sleeping in the fields; your face is pale and wasted from study, but not broken or scarred from encounters with the spear.]
8 In *Cortes de los antiguos reinos de León y de Castilla* (1861–1903, 3: 137–8), hereafter cited as *Cortes*.
9 Jurados were parish councillors, concejo or city council representatives without vote who defended the rights of the common people and preserved the public order.
10 See for example the 1484 *Ordenanzas reales de Castilla* by Díaz de Montalvo (1990), which cited laws already confirmed by Juan II. In theory, artisan and merchant positions were forbidden for all caballeros. In practice this order was not followed.
11 On the concept of class and social stratification, my analysis is in line with Wright's (2009) integrative approach.
12 On the influence of the debate on the nature of nobility on cancionero poetry within a larger theoretical frame, see Serverat 2001. See also Gerli's (1996) insightful study on Diego de Valera, which highlights performativity, as well as Di Camillo 1996.
13 On these urban elites ("patriciados urbanos"), see the useful overview by Sánchez Saus 2004. On cities and the distribution of power, see for example Asenjo González 1999, 2006. For some suggestions of how the everyday life and values of urban oligarchies are portrayed on literary works such as *La Celestina*, see Ladero Quesada 1990.
14 The poem is found in cancionero HH1, edited by Severin 1990: 21.
15 All references and quotes of Gómez Manrique's texts in this book are from the 2003 edition.
16 See also the broader and useful Beceiro Pita 2007. On the relation between administrative jobs and culture, see Sanz Fuentes 1990.
17 On this process in an Iberian and wider European context, see the important studies by Elias 2000 and Gomes 1998. See also Ladero Quesada 1999a: 154–244. On the role of the nobility and persons of various social groups in specific administrative offices, such as the royal Chancellery, see Montero Tejada and García Vera 1992.
18 For some key points on the (conflictive) relation of the monarchy and fifteenth-century politics, society, and culture, see the insightful studies collected in Nieto Soria 2006. and *La nobleza peninsular* 1999. See also Gerbet 1989, 1997.

19 On the growing importance of these and other "Crown officials" and the tight connection among many of the relevant posts, see the persuasive study by Ruiz García 1999. Also of interest are the conclusions reached by Cabrera Sánchez 2002 on the role played by education, social status, and political and administrative positions.
20 García Marín 1974: 133–4 and passim. For this issue in regards to the escribanos, see for example Arribas Arranz 1964a: 191–201.
21 Solana Villamor 1962: 19–20. On the possibilities of social mobility in the wider context of the development of fifteenth-century nobility, see Ladero Quesada 1996–7: 19–36. For insightful comments on the changes of the nobility during this period, see Rucquoi 2000: esp. 250–88.
22 Because the term "letrado" is given different meanings – from a general and literal "man of letters" to a loose "knowledgeable," to "university graduate," particularly in law – which can lead to confusion, I have avoided using it broadly here.
23 This is a well-known phenomenon. See, for example, García Marín 1974: 221–4.
24 On Iranzo see Carceller Cerviño 2000; also see Montero Tejada and García Vera 1992 on Iranzo and other nobles with roles in the royal Chancellery.
25 For this and other similar cases, see Carlé 1981, 1993. On Alfonso de Robles, see also Diago Hernando 1998–9. See also the studies collected in Quintanilla Raso 2006 for several aspects related to mobility into the noble ranks.
26 On the urban middle classes and the development of cities, see Asenjo González 1999, García de Cortázar 1994, as well as the volume *Las sociedades urbanas* (2003) and the bibliography cited therein; on urban oligarchies and a budding bourgeoisie, see Rucquoi 1995. Monsalvo Antón (1995: esp. 102–3) has highlighted the socio-economic diversity of urban elites as well as their relation to power. On the formation of a bourgeoisie, see the classic study by García de Valdeavellano (1991).
27 Asenjo González 1986; Bonachía Hernando 1988; Ladero Quesada 1992a; Montero Vallejo 1992: esp. 245–324; Valdeón Baruque 2007.
28 Cabrera Muñoz 2001; Collantes de Terán Sánchez 1980, 1992a, 1992b; Ladero Quesada 1999a: 165–9. Valdeón Baruque (1995: esp. 321–2) highlights the role of the conversos in the urban groups of power and the consequent double nature, both religious and socio-political, of the violent conflicts that targeted them. On this type of urban conflict, also see Mackay 1991.
29 For a useful overview of the conversos during the reign of Isabel, including observations on their social, economic, and professional status, see Rábade Obradó 1993, 2002, and the bibliography cited therein. See also Ladero Quesada 1992b. The documentation studied by Gómez Mampaso (1980) shows many conversos in artisan and small merchant occupations, including

tailor, shoemaker, and used-cloth merchant, as well as escribano, among others. For a comprehensive overview on the state of converso studies, see Perea Rodríguez 2008.

30 Among the studies on the topic, see the classic by Márquez Villanueva 1957, as well as Ladero Quesada 1992b and Lorenzo Cadarso 1994. Conversos had an important presence in city council government in Córdoba. See Cabrera Sánchez 1998: 126, 137–8, 149–52, and passim.

31 The role of the continos seems to have been loosely defined, as they were employed in a continuous fashion in the court and fulfilled various duties, often residing in cities where they held other political and administrative positions, presumably defending the interests of the Crown. On the role and function of the continos, see Montero Tejada 1999.

32 On the subject of urban oligarchies and power, see for example the studies in Asenjo González. 2009. On urban identities, see Val Valdivieso 2006.

33 See the important analysis in Val Valdivieso 1995.

34 Val Valdivieso 1996 has emphasized such diversity in her important study of urban society and socio-political aspirations; see also 2001.

35 See the insightful study by Rucquoi 1995. For Rucquoi, merchant and caballero are "two sides of the same social group" (360). Rucquoi has also noted the overlap of competencies with the conversos in the struggle for urban power (esp. 369). See also Contreras 1995: 689–92, and passim; Mackay 1986; Guerrero Navarrete 1986: 155–72 for Burgos; as well as Ladero Quesada (1990: esp. 99–102), who identifies *Celestina*'s Pleberio as a merchant caballero, and, to a lesser extent, Calisto. Even though the term "mercader" is customarily used in relation to large commerce, it is important to note, as does Collantes de Terán Sánchez (1992a) when discussing the traperos, the important commercial role of those in the textiles trade, particularly in Andalucía. For these issues in a wider social and political context, in the illustrative case of the city of Valladolid, see Rucquoi 1997, as well as Astarita 2005.

36 See for example the case of Gonzalo de Burgos, public escribano in Gran Canaria (Palenzuela Domínguez 2003: esp. 46–7, 85–7) or the escribanos in Valencia studied by Cruselles Gómez 1998: 169–86. On the group of merchants, see Collantes de Terán Sánchez 1992a. This trend would continue through the sixteenth century (Pike 1972).

37 [Your beloved daughters, / may you see them nicely wedded / with caballero husbands / and with honourable men of wealth, / with courtly merchants / and with rich burghers.]

38 [Now the customs of chivalry have changed into theft and tyranny. We no longer care about the virtuous caballero, but about the abundance of his riches. His cares used to be to fulfil great deeds, but now they are turned into

pure avarice. They are no longer ashamed to be merchants and to hold even more indecorous jobs, but they rather think that they are befitting. Their thoughts, which used to be solely about the public good, are now scattered with their great desire to gather riches through land and sea.]

39 For an overall profile of the regidores, who did not have to be university graduates (letrados or *bachilleres*), who owned cattle, vineyards, and land, had family in town, and were more involved than others in concejo discussions, see Monsalvo Antón 2003: 428.

40 For the case of Madrid, see Losa Contreras 1999: 624–8. González Alonso (2001) has laid out the main features of concejo reform under the Catholic monarchs.

41 The corregidor was a chief magistrate apppointed by the Crown, a city governor with civil, judiciary, and military competencies.

42 See recent views on the subject and bibliography in Val Valdivieso 2007, Sánchez León 1998, and Diago Hernando 2001.

43 An example is the case of the escribanos of Burgos studied by Guerrero Navarrete 1986: 200–4.

44 As an example, see Polo Martín 1999: 146 for a list of regidores in many important cities who also held jobs at court. For an overview of multiple office holding in the courts of fifteenth-century Castile, see Phillips 1978: 123–6.

45 The maestresala had among his duties that of overseeing the delivery of food to the lord's table and ensuring that it was not tainted. It was a sensitive position and given to those deemed most trustworthy.

46 For the office of secretario, see Bermejo Cabrero 1979. For a description of the types and competencies of notarios and escribanos (both *escribanos de cámara* and *escribanos públicos*), see Pino Rebolledo 1972: 19–29. For the period preceding the fourteenth century, see García Valle 1999. The terms notario and escribano were often used interchangeably (Pino Rebolledo 1972: 20). See also the comprehensive studies by Bono Huertas 1979, 1990, 1994; Corral García 1987; and Pascual Martínez 1981, 1983; the studies collected by Ostos Salcedo and Pardo Rodríguez 1995; Pardo Rodríguez 2002; Pino Rebolledo 1991; Polo Martín 1999: 318–61. For a comprehensive study of the role of secretaries through the early eighteenth century, see Escudero 1976: esp. 1: 3–40 for the figure of the fifteenth-century secretario. For the duties of the escribanos in Madrid, see Losa Contreras 1999: 357–9.

47 On Fernán Díaz de Toledo, see Bermejo Cabrero 1979: 190–8. On Fernán Álvarez de Toledo, another powerful converso secretary, member of the Consejo Real, regidor of Ciudad Real, as well as many other titles and brother-in-law of poet Juan Álvarez Gato, see Vaquero Serrano 2005.

48 Montoro aptly dedicated the poem "Hombre de rica familia" to Gonzalo de Hoces, veinticuatro of Córdoba. A bushel of wheat cost 73 maravedís in 1493, and on his voyages Columbus paid each ordinary seaman a salary of 22 maravedís a day or 666 a month (Morison 1942: 143). Celestina could buy about two litres of wine with 8.5 maravedís (Ladero Quesada 1990: 103n).

2. Escribano Culture and Socio-professional Contiguity

1 For insightful comments about the relation of escribanos and monarchic power, see Pardo Rodríguez 1992.
2 See, for example, the cases cited by Carpio Dueñas 2000: 308–14, 326–9.
3 For the text of the Seville ordinances, see Bono and Ungueti-Bono 1986: 44–56. For the *Pragmática de Alcalá* and its projections through many centuries and countries, see Rodríguez Adrados 1988.
4 See the document published in Pérez-Bustamante and Calderón Ortega 1983: 253. The document was signed and apparently written personally by the marqués, lacking the signature or otherwise authenticating traits typical of an escribano. It includes a meaningful narrative of the handling of official documents. Full text in Pérez-Bustamante and Calderón Ortega 1983: 252–5.
5 I have discussed some related issues in Gómez-Bravo 2005.
6 The pregoneros were hired by the city council for life with a right to a retirement pension. See the case of Burgos studied by Guerrero Navarrete 1986: 204–5.
7 An urban district linked to a parish, or ward.
8 More is known about escribano professional associations in the Crown of Aragon. See, for example, Cruselles Gómez 1998. Ferrer i Mallol (1974) has studied Catalan notarial practices in detail.
9 The Romeros were a powerful family of bureaucrats, starting at least with Sancho Romero I, who worked for King Enrique III. His son, Sancho Romero II, would follow and increase in importance as royal secretary and escribano. He appears as escribano under Queen Catalina's orders in 1411, 1413 (docs. in Esperabé Arteaga 1914–17, 1: 88, 96, 97, 98), and in 1416 (doc. in Pérez-Bustamante and Rodríguez Adrados 1995: 225–6). In 1418, he was the escribano who countersigned the queen's will, and in 1419, along with Martín González, he recorded Juan II's oath in Toledo. On the Romeros, see Cañas Gálvez 2008: 403–6. Sancho Romero also acted as escribano under the king in 1421 and 1431 (docs. in Esperabé Arteaga 1914–17, 1: 106, 107, 108, 109, 111, 119), also in 1419 (doc. in Pascual Martínez 1981: 164–9). He also appeared in 1416 in Valladolid as the king's escribano de cámara and notary public authenticating transactions and drawing documents for Santillana. See

the documents published by Pérez-Bustamante and Calderón Ortega 1983: 168–71, 176. Several bureaucratic functions were combined in Sancho Romero II's son, Diego Romero, who became not only Juan II's secretary and escribano de cámara, but also *escribano mayor de rentas, contador mayor de cuentas*, treasurer and *alcalde mayor* of Toledo, as well as member of the Consejo Real. Bermejo Cabrero (1979) briefly discusses his role as secretary (198–9, 255–6). Martín González, the other escribano whom Baena addresses in his poem, appears issuing documents under the king's orders in 1420, 1421, and 1426 (docs. In Esperabé Arteaga 1914–17, 1: 100, 101, 102, 111, 117). He also appears as the escribano of King Juan II in 1423 approving and ratifying the 1411 peace treaty with Portugal. The document is a twelve-page cuaderno (published by Almeida et al. 1961: 58–69). Martín González's hand and *signum* are clearly visible in plate n. 5.

10 The term "carta," as used by Baena in his poem, often has the meaning of a single piece of paper, in keeping with the Latin *charta*. In Fernández de Santaella's (1992) *Vocabulario Esclesiástico*, *pagina* is *carta* or *escriptura* (see entry *pagina*), while *pagella* (see entry) is related to *pagina* and *cartilla*; cartilla is equated to *carta chequilla*, or *cedula* (see entry *Pictaciolum*). See also the definition of "pages" in Palencia's (1967) *Universal vocabulario*: "Pagine. son memoriales. y es pagina cartilla: no pareçe nombre deriuado. paginas son foias aparte enlos libros assi dichas por que a vezes se escriuan." In another context, cartilla could also bear the meaning of 'reading primer.'

11 See for example the narrative mode used in the registers published by Pérez-Bustamante and Rodríguez Adrados 1995: 171–2, 174, and passim.

12 [I, Juan Alfonso, your escribano, / with great courtesy and reverence / to your high excellency, / kissing your hand, present this carta.]

13 [Fernand Martínez de Burgos, son of Juan Martínez de Burgos, escribano that was of that city, wrote it.]

14 He compiled prose and verse works in what has been termed the *Cancionero de Fernán Martínez de Burgos*, which survives fragmented and scattered in later copies (MN16, MN23, MN33, MN34, MN35, MN49) (Dutton 1990–1, vol. 2). See also Garcia 1979.

15 For example, the colophon to the *Cancionero del Duque de Gor*, a manuscript copy of selected works by Fernán Pérez de Guzmán (Fundación Bartolomé March Servera, Palma de Mallorca, Mss. B89-V1-13, MM2), differentiates between Pérez de Guzmán's work ("fizo e copiló") and that of Antón de Ferrera, *criado* of the Conde de Alba, who actually wrote it down ("escriuiólo"), finishing it ("acabóse de escriuir") on 1 March 1452. Description of the manuscript in Díez Garretas and Diego Lobejón 2000: 77–8. There

exists an early tradition of notarial signatures in literary works. Aside from the issues of authorship that concern the *Libro de Alexandre*, and which are a matter of heated debate among scholars (see the introduction to the edition by Casas Rigall 2007: 18–25), the explicit that concludes Ms. P reads like a notarial signature written in the appropriate terminology: "Si queredes saber quién fizo esti ditado, / Gonçalo de Berçeo es por nombre clamado, / natural de Madrid, en Sant Millán crïado, / del abat Johán Sánchez notario por nombrado" (725). This terminology is in line with that used by thirteenth-century notaries, studied by García Valle (1999: 32–77, and passim). A similar signature appears in Berceo's (1992) *Vida de Santo Domingo de Silos*: "Yo Gonçalo por nombre, clamado de Berceo, / de Sant Millán criado" (stanza 757).

16 Both appear in documents such as the one bestowing a number of privileges on the town of Buitrago in 1443. See the document published in Pérez-Bustamante and Calderón Ortega 1983: 272–7. In Santillana's documents there is also a "Sancho Garçia de Medina, escrivano del Rey e su notario publico," who is likely the same as the unidentified poet García de Medina, a few of whose poems appear in SA7. See the 1433 document published by Pérez-Bustamante and Calderón Ortega 1983: 195, 205, 206. Some of Santillana's and Diego de Burgos's poems may have been transmitted together, as evidenced by MT1, which contains the poems that both wrote for King Afonso V of Portugal. However, Santillana failed to include Diego de Burgos's works in the compilations of his own works. Burgos wrote a grand poem headed in SA10 by an introductory prose prologue (ID1709, 1710) upon Santillana's death. The poem, made up of 236 stanzas (or a few less depending on the copy) and found in HH1, SA10, MN55, 11CG, is understood to have helped him stay employed in the house of Mendoza under Cardinal Pedro González de Mendoza. The Burgos are documented as one of the most powerful families in the city of Burgos, with strong representation in the city's key government positions such as the regimiento (Guerrero Navarrete 1986: 160–4). Other cases of close collaboration between a noble and an escribano include Manuel Rodríguez de Sevilla, notary of the town of Benavente, who acted as copyist of books for the counts of Haro and Benavente (Lawrance 1984: 1093). The mid-thirteenth-century *Poridat de Poridades* had already declared a skilled and trustworthy escribano ("un escriuano sabio, et fiel, et entendido, et percebido"), one who was also of caballero status, to be of utmost importance for a noble (1957: 53–4).

17 See instruction by Torquemada in 1485, 1488, and 1498 (Jiménez Monteserín 1980: 110, 120, 128–9, 177, 178, and passim). See the same for a full account

on the Inquisition's modus operandi and its production, copying, disseminating, and archiving of documents. Also useful are González de Caldas 2000: esp. 225–473, and Rábade Obradó 1995b.
18. For example, those absolved repented heretics or "reconciled" by the Inquisition had their face marked with hot iron, according to the 1485 instruction by general inquisitor fray Tomás de Torquemada, if they were found guilty of fiscal fraud (text in Jiménez Monteserín 1980: 175).
19. ["from the lineage of Simeon arose the chancellors of the Jews, and those are the ones that the Holy Scripture of our church of Christ calls scribes in Latin, which is rendered as scribe, chancellor, or notary in the language of Castile."]
20. [Such pre-eminence had the kings and lords of Castile, that their Jewish subjects, bearing memory of their lords, were the wisest and most honourable Jews in all the kingdoms of their diaspora in four distinctions: lineage, wealth, goodness, and knowledge. And the kings and lords of Castile always found that all that we Jews have relating to the gloss of the law, as well as in their laws, rights and other sciences today, was all composed by the wise Jews of Castile; and through their doctrine today all Jews are ruled in all the kingdoms of their diaspora.]
21. [Furthermore, my lord, I saw many more of them at the house of the *relator* learning how to write than at the house of the Marqués Íñigo López learning how to joust. I can also assure your lordship that there are many more Guipuzcoans at the house of secretaries Fernán Álvarez and Alfonso de Ávila than at your own house or that of the constable, even though you come from the same land.]
22. Text in López Martínez 1954: 385. The converso groups or confederations can refer to converso societies such as the confraternity of the Santa Trinidad in Barcelona studied by Madurell i Marimón (1958). Garci Sánchez, jurado of Seville, gives notice of two clashing cofradías in Toledo, one constituted by Old Christians and another by conversos. King Enrique IV forced peace between the two and had them merge into one larger new cofradía in April 1465. The king himself joined as a member and gave the newly formed mixed society a handsome sum of money (text in Mata Carriazo 1953: 50).
23. For an interpretation of some of the aspects concerned in the virulent poetic exchanges among conversos, see Gómez-Bravo 2010.
24. [Rather, my lord, you should give / a great part of your wealth / to many poor hidalgos / that are in your kingdom / because, my lord, you must believe / that in time of need / these will be the ones who will help you.]
25. [but there is no longer mention / of he who keeps virtuous / only, my lord, of anyone who has / his bag well stocked.]

26 Juan Poeta is documented as escribano ("scriptori et officiali in regia dohana" and "confector librorum dohana et secrezie") in the court of Alfons V of Aragon. On the various details of his elusive life, see Costa 2001: 144–52.
27 ID1299 is a poem dedicated to don Sancho de Rojas, Bishop of Palencia, and should help qualify his traditional characterization as a low-class, jocular troubadour. He also writes of himself as a caballero of fortune, "cavallero afortunado" in ID1325 (v. 13), at the same time that he reminds the Condestable don Álvaro de Luna, to whom the composition is dedicated, that he spent his childhood and youth in court (vv. 37–9). This was also a prolific social group for Provençal poetry (Burgwinkle 1999). On Villasandino and his social status, see Mota Placencia 1990: esp. cxxxvii–cxli.
28 [The poor caballero has no other path to prove himself but that of virtue, and be affable, well-bred, courteous, self-restrained, and diligent, but never proud, arrogant, or a rumourmonger, and, above all, be charitable, because by giving two maravedís to the poor gladly, he will prove that he is as liberal as the one who gives alms at the toll of the bell. And seeing him so adorned by the said virtues, there will be none who, even without knowing him, will not value him and know him to be of good lineage, as it would be a great wonder if he were otherwise, because praise was always the prize of virtue and the virtuous cannot but be praised.]
29 According to Baena, the art of poetry is a God-given gift ("por graçia infusa") that can be perfected through study and courtly experience and therefore marks a noble elite "noble fidalgo, e cortés" (1993: 7–8). Similarly, as Garcia de Resende states in his *Cancioneiro Geral* (1990–8) the *arte de trovar* is necessary at the courts of great princes ("nas cortes dos grandes princepes" 1: 10). On the function of poetry as elite practice, see the insightful studies by R. Green 1980, Johnston 1996, and Weiss 1990. For the understanding of poetry in the *Cancionero de Baena*, see Weiss 1990: 25–54.
30 For a detailed study of Santillana's understanding of poetry, see Weiss 1990: 165–228. All citations of Santillana's (1988) texts refer to Gómez Moreno and Kerkhof's edition.
31 [much like those who are descendants of noble and great old ancestors are called generous, likewise the intellects that are accustomed to the solitude and care of the other old and sanctioned sciences are called generous and noble. What the first have naturally, the others reach through learning. It is intellects so disposed that poetry nourishes and instructs.]
32 Discussion in Gómez-Bravo 2010.
33 A mozo de espuelas walked in front of his master's horse.
34 [if virtues are found / in the poor and the lowly / ... / if he did the better work / let him be considered the best.]

35 On the important function of the corregidores, see Bermúdez Aznar 1974; González Alonso 1970; Lunenfeld 1987; Rufo Ysern 1991.
36 For the lineage of the Manriques with details on the life of the poet, see Montero Tejada 1996. On the role of lineage and kinship among Castilian nobility, see Beceiro Pita and Córdoba de la Llave 1990. For details on Gómez Manrique's biography, see the introductory study to Manrique's works by Vidal González (2003). For the important role of lineage for the house of Mendoza, see Sánchez Prieto 2001: esp. 227–36. The significance of having a mastery of written culture for posts of a chiefly political, judiciary, and administrative nature such as that of the corregidor was highlighted by Alonso Ramírez de Villaescusa, himself a corregidor and probably a converso, in his 1493 *Espejo de corregidores y juezes*, a manual for corregidores and judges. Ramírez de Villaescusa strongly cautions about military caballeros ("caballeros de armas") who lack knowledge of letters ("letras") and practical cases ("de experiençia de causas no tienen cognosçimiento ni saber," text in Pérez Priego 1997: 1173). Between 1485 and 1490, Ramírez de Villaescusa overlapped with Gómez Manrique in Toledo, where Villaescusa was sent as judge concerning property confiscated by the Inquisition (Hernández Gassó 2007).
37 For a brief overview of the documentation supporting this standpoint, see Márquez Villanueva 1960: 164–5.
38 [the treason was known and discovered and Gómez Manrique, who was *corregidor* by the king in the city at the time, arrested some conversos who were part of the plot, and he found out the truth of what they had planned … And Corregidor Gómez Manrique, seeing that by executing so many people the city would become depopulated, decided to give them fines instead.]
39 For an in-depth study of the model of control of public officials instituted by Isabel and Fernando described by Pulgar, see Garriga 1991.
40 A full account of Santillana's life and career, along with the transcription of important documents, can be found in Pérez-Bustamante and Calderón Ortega 1983.
41 For the biography and professional details on Álvarez Gato, see Márquez Villanueva 1960: 13–79; and Pescador del Hoyo 1972–3, as well as Gómez-Bravo 2011. Parts of the study on Montoro and Álvarez Gato presented in chapters 2 and 9 have been published in Spanish in Gómez-Bravo 2010 and 2011 respectively.
42 Receiver, with duties for management of the concejo's rents.
43 On these practices, see Losa Contreras 1999: 175–81. On the impact of the group of "caballeros y escuderos" in concejo meetings as recorded in the *Libros de acuerdos*, see Losa Contreras 1999: 177.

44 For information on the church of San Salvador, see Vizcaíno Villanueva 1991–2.
45 The 1576 *Relación* concerning Pozuelo de Aravaca as part of King Felipe II's efforts to collect detailed information of Spanish towns confirms the hidalgo reputation of both the poet and his brother: "por tales hijosdalgo les vieron tener e reputar" (Viñas Mey and Paz Remolar 1949: 484).
46 *Libros de Acuerdos* 1932–87, 2: 218–19. Although she does not discuss Álvarez Gato's role, Rábade Obradó (1996) has provided a vivid depiction of the context surrounding the early years of Madrid's inquisitorial tribunal.
47 See chapter 6. The professional line between artisan, merchant, and shopkeeper is more fluid and continuous than rigidly divided, with some occupations and activities overlapping (Navarro Espinach 2006: 163 and passim). On the profession of the *aljabibes*, which included trade in used clothing as well as the ability, similar to that of the tailors or *roperos*, to make new clothes, see the city ordinances of Córdoba published by Ramírez de Arellano 1900b. The overlap of the two professions would explain the double denomination of *aljabibe* and *ropero* applied to Montoro. According to the ordinances for the *aljabibes*, those seeking to also work as *roperos* needed to show greater economic solvency. On the subject of Montoro as a member of the budding merchant and professional "bourgeoisie," the medianos, see my study (Gómez-Bravo 2010). We have evidence of Montoro's wealth through his will, which was published by Ramírez de Arellano 1900a: 488–9.
48 This membership was first proposed by Gerli 1995: 270 in an important article. See also Gómez-Bravo 2010. On the Cordoban caballeros de cuantía or de premia, see Cabrera Sánchez 2003. On the medianos and this caballería popular, see, for example, Pescador 1961–4, Powers 1988: esp. 93–111, and T. Ruiz 1994: 237–48. Further bibliography in Gómez-Bravo 2010.
49 See a useful overview of the documentation in Costa 2001: 17–20.
50 [She is a relative of the Ropero, a compatriot of Seneca, Lucan, Martial and Avicenna. It is in the soil.] A fifteenth-century testimony of a similar belief in the goodness of Cordoban soil for the production of intellectual talent is found in the description of Córdoba penned by Jerónimo, canon of the Real Colegiata de San Hipólito in Córdoba (Nieto Cumplido 1973: 44, and passim).
51 [Man of low family / of the lineage of David / tailor of simple craft / but no Roland in combat.]
52 On the documentation that shows Román as royal contino, see Mazzocchi's (1990) introductory study to Román's *Coplas de la Pasión* (15, 22).
53 On the twenty-four Cordoban aldermen or veinticuatros, see Cabrera Sánchez 1998: 101–26.

54 [Since it does neither increase wealth / nor does writing poetry give any advantage / we adore you, thimble / let us give you thanks, needle.]
55 In the manuscript, the reading of the word "indino" is unclear; it may also be read as "iudino," or Jewish.
56 It is important to note that there may be a family connection between Baena and Antón de Montoro. Baena may have lived part of his early life in Seville. There is a Juan Alfonso de Baena documented as living in that city in 1408 in the documents extracted by Collantes de Terán Delorme 1972–80, 1: 262, and a Juan Alfonso, escribano, listed as working in Córdoba around the same dates (1406–22) (roster provided by Ostos Salcedo 2005: 222, and passim). It is also possible that the poet may be identified with a Juan Alfonso, royal escribano working under King Enrique III, who was in charge of drawing a document on 28 June in Salamanca around the year 1397 and who signs as "Johanne Alfonso" (see a photo of the document in Romero Tallafigo et al. 1995: 166–7, plate 35). This would be one of the earliest documents known about the poet, then a young man in his early to mid-twenties, having been born ca 1375 according to current estimates. Baena's first datable composition is one dedicated to the death of Enrique III on 25 December 1406 (ID1180). Of possible relation to the poet is Rodrigo Alfonso de Baena, a tailor, who, along with his wife, is recorded as having bought a vineyard in Córdoba in 1452. The document is extracted in Ostos Salcedo 2005: 218. For a review of other points of Baena's biography unearthed by Nieto Cumplido and others, see the introduction to the edition of the *Cancionero de Juan Alfonso de Baena* (1993: xiii–xviii) and the bibliography cited therein, as well as Gerli 1995: 270.
57 For an overview of the documentation on Mena's biography, see for example the introductory study to Mena's (1986) selected poems by Azáceta (16–22).
58 The document was first published by Rada y Delgado (1874: 300n) and later in Fuentes Guerra 1955: 99–104. Burgos was the seat of the powerful Cartagena-Santa María converso family, of which the escribano who drew the document, Juan Sánchez de Santa María, appears to have been a member.

3. Pervasive Papers

1 [Many are the temples and edifices that past kings and emperors built, of which there remains no stone that we can see but there are writings that we can read.]
2 See, for example, the use of widely disseminated letters and other texts for legitimizing political propaganda by Queen Isabel in Carrasco Manchado 2006: 101–51, 202–53, 349–420, 475–538, passim.

3 The situation is reminiscent of that described for England. See, for example, Knapp 2001.
4 On the history of paper as writing medium, see, for example, Bloom 2001 and Valls i Subirà 1978–82, particularly vols. 1 and 2.
5 See the study on Barcelona booksellers and their paper trade by Madurell i Marimón and Rubió i Balaguer 1955: 14*, and passim. There is ample evidence from inventories and other primary sources. See, for example, doc. 177 in Madurell i Marimón and Rubió i Balaguer 1955.
6 For an interpretation of the function played by these inscribed papers see Gómez-Bravo 2013.
7 For a discussion on these developments during the period and its implications to cancionero poetry, see Gómez-Bravo 1999a, 1999c and the bibliography cited therein.
8 [For lack of time, I don't speak any further of these perjuries, for the writing of which ten hands of paper would not be enough.]
9 [Reply by don Joan de Mendoça blaming two of the Almirante's servants for the stanza he received from him, as one of the servants was a very good penman and the other an even better poet: Of this stanza that touches me / you can only claim ownership of the paper. / I hear Gabriel's voice / I feel Coca's hands / It is not hard for you to win / as I cannot prevail / with three (or at least two and a half) writing against me […] Your work shall be in vain / you will get no benefit from it / as you want to go with paper / where your hand cannot reach.]
10 [woe is me for I used to / write devout things / and now love with its continuous struggle / commands me without cheerfulness / to write sorrowing things […] I make days of my nights / my paper weeps blood.] This image may be actually referring to the documented practice of using one's own blood to write (painful) letters. One such case, recounted by chronicler Fernández de Oviedo, involves a youth who, in obeying his lady, slashed his finger to draw blood with which to write to her (Avalle-Arce 1974: 624–5).
11 [the paper is left scribbled / feelings battle / within the paper's field; / the pen appears in the hand / between hope and fear / it moves, takes heed and hustles; / over this fight love / has a safe fortress.]
12 [a long account could be made / but I hold up the pen / so as not to touch the paper.]
13 [Which tongue will tell / or which hand will write / my painful cares?]
14 [If all the land were turned / into white Tuscan paper, / and the rivers were transformed / into ink with which to paint / a pain so inhuman, / such materials / would sooner be spent / before half of my troubles / and unequal torments / could be told.]

15 [With a heart much unsettled / and a trembling hand / I must go on writing.]
16 I understand the paratext to have a more positive relation to the text than that envisioned by Genette (12).
17 [One day when they played *cañas*, Juan Álvarez threw these stanzas wrapped around a rod onto a roof that overlooked a window where that lady sometimes stood.]
18 On Fernando de la Torre's game, see Marino's (2006) insightful study. On Pinar's, see Sanz Hermida 1996.
19 For an overview of the appointment of public officials in this manner, see Polo Martín 67–80.
20 [Vázquez de Palencia against fray Íñigo on his poem Vita Christi, writing (these stanzas) to his lady friend because she sent for the poem and it was given to her by his servant without his permission and (the stanzas) say thus.]
21 [Other [stanzas] because a lady wrote to him asking to have a letter that she had written to him returned to her, and he sent it along with one of his and both letters talk to the lady saying thus.]
22 [a caballero sent a letter to a lady and the address read thus: No ugly lady / may open or read me.]
23 [what was inside is: You are not her to whom I come / therefore don't read me / you lady are not heeding / the address written on me / close me again / and don't let anyone come near me / because she whom I came to seek / must not be here.]
24 [To the most lusterless lord / whom it is a marvel to see / from small to great / a grandee of Castile.]
25 [The closing of the letter, which was closed with a Merced coat of arms.]
26 For the practice of publication in manuscript form, see the classic study by Root 1913. For the use of relevant terminology (*editio, edere, emittere in publicum*), see Bourgain 1989.
27 [Stanzas that Afonso Valente made in Tomar to Garcia de Resende without forwarding them to him.]
28 [Reply by Garcia de Resende following the rhymes to all the stanzas by Afonso Valente and which he found without Valente ever sending them to him. *And they are copied out of order so that he may obtain the originals.*]
29 [Juan Alfonso tell me if it would please you / to give me the stanzas and poems that I made.]
30 [Yet a second time I demand / that you give me those poems at once.]
31 For a view of the various formats and supports used in legal documentation, it is useful to look at the Bancroft Library's collection of beautifully digitized Catalan documents, posted online at http://sunsite.berkeley.edu/catalan/.
32 See the beautiful reproductions of documents held at the Alba Archives in Varela 1987. See also the introduction to Colón 1992: 81. A comprehensive

view may be gleaned from many other document collections, such as that held in the Instituto Valencia de Don Juan, described and catalogued by López Pita, Luis López, and Ser Quijano 2002. For the uses of writing and documentary practices in narratives of exploration and conquest, see González Sánchez 2007.
33 [He talks with these stanzas and makes them messenger and message so that they go and encounter the lady and divulge that which he did not dare tell her.]
34 Some of these forms of dissemination were already in place in Provençal lyric through groups collecting, glossing, and circulating poems. See Poe 2000.
35 On King Pere's poetry, see also the brief introduction by Cluzel 1957–8: 362–9.
36 Text in Sesma Muñoz 1987: 85–7, 125–7.
37 [Motes that the Almirante de Castilla sent to the gentlemen and ladies from one of his towns and they say that they were found written in an inn.]
38 [Third law. On the dishonour that a man does unto another by way of poems and rhymes. People slander and dishonour one another not only through words but also through writings by making cantigas or rhymes or bad poems of those they wish to slander. This they do at times publicly and at times covertly, throwing those bad writings into the houses of the great lords, or in the churches, or in the public squares in cities and towns so that everyone can read them.]
39 For forms of dissemination of slander through later centuries, see Castillo Gómez 2006: 229–37.
40 See for example the case of the *mesón* advertised for lease in this manner in 1477 and 1498 in Toledo described in Gómez-Menor 1970: 30–1, doc. 2.
41 Text in Medina de Mendoza 1853: 183. The document was publicly posted on the door of the Church of Santa María in the town of Colmenar de Oreja, where the court was residing at the time, as testified by the notary, who also acted as one of the witnesses. See the whole document in *Memorias de don Enrique IV de Castilla* (1835–1913: 573–8). The document contains many offers to provide copies to any interested parties (esp. 577–8).
42 For instance in Carrillo de Huete's *Crónica del halconero* (1946: 154), and in Lope de Barrientos's *Refundición de la Crónica del Halconero* (1946: 150).
43 Examples in Orejudo 1993: 25, 164, 168–9, 172–3.
44 See further examples in Castillo Gómez 2006: 209–12. The classic case, of course, is Martin Luther's posting of his 95 Theses in 1517.
45 The letter is preserved at the Archivo General de Simancas, Registro General del Sello, fol. 774 (see García de Salazar 1967, 1: xv; fols. 237–9 for photo of the document). For an example of a letter from the Inquisition in Toledo posted on a church door summoning suspects, see the text published by Weiner in his edition of Horozo 1981: 102.

46 This is the case, for example, of the tenth-century original founding charter of the monastery-*infantado* of Covarrubias. This document, as it stands today, bears piercing marks where nails were placed to affix it to the board for its public display. The parchment was trimmed to the size of 750 × 520 mm to fit the board. It bears further damage due to the various times when it had to be removed and presented as evidence in litigation before the Royal Chancery in Valladolid. It was later folded, presumably for the purposes of archiving it, which caused further damage. This charter, like others, was copied singly or in various cartularies in order to preserve its important content from destruction. See the text and study in Zabalza Duque 1998: 396–407, along with other revealing examples. Other examples can be found in many of the documents housed at the Cathedral in León (photos of some in Estepa Díez et al. 2002: 304–7). Some of Columbus's documents preserved in the Alba archives have piercing marks in corners and/or borders, perhaps pointing to a mode of archiving or posting. Reproductions in Varela 1987.

47 (*Libros de acuerdos* 1932–87, 1: 325).

48 See Castillo Gómez 2006: 220–4. This public display was also for the purposes of settling disputes over prices or unreasonable monetary demands from officials. See for example the case documented by the meeting minutes in Guadalajara published by López Villalba 1997: 188–90. Further, the *Ordenanzas reales de Castilla* prescribe the use of such tablas for judges, their escribanos, and other officials (Díaz de Montalvo 1990: fol. 70v). The same text appears in the 1480 *Ordenamiento de las Cortes de Toledo* (*Cortes* 1861–1903, 3: 170).

49 See the text in *Cortes* 1861–1903, 3: 193–4. Cities kept careful records of the publication by *pregón*. See, for example, the *Libros de acuerdos* 1932–87, 1: 38, 118, 139, 248, 299, and passim.

50 [but your consideration and your goodness have been such that, as a remembrance of our exchanges ... you kept me in a little cell in your memory affixed by the nails of love.] The allegory of the house as that of human consciousness and of the private chamber as the locus of innermost thoughts, as well as the place for private writing and reading, had already been used earlier by Portuguese King Duarte in his *Leal conselheiro*. See Gómez-Bravo 2005.

4. The Hands Have It

1 The role of hands in such ceremonies is amply represented in manuscript illuminations. See some telling examples in Gudiol i Cunill 1955: plates 165–70, 172–3, 175, 178–9, 181–6. On the requirement of touching the

Gospels while being sworn into office, see García Marín 1974: 225–8. On the changing role of the relation among voice, gesture, and written record in similar ceremonies, see Beceiro Pita 1994.
2 Text in Sánchez Prieto 2001: 191–2. Additional formulas for the swearing in of a post requiring both tactile and oral components, in *El libro del juramento*. Text in García Marín 1974: 367. For an overview of the ceremony of swearing in of public officials, see Pérez-Bustamante 1974.
3 Both illuminations in the edition of the Alba Bible (1920–2) between pp. 58 and 59.
4 *Fuero de Uclés, Fuero de Soria, Fuero de Navarra, Fuero general de Navarra; Fuero de Teruel*. Photos of these illuminations in Delclaux 1973: 157, 169, 205. See also Silva y Verástegui 1988: 141–50.
5 See the document of the complex ceremony recorded by escribanos in Pérez-Bustamante and Calderón Ortega 1983: 133–42. Santillana performed the same swearing-in ceremony several times in his lifetime. Description in the documentation published by Pérez-Bustamante and Calderón Ortega 1983: 155, 157, 276, 281, 310.
6 Examples of fourteenth-century documents to this effect in Pérez Chozas et al. 1932: 301–3, 328.
7 Prescribed text to be recited by the accused in such cases as instructed by Diego de Deza in 1500, in Jiménez Monteserín 1980: 124, 125.
8 For a relevant contribution to the study of maniculae, albeit centered on a later period, see Sherman 2008: 25–52.
9 Photo in D'Ancona and Aeschlimann 1949: plate CXXX. On the evolution of the iconography of the Annunciation, see Franco Mata 1998: 175–6; see also plate 81 for a representation of Gabriel and a roll in the scene of the Annunciation in a frontispiece of the Cathedral of Burgo de Osma (Soria, ca 1290–1300).
10 Lucas of Tuy's *Vita Sancti Martini* constitutes chapters 53–75 of his *De Miraculis Sancti Isidori*. In Viñayo González and Fernández González 1985: 24–5.
11 [But don't think, my good people, that those titles that they put before prayers are true, nor are they devised by the saints. Rather, they are the work of lying hypocrites or preachers, saying that if one carries that prayer with him, one will never die by fire or drown in water, as well as many other useless and deceitful things.]
12 On literacy and manuscript production, see Sánchez Mariana 1988.
13 The rise of the signature in royal and noble courts, its evolution from the earlier signum, as well as its relationship to a similar phenomenon in painting have been studied by Jeay 2001.

14 [The handwriting could also vary due to the sickness or old age of the escribano, for a man writes in one way when he is young and healthy and in another when he is old and sick.]
15 See for example the text of the cartel submitted by the Count of Cortes to the citizens of Pamplona, in Orejudo 1993: 172–3. Covarrubias (1998), under the entry for "Carta," defines cartel as both the paper that is posted announcing of the terms of a joust, at whose bottom the participants had to sign, and a slandering libel posted in secret (313).
16 [And let the carteles stop because it is a shameful thing for two caballeros to waste ink and paper without fighting with their hands.]
17 The first and last of the four volumes are kept in the Sección Osuna of the Archivo Histórico Nacional (Mss. 3406-1, 3406-2), one covering the years 1504 through 1506 (published by Szmolka Clares et al. 1996) and the last with documents dating from 1513 through 1515 (published by Moreno Trujillo et al. 2007). All references here to the first manuscript are to volume 1 of the Szmolka Clares (1996) edition. The second and third volumes, containing Tendilla's correspondence from 1508–13, are held at the Biblioteca Nacional in Madrid (Mss. 10230 and 10231) and were published by Meneses García (1973–4). Additionally, García López (1995) published Tendilla's letters from 1497. Moreno Trujillo (2002) has highlighted the crossover between personal and official register in Tendilla's manuscripts. For some insights on the procedures followed by Tendilla in regards to document production and transmission, see the introductory study in Moreno Trujillo et al. 2007: esp. 43–75.
18 [Please forgive, most magnificent master Girónimo, that this letter is not written in my hand, because I am ill with a cold, and I remain much like a brother to you.]
19 [I, my lord, do not ask forgiveness of your mercy for not sending this letter written in my hand, because what I write is but nonsense, as anyone can plainly see; and also, because my head is not fit for it.]
20 [I have stopped writing but not serving, the cause being that my handwriting is bad and that I am ashamed of it.] Letter to don Pedro Venegas, dated 27 June 1515. Published in Moreno Trujillo et al. 2007: no. 1267.
21 [To don Jorge Manrique requesting that he favour a work of his that he sent for his perusal: Noble, superior gentleman / whom my desire serves / I ask that you bestow upon the work in which my time was wasted / such favoured favour / as to adorn the ugly, / and taking pity on my pain / my noteworthy caballero / may you deceive with such deceit / as to gild tin with gold / doing what a silversmith could not.]

22 Santillana signed the 1455 codicil to his will, marking every single page of the eight-leaf paper cuaderno. See the transcription of the will and codicil in Layna Serrano 1942, 1: 316–33.
23 Two such manuscripts are Lucena's *De vita beata* (BNM Ms. 6728) and Alfonso de Palencia's 1488 copy of his *Diccionario latino-castellano* (Escorial, Ms. f.II.11) (Sánchez Mariana 1996: 215).
24 [Because this is written in my own hand, don't marvel at the bad handwriting.]
25 I use the term "reproduction" to mean the replication of a text in any medium, as used by R. Williams 1995.

5. Papers Unite

1 For an excellent review of the different means of material support of texts throughout history, including tablets, rolls, booklets, and codices, see Sirat 2006.
2 Burning would have worked better for paper or wooden tablets. For the lessened effect of heat on parchment, see Reed 1972: 316–18.
3 [I beg that this letter of mine and all the other ones that I have written to you be burned by you or be kept in a chest of which you may guard the key, so that none may ever see them, and may be returned to me when God pleases that I may see you again.]
4 [I gave his Majesty an account of the content of Pedro Treviño's book, and he commanded me to burn it; I threw it in the river Tajus tied to a stone.]
5 For example, 92, 104, 136, 144, 161, 193, 237, 286, 346, 352, and passim.
6 [as I already wrote about everything to Your Majesty when I left for my discoveries, and I left the wrapper/folder in the Isabela so that Your Majesty would be well informed if the ships that were expected came and went before my return ... many particular things that were necessary, about of all of which and of their instruction I sent Your Majesty a copy in the same wrapper/folder.]
7 [The wrapper/folder for the cards should be done thus: A parchment skin about as big as a sheet of paper on which the following should be written. And the back of the said wrapper/folder should be of the same colour as the back of the cards.]
8 In their edition, Dutton and González Cuenca (1993) propose the emendation of "enboltorio" to "emboltero" in order to restore the rhyme, interpreting the term as "carpeta" or binder (711).
9 Here, the term "roll" refers to being disposed vertically and "scroll" as written and read horizontally. However, this assigned meaning sometimes

does not work well in correlation to visual representations of the medium in art. Further, Spanish terminology does not always differentiate between the two, as is obvious from the citations in this section.
10 On the weight of the rabbinical tradition on the illuminations of the Alba Bible, see Nordström 1967.
11 There are of course individual books that have survived in scroll format. One is the Book of Esther held at the Archivo Histórico Nacional, Códices 1423, from the fourteenth or fifteenth century, in Hebrew.
12 The *Biblia romanceada* (1994) of the Real Academia de la Historia, Ms. 87 (translated 1300–1400; copied 1450–1500) has "proceso de libro" or in shortened form "proceso" (fol. 121v, and passim). The *Biblia ladinada* (1995) from the Escorial library (Ms. I.-i.-3) (translated 1300–1400; copied ca 1425–ca 1450) offers "rollo de libro" (fol. 281vb, and passim).
13 See the thorough study by Gamble 1995: esp. 42–81, and the bibliography cited therein; the classic study by Roberts and Skeat 1983, as well as Blanchard 1989, and Clemens and Graham 2007: 250–8, and the studies collected in Martin and Vezin 1990. See also Clanchy 1993: 135–44.
14 For an overview of the scholarship on this topic and its main implications for the Christian tradition, see Stallybrass 2002.
15 See, for example, the illuminations depicting Hebrew codices in Spanish manuscripts held at the British Library, published by Narkiss 1982: 88–9, 90, 99, and passim. For the various formats used in medieval Hebrew manuscripts, see Sirat 2002.
16 Held at the Archivo de la Corona de Aragón, Barcelona, sign. Pergam. Jaime I, measuring 940 × 635 mm. See Delclaux 1973: 144–5.
17 The Newberry Library houses the same work in codex form (Ms. 29). On these manuscripts, see Clemens and Graham 2007: 255–7.
18 See Robbins (1939) for the use of such rolls in medieval England for congregational purposes. As Robbins points out, the *Arma Christi* roll was, in contrast, copied into small book format when intended for private devotional practices (416–17). A related example may be found in the fifteenth-century roll from the Newberry Library, Ms. 32.
19 See, for example, the insightful study by Kelly (1996). Of relevant importance are also the Litany Rolls at the Beinecke Library, Ms. 810 (ca 1300 from Milan).
20 Roberts and Skeat 1983: 51–2. As an example of bull format, see description and photographic reproduction of the bull sent by Martin IV to Pere III of Aragon (el Gran), which measures 895 × 648 mm in *Exposición antológica* 1959: 53); or the bull of Alexander VI giving Isabel and Fernando their "Catholic" title, measuring 445 × 590 mm (described in *Exposición antológica*

1959: 200–1). Other documents probably intended for public display purposes also had very large formats. One such document is Isabel and Fernando's marriage charter given in Zaragoza on 12 January 1469, which measures 640 × 450 mm (*Exposición antológica* 1959: 195).

21 This was already noted, with further examples taken from archival evidence, by Arribas Arranz 1948: 5–6. See also Clemens and Graham 2007: 160–1.
22 Examples of such texts in Ser Quijano (1998): document 3 in parchment (measuring 287 × 390 mm) (48–50), or document 4 in parchment (332 × 500 mm) (50), and passim.
23 This roll is dated 10 February 1452, commissioned by the administrators of the Pía Almoina and contains documentation on the properties belonging to the Cathedral in Lleida signed by notary Martí Benet. It is written in Latin and Catalan and measures 610 cm long. I would like to thank Ana Carles, of the Arxiu Capitular de Lleida, for her help with this manuscript.
24 Arribas Arranz (1948) has studied legal rolls. On the monastic rolls, see Petrucci 1998: 49–50. Dufour (2009) has published an illustrative study, complete with transcription and images of the "rouleaux des morts."
25 Description of several of these documents in the inventory in Gil 1989: 159–60.
26 [*Cartapel*: the long writing that joins sheet to sheet and does not turn pages, like the edicts that are fixed on the doors of churches, tribunals, and public places.]
27 On the tiras, see also Arribas Arranz 1948.
28 Arribas Arranz 1948: 8–9. See also Ostos Salcedo 1998. The mandate appears clearly ariculated in Díaz de Montalvo's (1990) *Ordenanzas reales,* where the *palmo* is the unit used to assess the price in the preparation of procesos (fol. 78v, Libro 4, título XVIII, ley 12). The *Fuero de Briviesca* (1313) had already laid out similar laws for the escribanos, specifying that each *palmo* had to contain at least twenty-three written lines (BNM Ms. 9.199, fols. 59v–60r).
29 The regulations were repeated in later legislation, such as the comprehensive *Repertorio universal* by Hugo de Celso (2000: fol. CXXVIV).
30 For a description of the register see Bono's epilogue in Pérez-Bustamante 1985: esp. 64–5, and plates for fols. 3r–v ("hoja adjunta"), between pages 88 and 89.
31 Mary and Richard Rouse (1991) interpret the extant evidence as showing that literary and other rolls were by design ephemeral (26); but see Paden 2004–5 and 2011.
32 Under the entry "Phariseus" in his *Vocabulario eclesiástico*, fifteenth-century author Fernández de Santaella (1992) distinguished the Pharysees from other Jews, among other characteristics, because of their use of "rotulos," meaning

the *tefillin* or phylacteries. The same author defined *Filaterium* as a "cedula o cartilla" with the commandments that the priests wore on the forehead or the chest (see entries).

33 For the text and illuminations of this cartulary, see Fernández Conde 1971. For the iconographical depictions of rolls and also of the codex format in the life of San Martino, see Viñayo González and Fernández González 1985.
34 Published by Prieto Escanciano 1997. Vol. 1 contains a facsimile of the manuscript.
35 Relevant remarks in Silva y Verástegui (1988: 15–22) in relation to Navarrese cartularies.
36 See the interpretation of the illuminations in this manuscript in the study by Lacarra Ducay 1989. For a useful overview of Canellas's contribution to notarial practices, see Bono Huertas 1979, 1: 269–75. On the content and format of early legal documents such as donations, see Lacarra Ducay 1993: 74–5.
37 See the illuminations of the Getty manuscript in the lavish 1989 edition.
38 For example, fols. 22r, 24r, 29v, 31v, 32r, 36r, 41v, 62v, 72v, 78r, 83r, 83v, 84v, 85v, 95v, 97r, 113v, 114r, 126v; 133r; 146r, 185r, 193v, 205r, 205v, 212r, 212v, 217v (where the booklet in the judge's hands may represent the sentence which the roll written by the scribe is intended to appeal), 257v, 276r.
39 Written records show that sentences were often recorded on booklets or cuadernos. See, for example, the record of a 1439 legal sentence ("carta de sentençia") in the diplomatic collection of Santa Catalina del Monte Corbán. The cuaderno, as the text refers to the document, is eleven pages, the two last left blank. Copy of the document in Toro Miranda 2001, 1: 309–23, quoted passage in 323.
40 John G. Johnson Collection, Cat. 759. Photo and description in the online catalogue.
41 For an excellent study of this piece (including a photo on page 359) and the medieval iconography of Christ, see Fernández González 2001.
42 Photo in Domínguez Bordona 1929: 125. The same iconography is found in the Crucifixion miniature in the *Misal Rico de Isabel la Católica*, also belonging to the Royal Chapel in Granada, copied by Flores, escribano. Photo in Yarza Luaces 2005: 84.
43 Sirat 2006 emphasizes the transition from a sloping to a horizontal surface through the Middle Ages and later centuries, with the effects on the use of writing instruments and the anatomical implications of writing (412–22). Although the work focuses on Renaissance Italy, it is useful to see the thorough introduction to the materials and spaces for writing in Thornton 1997.

44 Photo and commentary in Planas 1998: 138–40, as well as Parkhurst 1941: 300, fig. 33.
45 Photo in *Quand le livre* 1992: 53. For similar depictions of an author and his writing instruments as well as varied material supports, see the fifteenth-century translation of Avicenna's *Trattato di medicina* by Gherardo Cremonese, at the Biblioteca Medicea Laurenziana, Ms. Gadd. Reliqui 24 (fol. 1r). Photo in Morandini et al. 1986: 166–7.
46 Photo in Borchert et al. 2002: 127, plate 148; Sanchis y Sivera 1929: 25, plate 44; and on book cover.
47 Of relevance is the seminal work by Saenger 1989, 1997. I have discussed practices in fifteenth-century Iberia in Gómez-Bravo 2005.
48 For this latter roll, see A. Taylor 1991: 68–9. For some key insights on medieval and, more specifically, German rolls see Rouse and Rouse 1991. For a good overview, see Beltrán 2009.
49 Ferreira 1986: 65. For a full study and description of the manuscript, which measures 340 × 460 mm and is folded at the middle as part of the original format, see Ferreira 1986: 60–73.
50 For an accurate evaluation of this terminology, see Gonçalves 1994.
51 [This [monk] knew how to read / little, as I heard it told, / but he knew how to love well / the Virgen, who has no equal; / and therefore he went on to compose / and to put together five psalms.]
52 [This is the prologue to the Cantigas de Santa María, pointing out the things that are necessary for poetic composition: Because poetry is something in which lies / understanding, that is why whoever makes poetry / should have it, along with a good deal of reason / so that he may understand and know how to express / what he intends and pleases to say / because good poetic composition is thus done.] The verb "dizer" may be ambivalent, meaning both to say or to express and possibly to write according to rhetorical rules (*dicere*), following the poetic terminology of the time. On this topic, see Gómez-Bravo 1999c. For an insightful study on this miniature, see Cómez Ramos 1987.
53 For the interpretation of the iconography of King Alfonso as author and its relationship to the king's specific modes of authorship and the operation of his scriptorium, see the different contributions to the volume edited by Montoya Martínez and Domínguez Rodríguez (1999) and the copious bibliography cited therein. For the relationship between authorship and power in Alfonso X, see Ruiz García 2006.
54 Blecua's edition replicates the layout and also contains photos of the illuminations. See Weissberger 2005 for a knowledgeable discussion of the use of miniatures in this cancionero in the interface of political propaganda

and personal advancement. Additionally, it should be noted that also in the fifteenth century, Rogier van der Weyden painted the portrait of Philip the Good, Duke of Burgundy, holding a perfectly rolled written piece of paper in his hand (photo in Borchert et al. 2002: 174, pl. 199 and 247, pl. 62), a motif used by other important figures such as Emperor Maximilian I (photo in Borchert et al. 2002: 186, pl. 205) (also used in portraits by Jan van Eyck, Pietro Perugino; photos in Borchert et al. 2002: 202, pl. 223; 205, pl. 228, 210, pl. 237).

55 Currently in the Museo del Prado, Madrid, on a ten-year loan from the Infantado family. See the useful Bosch 2000: 86–8.
56 Santillana mentions in the codicil appended to his will his orders for the purchase of this image in Medina del Campo. Text in Layna Serrano 1942, 1: 328.
57 For the common use of this type of hand during the period, see Sánchez Mariana 1988: 322.
58 There is much written on this subject. See, for example, Lalou 1992; Rouse and Rouse 1989, and the bibliography cited there. For a recent and useful general introduction to various kinds of material supports, see Clemens and Graham 2007: 3–17. Slate tablets (*pizarras*) were commonly used in Visigothic Spain; see the volume edited by Velázquez Soriano and Santonja Gómez 2005. Lead is the support for the famous lead books of Sacromonte (see, for example, Barrios Aguilera and García-Arenal 2006). During the inquisitorial process against the Arias Dávila family, a witness mentions some brass sheets ("láminas de latón") with the name of Adonay written on them that his father gave to Elvira, one of the accused and the wife of King Enrique IV's contador mayor. Witness's account in Carrete Parrondo 1986: 115, no. 220. Four stanzas of the Spanish epic *Poema de Fernán González* were found copied on a terracotta shingle in Villamartín de Sotoscueva (Merindad de Sotoscueva) (Burgos), dated in the fourteenth century.
59 Torre y del Cerro 1968: 34. The differentiation between libro and cuaderno operates throughout inventories, as for example those published by Torre y del Cerro 1968: 17–18. These inventories also mention cuadernillos (Torre y del Cerro 1968: 5).
60 [Some cuadernos tied with hemp string that have "Revelaciones de Santa Brígida" written on top] (in Torre y del Cerro 1968: 32). For photographic reproductions of many documents in booklet or single leaf form from Isabel's reign, see, for example, Checa Cremades et al. 2004.
61 Torre y del Cerro 1968: 17–18, and passim. A bound book is often specified as such, as in Gonzalo de Baeza's expense book: "Vn libro de molde, enquadernado, para leher, que costo 434 mrs" (text in Torre y del Cerro and Alsina de la Torre 1955–6, 1: 380).

62 Whether a jewel-like book, or cuadernos accompanied by a single illuminated leaf ("hojuela dorada"), the inventories specify the medium. See, for example, Torre y del Cerro 1968: 24–5.
63 [fourteen paper cuadernos, in quarto size, printed, which is a *Confisional* by el Tostado.]
64 For example, doc. 177 in Madurell i Marimón and Rubió i Balaguer 1955.
65 There is much written on the subject; see for example Bataillon 1989. On booklets and lyric compilations, see H.W. Storey 1993: esp. 111–70, 341–419. Saenger (1975) has pointed to the lack of applicability of the *pecia* model to the production of some court codices (406). Alfonso X (2004) had regulated the loaning of pecia (quaderno) in his *Siete Partidas*: "E el exenplario dar por quaderno que prestare a los escolares para escreuir o para emendar sus libros" [The *exenplario* should give for each cuaderno that he lent to the students for them to write or to amend their books] (III, 1r).
66 He was the son of the famous Fernán Díaz de Toledo, who held various important administrative positions under Kings Juan II and Enrique IV (Benito Ruano 2004: 23–4).
67 See the study of this manuscript by Arribas Arranz 1964b. Many of the documents prepared by scribes for official business are preserved in both their original support and in a later copy on a different medium. For illustrative purposes, it is useful to look at extensive collections of copies of single documents or their originals preserved in booklet or single leaf form such as the ones housed at the Instituto Valencia de Don Juan. Description and transcription of the documents in López Pita et al. 2002: for example, vol. 1: 309–25; vol. 2: 90–127, 127–34, 149–60, 165–6, 166–8, 169–72, 214–15, 216–26, 227–8, 294–6. A similarly executed register is that of late fifteenth-century Aragonese notary Rafael Osón in the Archivo Histórico de Protocolos in Zaragoza (Cabanes Pecourt and Pueyo Colomina 2001).
68 Text of the formulary and study in Cuesta Gutiérrez 1947; comments in Arribas Arranz 1964b: 19–20.
69 See an example in the documentation concerning the town of Laredo in 1391: "E yo Pedro Perez que fui presente a todo lo que dicho es ante los dichos alcaldes ... fiçe escrevir este testimonio en este quaderno en catorze fojas de papel con esta en que ba el mio signo y ban cossidas con filos de lino blanco e en cada una foja e en cada una plana en son derecho escripto mio nombre e por ende fize aqui este mi signo en testimonio de verdad. Pero Peres." [And I, Pedro Pérez, witness to all that has been said in front of the said mayors ... had this testimony written in this cuaderno of fourteen paper leaves, including this one that bears my signum, and they are sewn with white linen threads and in each leaf and on each page on the right my name is written; and therefore I made my signum here in testimony of the truth]. In Cuñat Ciscar 1998: 154–5.

70 The usefulness of the cuaderno format for the issuing of specific regulations that were subject to constant revision is clearly explained in one of the cuadernos: "E llamole cuaderno porque … han de subceder los cuadernos viejos e nuevos de todas las rentas" [And I call it cuaderno because … the old and the new tax cuadernos will succeed each other]. Text in Ladero Quesada 1973: 362.

71 "Escriuen los onbres en sus quadernos por remembrança de las cosas que les deuen. E otrosi lo que ellos deuen & a las vezes escriuen verdad & a las vezes el contrario por oluidança o maliciosamente" [Men write in their cuadernos in order to remember what others owe them; and likewise what they themselves owe. And sometimes they write the truth and some other times they do not, due to forgetfulness or out of malice] (III, fol. 67v).

72 Many other notarial cuadernos have survived. There exist, for example, eighteen by Juan de Valladolid, escribano of Granada in 1502–4 (Moreno Trujillo 1995: 77).

73 For a review of the terminology used to refer to books, booklets, and loose leaves, see Bataillon 1989, Dolbeau 1989, Guyotjeannin 1989; also Rizzo 1973. Indeed, Queen Isabel's inventory lists *librillos* consisting of only four leaves, ten leaves, or other numbers (Torre y del Cerro 1968: 6 and passim), as well as cuadernillos and *cuadernicos* with a similar number of pages (Torre y del Cerro 1968: 380 and passim). Juan de Flores (1981) modestly refers to the circulation of his short prose work *Triunfo de amor* as a cuaderno (which he fears will later be burned) and therefore as a not quite perfect product (177). This "booklet" is seventy paper folios long, and 220 × 150 mm in the Biblioteca Nacional manuscript (Ms. 22019). Passage quoted on fol. 70r.

74 Full facsimile in Menéndez Pidal 1914.

75 [*Cartapacio*: manuscript book in which to write diverse matters and subjects; also the cuaderno on which one writes as one's master lectures.] On the suitability of using the term "cartapacio" in cancionero studies, see Beltrán 2006.

76 For Palmireno's and Vives's recommendations on the use of cartapacios, also supported by Erasmus, see Gallego Barnés 1982: 97–8. For Vives (1971), see his treatise on education, *De tradendis disciplinis* (xxxviii–xxxix, 108).

77 A 1463 inventory lists a paper cartapàs with parchment covers: "hun cartapàs de paper cubert de pergamí, qui és lo Procès de la donzella." In Madurell i Marimón1974: 87.

78 [One called Repertorium in his hand with some poems of little value in the form of a cartipas.]

79 [in this and in writing my cartapacios I spit all my venom and find repose, as I would burst if I didn't.]

80 Cantera Burgos 1952: 200. Converso bureaucrat Alvar García de Santa María, brother of Bishop Pablo de Santa María, was royal escribano, chronicler to Juan II, member of the Consejo Real, secretary of the concejo of Burgos, and contador to Prince Juan of Aragon, among other posts. On this important figure, see Cantera Burgos 1952; and Cañas Gálvez 2008: 402.
81 For a good study of Pérez de Guzmán's cancionero dedicated to Alvar García de Santa María, see Díez Garretas and de Diego Lobejón 2000.
82 See the observations on this manuscript by Plaza Cuervo 1995.
83 Marcos Álvarez 2005. The current binding dates from the nineteenth century.
84 See the excellent study by Martos 2005, who also cites previous studies on the topic.
85 For the editorial phenomenon of the "pliegos sueltos" in a wider European frame, see, for example, the studies in Chartier and Lüsebrink 1996, as well as Beltrán 2005–6.

6. Paper Politics

1 [And how shall I judge your intention, which was to copy my rough admonishments and my simple consolations into a durable support in the pages of a book, where past things become present and forgotten deeds are reduced to memory, of which either glory or shame may follow for those past and example for those present?]
2 Marín Martínez et al. 1997, 2: 330.
3 For the processes of document production in Castile during the administration of Isabel and Fernando detailed here, see the classic book by Martín Postigo 1959.
4 For a detailed explanation of the nature of the carta misiva and cédula real, see Marín Martínez et al. 1997, 2: 330–3.
5 According to *BETA*, there are around 4,000 literary manuscripts of all sizes, but extraordinarily few of them have identified scribes. This is different from documents, which particularly in the fifteenth century are signed by all those involved in their production.
6 Published by Torre y del Cerro 1954.
7 For a helpful introduction to the format of municipal libramientos, see Pino Rebolledo 1991: 82–103. City council payments were handled in a similar manner. See the case of Guadalajara as revealed in the text of its cuadernos, published by López Villalba 1997: 90–1. For meaningful parallels with the handling of petitions to the pope, see Linehan and Zutshi 2007: esp. 999.
8 Torre y del Cerro 1954: 15; and sample entries in pp. 58–9, 82–8, and passim.

9 See for example the modus operandi of the escribanos in Madrid studied by Pérez-Bustamante and Rodríguez Adrados 1995: 194–8. The 1427 ordinances of the city of Guadalajara specified that the escribano record the expenses of the concejo in his register and sign his name on the back of documents ordering money withdrawals (text in Layna Serrano 1942, 2: 526).
10 Rodríguez Adrados 1988: esp. 542–639; Bono's epilogue in Pérez-Bustamante 1985: 60–5; and Bono Huertas and Ungueti-Bono 1986: 37–43.
11 See, for example, the cuadernos of the escribanos in Seville extracted by Bono Huertas and Ungueti-Bono 1986: 99, 111, 123, 127, 130, 135, 136, 137, 138, 145, 233 (which includes a longer specification referring to a "pedaço de papel metido en este libro" [piece of paper slipped into this book]), 304 and passim; also documents 75 and 76 in Piergiovanni 1994: 618–19, and Ostos Salcedo 1994.
12 Such inclusion is indicated in the text by, for example, "sunt in bursa" (Madurell i Marimón and Rubió i Balaguer 1955: 188, doc. 22, doc. 95).
13 "Delos libros dela camara // Estauan enla camara del prinçipe don Johan, mi señor, quatro libros para la rrazon e cuenta della, y eran estos. Vn borrador de trezientas hojas, el qual tambien se llama manual e algunos lo dizen diornal: eneste libro a la jornada (quiero dezir ordinaria mente o arreo), se asientan todas las cosas que entran o salen enla camara, e aqueste libro es la llaue e padre e rregistro de todos los otros libros dela camara. Tienele a cargo el moço de camara que tiene las llaues della, e él le escriue de su mano, e eneste el escriuano de camara firma lo que se trahe a la camara en cada partida, e lo que sale dela camara asi mismo, o lo firma e rrubrica en fin de cada plana dela hoja" [On the books of the chamber // There were four books in the chamber of Prince Johan, my lord, for accounting purposes and they were these: A rough book of three hundred leaves, which is also called manual and others call it *diornal*. In this book are noted all things that enter or leave the chamber daily, and this book is the key and father and register of all the other ones in the chamber. The chamber attendant who has the keys to the chamber has it in his charge and he writes in it in his own hand. The escribano of the chamber notes what is brought into the chamber daily and likewise on what comes out of it, or he signs and rubricates it at the bottom of each page] (34–5).
14 For the professional and personal use of documents such as the albalá (*albarà*), memorial, and *recordança* by merchants, see Madurell i Marimón and Garcia i Sanz 1973: 130–43. The different formats of single documents can be seen in photographic reproductions, such as those in Gimeno Blay 1985. The plates for docs. XIIA-B and XIIC-D are photos of two original albalaes ("albarà") in paper with fold marks and address on the verso. The

first measures 194 × 72 mm, the second, 220 × 93 mm, and are dated 1366 and 1368 respectively. There are documents such as XIV (from 1392) that resemble the physical layout of some cancioneros, both in the visual disposition and in the wording of the rubrics. A similar document is XVI, dated 1402.

15 Both bills were, if we are to trust the texts, issued in the same month but within a year from each other, which meant that the predicament of having to find a hard-to-rhyme date (March) had already been solved the first time around (in rhymes *marzo-estarzo*). In order to appreciate Montoro's familiarity with the formulas used in legal language and his ability to turn it into rhymed verses, it is useful to compare the wording of the first poem with an actual receipt (carta de pago) issued in Valladolid some years later: "Conosco yo Antón Enríquez, limosnero del rey, nuestro señor, que recebí de Luiz González, mayordomo desta villa de Valladolid, mil e dozientos maravedís del yantar [que?] ha de dar al rey. // Y, porque es verdad, ago lo presente de mi mano. A VI de Agosto de 1508 años. // Fray Antón Enríquez, limosnero" (signed and rubricated) (published by Pino Rebolledo 1991: 85). The full text of ID3035 (MN19) is: "Con'os yo el de Montoro / que de vos persona franca / Obispo de Salamanca / recebi diez piezas d'oro / en este mes del estarzo / que no lo niegue ninguno / a diez y siete de Marzo / año de sesenta y uno."

16 It would be useful to compare the format and language of some extant legal documents, such as the "Libramiento a favor de Diego de Salcedo" from Christopher Columbus's personal papers, which has irregular edges roughly measuring 215 × 90 mm (reproduction in Varela 1987: doc. XXI) with poetic libramientos by Montoro or Baena.

17 For information on legal documents dealing with incarceration and requests for release from prison, among other forms of legal paperwork associated with imprisonment, see Bono Huertas 1979, 2: 69–70. For an earlier discussion of these and other texts presented in this chapter, see Gómez-Bravo 2004.

18 [cloak, smock and doublet / cost one thousand maravedís / pants, buskins, / hat cost two hundred. / Great lord, it is not two million / so let me know what you say.]

19 [but if it were disagreeable / I ask that the burning fire / be its due prize / or let it be torn with the torn / orders of payment.]

20 [Another work by the same to Diegarias, contador of the king because he did not want to accept a libramiento, but tore it to pieces telling the bearer that he would never accept another if the poet did not write a poem to him.]

21 See, for example, the provision in the *Ordenamiento* of Madrigal in 1476. Text in *Cortes* 1861–1903, 3: 27–8.

22 [The address on the albalá: To you Madam María / named Carachula / Juan de Tapia is the man / who sends you this albalá.]
23 See the description of an example of such practices in Ahumada Batlle 2003: 43–4, and passim.
24 On the physical make-up of letters and single documents, and the continuities of their material and inscriptional traits through the centuries, see Sirat 2006: 263–6, and passim.
25 Bofarull y Mascaró 1847–51: 154. This practice is commonly identified with document processing in the Chancellery. For the complex bookkeeping system in place at the court of King Pere IV el Cerimoniós, see Gimeno Blay 2006.
26 Torre y del Cerro and Alsina de la Torre 1955–6, 1: 7–8. This recalls the *modus operandi* of paper processing by the Roman curia, where granted requests were marked in the margin of the document by a "Fiat." See an example in the document described in *Exposición antológica* 1959: 201, where Cardinal Cisneros beseeched Pope Alejandro VI to found the Colegio of San Ildefonso in Alcalá in 1499.
27 Copies of royal letters and other documents were customarily archived together in the same threaded book (for example documents in *Libros de acuerdos* 1932–87, 1: 57, 64, 68, 105, and passim; and subsequent volumes).
28 *Relación de tripulantes del segundo viaje de Cristóbal Colón y sueldos adeudados*, Archivo General de Simancas, Contaduría Mayor de Cuentas, 1ª Época, legajo 98. I would like to thank head archivist Ángel J. Moreno Prieto and Isabel Aguirre, both at the Archivo General de Simancas, for their help with the Simancas manuscripts.
29 See the study by Floriano 1950, 2: 232–43, esp. 233, 242–3; also García Leal 2000.
30 On the rise of the cartulary and the issues that it poses regarding documentation, archives, and memory, see the excellent study by Geary 1994. See also the classic and insightful study by Clanchy 1993: esp. 44–80, and the useful general introduction by Clemens and Graham 2007: 222–39. See also the documents described by Ubieto Arteta 1991.
31 See *Los Reyes bibliófilos* 1986: 22. See also Marín Martínez et al. 1997, 2: 253–6. In traditional diplomatics, these terms are often used to refer to copies of documents given to a monastery or another institution often in relation to donations. In contrast, the term registro or register is understood as the collection of documents issued by an institution. Although such conceptual differentiation of extant codices is useful, it may not adequately mirror period practices. There are codices such as the *Registro de Corias*, which is in intent similar to a cartulary or tumbo but is in fact envisioned as a register since its

inception in the thirteenth century and was indeed thus called by the scribe, who heads the compendium ("codicem") of the documentation regarding the properties belonging to the monastery ("legis noue ac ueteris instrumenta") with an "Incipit prologus registri Coriensis." In Floriano 1950, 2: 3. Relevant comments in Floriano 1950, 2: 235–6.

32 Asenjo González 2006: 148, with bibliography. See the related phenomenon in other parts of Europe studied by Rosenwein on the establishment and consolidation of Cluny's property. On the forces behind the creation of cartularies, see Bouchard 2002. See also the comments on the driving forces behind the *Libro de las Estampas*, *Libro de los testamentos*, and *Tumbo A* in Boto Varela 1995: 63–4.

33 See Clanchy's (1993) discerning comments on document forging (318–27), as well as Hiatt 2004, and the documents studied by Ubieto Arteta 1991.

34 On forgery practices both in monasteries and possibly by nobles, see Zabalza Duque 1998: 16–17, and passim.

35 For a quick overview of the developments in law codes of the thirteenth century, see Ubieto Arteta's contribution to the introduction of Canellas's *Vidal Mayor* (1989, 1: 16–17). On the *Usatges*, see the useful overview by Font Rius 1988: lxxvi–lxxix.

36 Text in Miquel Rosell 1945, 1: 1. A similar rationale appears in the preface to the *A* cartulary of the Cathedral of Santiago de Compostela (*Tumbo A de la Catedral de Santiago*), whose compiler affirms the need to save original documents from oblivion, decrying their level of neglect and old age. For this reason, he set out to copy them all in one book, almost like in one body ("illa omnia testamenta translatarentur et translata in uno libro, quasi in uno corpore"), so that those that were scattered and lost may thrive when preserved as one ("et alia, que huc et illuc sparsa, amitti solebant, cum aliis in unum coniuncta, ualerent melius reservari") (Lucas Álvarez 1998: 47–8). A similar thrust has been indicated for other cartularies, such as the *Cartulario de la Catedral de Sigüenza* (Gutiérrez García-Muñoz 2002). On the subject of cartularies as archives and memory, see the lucid study by Clanchy 1981.

37 For the text and illuminations of this cartulary, see Fernández Conde 1971.

38 [And we say that a register is a book done for the remembrance of the letters and privileges that are made. And it is important because if a privilege or a letter is lost or torn or the writing is destroyed due to age or anything else, or if any doubt regarding it should arise because it may be damaged or due to any other reason, the lost ones may be recovered and the old ones renewed through the register.]

39 On the relevance gained by the register in the late medieval and early modern periods see Maravall 1986: 472–8; and Petrucci 1995: 145–68.

40 Archivo General de Simancas, Diversos de Castilla, legajo 1, fol. 51; Martín Postigo 1959: 291–2.
41 [In addition, their Majesties made an ordinance of the rights held by the escribanos in their Kingdom when drawing up extra-judiciary documents. First they ordered that each of the escribanos have a book in which to write in detail the notes of the documents that came before him and needed to be prepared, declaring the names of those who granted the said document, as well as the day, month, year and place or house where it was given, and what was granted, specifying all the conditions or pacts and clauses or relinquishments that all parts entered. And as soon as such notes were written, they were to be read to all parties involved in front of witnesses, and make all of them sign their names, and give them the aforesaid documents signed with their names and scribal signs, without taking away or adding a word to what they have in their registers.] For the guidelines for Catalan notaries, see Durán Cañameras 1955. The specific regulations for other cities, such as the ones cited for Seville below, follow exactly the same pattern. For an excellent study of medieval notaries, see Bono Huertas 1979.
42 Martín Postigo 1959: 166, 291–5. Personal references to the person/s involved in the production of a document became increasingly frequent toward the fifteenth century (García Valle 1999: 18, and passim).
43 On this general topic, see Arribas Arranz 1964a: 219–21; also Pino Rebolledo 1972: 28–9.
44 The letter, dated 28 May 1492, is copied in the register known as the *Tumbo de los Reyes Católicos*, tomo IV, fol. 38r–39r (doc. 26) (Archivo Municipal de Sevilla) and published in Fernández Gómez et al. 1993: 76–8, with a colour photograph of the document.
45 [bound book in parchment in which to write all the privileges held by that city and its lands as well as all the legal sentences that have been decided in its favour.]
46 [Letter instructing to make a book of letters in large format in paper and another book of the privileges and sentences in parchment.]
47 Pino Rebolledo 1991: 45. On this type of "document custody" see also Polo Martín 1999: 350–1.
48 [A book in paper, half of it written and the other half blank, which begins in the first page with "Jhs. o pater mi."]
49 [This testimony was sent from the Inquisition in Aranda and is located in an envoltorio with other testimonies that have been copied in the register.]
50 Blank books were also used for the purposes of poetic anthologizing in other lyric traditions, such as Provençal, French, and Italian (Beltrán 1995b and the bibliography cited therein).

51 Beltrán 1999b: 29. Beltrán studies MN6 and other cancioneros that allowed for successive additions to their contents, often by leaving blank pages interspersed in the manuscript or by adding quires. For in-depth study of BM1, see Martí 1997, 1998; and Martos 2003.
52 "libro de traslados de cartas y otro de mis privilegios en una barjaca de cordoban colorado con su çerradura de plata" (*Carta a Nicolò Oderigo*, in Colón 1992: 520-1).
53 See the insightful study by Varela 1992: 159–64. Gorricio served as Columbus's personal scribe and assumed the task of writing a title or docket upon receipt of thus unmarked documents (Varela 1992: 161). On the archive created by Columbus in the monastery of the Cartuja de la Cueva in Seville, see Serrano y Sanz 1930, and Gil 1989.
54 All references here are to vol. 1 of the edition by Szmolka Clares.
55 For the use of "envoltorios," see chapter 5.
56 [For the said Juan de Castilla, sent with the said Gamarra]; [Another letter of a tenor similar to the aforesaid one for the said Juan de Castilla to ask him to join Pedro de la Dueña and take roll of the horsemen in Almuñecar. / Done on the same day.]
57 Spanish mercantile professional groups, such as the Consulado de Burgos, had similar bookkeeping practices. On this topic, see Basas Fernández 1963: 92–6.
58 [When a girl is born, I have her name recorded in my register. And I do this in order to know how many escape my net. What did you think, Sempronio? Was I supposed to earn my living out of nothing? Did I receive an inheritance? Do I own houses or vineyards?]
59 Petrucci 1995: 169–235, esp. 180-9. See also Sirat 2006: 475–86, esp. 479. For Salutati's notarial *protocollo*, see Petrucci's (1963) edition.
60 [And it was put in a figurative manner here, as it appears in this first register or original, so that he who had the task of copying it in a nice hand could copy it in the same way, in case the opportunity arose that it could eventually be presented to the King of Navarre, for whom it was commenced and finished.] A first draft of the work was finished in 1428 and the introductory letter was added ca 1429. See Cátedra's introductory study to the edition of Villena's work (1994: xvii–xx).
61 [Some days past Alvar González de Alcántara, of the family and service of the house of Prince don Pedro, most illustrious Duke of Coimbra, your father, on your account, my lord, asked me to send my poetry to your magnificence. And truly, my lord, I would like to please your noble person in more important, even if more laborious, matters, because these works – or at least the majority of them – do not deal with such topics nor is their form good

enough to seem worthy of a memorable register [...] But, most virtuous lord, firm in my low estimation of my works, but committed to fulfilling your wish, I had my poems searched for everywhere, and through the books and cancioneros of others, and I had them copied – in the order in which I composed them – which I send you in this little volume.]

62 The extent to which these two compilations resemble Santillana's first compilation is subject to debate. For an overview of Santillana's manuscript tradition and bibliography, see the introduction to his lyric in the edition by Pérez Priego 1999: esp. 75–85.

63 [And because his intention was that all of his works would be printed, he wrote a great number of them in his own hand and gathered them in a book, and he would have gathered his complete works if death had not finished him.]

7. Books as Memory

1 Clanchy (1993) emphasizes the impact of record proliferation and accumulation on an increase of literacy in the twelfth and thirteenth centuries. For the study of the implications of the growth of literacy toward the fifteenth see Saenger 1997. For further bibliography and considerations on this growth in relation to fifteenth-century cancionero poetry, see Gómez-Bravo 1999a, 1999c.

2 On the relationship between literacy and document production in early societies see Goody 1986: esp. 92–9, and 159–65.

3 New York, Hispanic Society, Ms. HC:NS 7/1, containing documents from the twelfth to the fifteenth centuries. See *BETA* manid 3761. Also see García y García 1963: 538–40. The *Libro memoria* is document 1, p. 539. I would like to thank Irene Zaderenko for this reference.

4 One example is the inscription from the tomb of Juan Álvarez, archdeacon of Mayorga, in the Cathedral of León transcribed by Franco Mata 1976: 464.

5 In a long poem studied by Lawrance (1981), Baena displays an impressive list of readings, which include Alfonso's *General Estoria*, among others, and which highlights a thorough acquaintance with juridical texts.

6 On Baena's emphasis of his compilation as history, see Beltrán 2001 and Weiss 1990: 49–50.

7 Upon receiving a letter from fray Tomás de Torquemada on 27 July 1489, during the inquisitorial process against the Arias Dávila family (text in Carrete Parrondo 1986: 148n250).

8 [And Scipio having heard these words he carefully noted them in his memory book in order to be able to repeat them in front of the Roman senate.]

9 [another blank book of white Pisan paper bound in paper boards covered in green leather.]
10 [Among many other virtues, [the king] had a very special one. He took such care of those who served him well that without waiting to be asked, he would grant a favour. And he used to secretly carry a book that was written in his own hand and about which nobody knew until after his death. In this book he had written the names of all those to whom he owed the most gratitude, specifying in what amount, all written in separate sections that read: "Such and such services were rendered to me, I must remember when there is an opening to reward them." And when a post would open and someone came to petition it, he would say "I have already granted it." And then he would secretly look in his book and see the names of those qualified people and to those whom he was most obliged, and he would grant it. At times, when those people had travelled away from the Kingdom in his service, he would call for them, and this amazed many. This was singular virtue because of which the good held great expectations for rewards. And this book is now in my possession. Furthermore, he had another secret book in which he had written the names of all worthy men in order to use their services according to their skills and rank. Some he destined to be leaders in big enterprises and some for others with lesser responsibility, some he meant to name ambassadors, or special envoys, and similarly for all the necessary posts and affairs.]
11 [In their administration they placed great care, as they did in their choice of persons destined to fill important positions in government, justice, war, and treasury [...] and in order to be ready to make wise choices they had a book, which held the record (*memoria*) of the names of those men with more skill and merit for the positions that became vacant.]
12 [Five memoriales, all written in our lady the queen's hand, may she rest in holy peace, one being a folded leaf written on one page and the other is ... on one side one of the leaves and the other leaf, which is cut, has up to thirteen lines written on it, and the other three leaves are all written in her majesty's hand. They are all unimportant memoriales, just stating various provisions that the queen made for different people at that time.]
13 Pérez de Guzmán is referring to the various books of lineages, such as the Libro de las Generaciones; Liber regum; *Livro de Linhagens* by Pedro de Portugal, Count of Barcelos; to which later would follow Diego Fernández de Mendoza's *Linages de España*; Pedro Gratia Dei's *Vergel de Nobles de los linajes de España* and *Genealogía y blasón de los reyes de Castilla*; and chronicles of a similar bent. For a good illustration of the forces behind the creation of these family genealogical books, see Dacosta's (2007) study on the Ayalas.

14 [The truth and certainty of the origin and birth of Castilian lineages cannot be accurately known but for what remained in the memory of the ancients. For there was and is always in Castile little interest in keeping historical records, which causes great harm. In regard to this, one finds in historical narratives many remarkable customs, two of which I shall recount. First, during the time when the Jews had kings, they used to have books telling the notable events of the year called annals, and kept them in bookcases (*armaria*) and boxes in the Temple, and they also had registers of the noble lineages. This lasted until the time of King Herod the Great, who had all the books burnt fearing to lose the kingdom. Incidentally, there was no other tyrant more afraid of losing his kingdom, and for this reason he had those writings burnt and even had the innocent children killed, which was singular and extreme cruelty. The second custom at that time, as we read in the book of Esther, was that King Ahasuerus of Persia had a *book of the services* that were performed for him, and also *of the rewards* that he had given in return, which were undoubtedly notable and praiseworthy acts. However, there is little concern about keeping memory of the noble lineages and of the services performed for the kings and the republic in Castile, and, to tell the truth, it is not even necessary because he is most noble who is most wealthy. Therefore, what need is there to consult the *book of lineages*? For in wealth we find nobility. Furthermore, it is not necessary to create a *written memory of rendered services*, for kings do not give rewards to those who best serve them or those who act most virtuously, but to those who please them and agree with them. It would then be excessive and superfluous to put into writing those two deeds, wealth and flattery.] Emphasis added. Santillana (1988), in the *Proverbios o Centiloquio*, recounts the same anecdote in his gloss on Ahasuerus (226).

15 [And about the Catholic Queen Isabel it is said that when she reigned with King Fernando, her husband, a paper fell from her sleeve in which she had written in her own hand: 'The town crier position must go to so-and-so because he has the best voice for the job.]

16 "Cinco libritos para escribir memorias, el uno dellos de hueso blanco e los otros dos de cuerno e las cubiertas historiadas, y el de hueso blanco tiene un escudo de las armas reales, y el otro está desencuadernado y fáltale media cubierta" [Five little books to write memories, one in white horn and two others in horn with historiated covers, and the white horn one has a shield with the royal arms, and the other is unbound and is lacking half a cover]. These are listed in the inventory of the books under the charge of the queen's camarero, Sancho de Paredes, published by Clemencín (1821: 479). Probably because of their material similarities, listed right before these in the inventory

we find two drawing books made out of boxwood, one of which was black and had nine boards: "Dos libros de dibujar de box, uno negro de nueve tablas con su cerradura e cabo de latón" (published by Clemencín 1821: 479).

17 See a description of these books used by sixteenth-century nobles in Bouza Álvarez 2003: 48–65, and Castillo Gómez 2006: 61–70. For the use of writing tablets and their relation to parchment notebooks in the Classical period, see Roberts and Skeat 1983: 15–23. According to Alfonso X's (2004) *Siete Partidas*, wills could be written on parchment, paper, wax tablets, or any other suitable writing medium: "Ley dozena. en que cosa puede ser escripto el testamento. En pargamino de cuero o de papel o en tablas quier sean con çera o de otra manera o en otra cosa en que se pueda fazer escriptura" [Law 12: on what thing a will may be written. A will may be written on parchment, on paper or on tablets, be they wax or otherwise as long as they can be written upon] (VI, 3v).

18 [Two drawing books, one black with nine boards, lock, and stump, and the other white with nine boards and lock and brass stump.] Listed in Torre y del Cerro 1968: 294. Under "Cosas menudas" (minor things) of the queen we also find two little wooden books for drawing with horn covers ("Dos librillos de madera para debuxar con vnas couerturas de cuerno," listed in Torre y del Cerro: 1968: 370), as well as two similar ones, one with nine wooden tablets and horn covers ("Vn librillo para debuxar que tiene 9 tablas que se apresçio con sus coberteros de cuerno en rreal e medio. Otro librillo de la manera del suso dicho, que se apreçio en vn rreal" listed in Torre y del Cerro 1968: 380).

19 [Wax tablets were in other times used by the consuls in Rome, and today are used by many. In them they entrusted to memory the things that they did not want to be publicly known, as they promptly eliminated them.]

20 [I never saw a laureate poet as punctilious as you are. With wax tablets you reason with me; you inscribe what I say, if I laugh, if I play, if I joke around with you. I am almost tempted to make you a poem, but I am afraid that you will write it down. To snatch a word and to gnaw it is more appropriate for a dog than it is for a man.]

21 [Copy of a letter of privilege that King Juan II gave to a nobleman // I also allow that you may feignedly enter the church and other sacred places without any devotion, carrying, instead of the books of hours or the Psalter, your book and memory of the rent and tax collections in your charge, pretending to pray the penitential psalms, as many members of the said generation of marranos do and is their habit.] The common currency of these books is witnessed, for example, by a 1469 memorial dealing with record-keeping matters of the Treasury and which requires that cloth merchants keep

accounting books and show them to tax collectors or other public officials every month (text in Ladero Quesada 1973: 351).

22 As an example, the size of the *Tumbo de Santiago, tomo E* measuring 450 × 325 mm (Archivo de la Catedral de Santiago de Compostela, Pasillo 634, *BETA* manid 4240) may be compared with books of hours and prayers such as that written by Sister Constanza measuring 166 × 112 mm (BNM 7495, *BETA* manid 3082). Sirat (2002: 54 and passim) studies small-format Jewish books.

23 "Movime tanbién a la copilación destas obras por verme ya llegar a perfeta edad y perfeto estado de ser vuestro siervo, y pareciome ser razón de dar cuenta del tiempo passado y començar libro de nuevas cuentas" [I was also moved to compile these works upon seeing myself in the perfect age and in the perfect state of being your servant; I thought it right to give account of the time past and to start a book of new accounts] (fol. 9v).

24 [and although to another are presented / here part of my quarrels, / when tallying the account / the sum of them all / is assigned to your name.]

25 [The glorious Messiah / requesting his registro / with all his hierarchies / ordered that, in our days, / a copy be seen / elected for Christ, / anointed for being a prophet / as caballero and king.]

26 Covarrubias (1998) states in his *Tesoro de la lengua*: "Memorial. La petición que se da al juez o al señor para recuerdo de algún negocio" [Memorial: petition that is given to a judge or the lord as remembrance of a particular business] (798).

27 Text in Garrido Atienza 1992: 326–8. For a brief introduction to the administrative memorial see Ladero Quesada 1973: 327–79. See also the abundant references to the memorial as a sheet or sheets of paper that get sent along with letters in the register of the Conde de Tendilla (1996) published by Szmolka Clares et al. For abundant and erudite information on different forms of documents issued by the Treasury, see Ladero Quesada 1973; and Ladero Quesada 1999b. This type of memorial would be used for the next few centuries by anyone seeking emoluments or positions from the Crown. On the origin of the term "memorial" in relation to memory, see Guyotjeannin 1989: 123.

28 [And on the capture of the king there is this difference between the memorial of the count and that of the Alcaide de los donceles, because the count says that when the Moors fled that the King of Granada was last.]

29 Pérez de Guzmán (1965) similarly reiterates the nature of his work in other places when he states that his intention is not to write a long history ("proceso de estoria"), but a memorial or register ("mas un memorial o registro" 10). For important insights on the idea of memorial and register in Pérez de Guzmán and other authors, see Folger (2003).

30 See also Alfonso de Palencia's (1967) definition in his *Universal vocabulario* under *registro*: "Registro. como memorial delas cosas acaesçidas" [Register. Like a memorial of things that have taken place.]
31 Miralles 1992: 11–13. See also the useful introduction and notes to Miralles 1999. Parts two and three are, significantly, entitled "Libre memorial breu de memories" [Short memorial book of memories] and "Recordancia e memoria" [Remembrance and memory] (Miralles 1992: 73, 89), respectively.
32 See Mata Carriazo 1953; also Sánchez Saus 1995.
33 [where they stole not only what is mine, but the registros with the procesado that I had written about it, given that, due to human weakness, memory has a greater share in forgetfulness than an abundance of remembrance.]
34 See the instruction for these judges in relation to their books by Diego de Deza in 1504 (text in Jiménez Monteserín 1980: 137).
35 For a description of the municipal memorial, see Pino Rebolledo 1972: 75–9, and the plate on 79.
36 For Barcelona, see Madurell i Marimón and Rubió i Balaguer 1955: 32*, and docs. 374, 380, and passim. See also the important study by Batlle 1981: esp. 17. Converso booksellers such as the Corrós family supplied blank booklets, ink, and pens to government departments such as the Royal Treasury and the Consolat del Mar (Madurell i Marimón and Rubió i Balaguer 1955: 18, 308–9n, doc. 177).
37 See for example Branca 1986a; Cicchetti and Mordenti 1985–2001; Mandingorra Llavata 2002, and the bibliography cited therein. For the relation between memory books and merchant culture, see Del Treppo 1972: 757–827. See also Del Treppo 1976: esp. 475–87 for useful information on the format and content of merchant account books.
38 A register and memorial (*ricordanze*) manuscript of this type is held at the Newberry Library, Mss. +27. Melis (1962) provides a useful study of merchant documentation, including ricordanze, blank books, and other types of documentary support (esp. 5, 357–90). Lapeyre (1955) surveys the "livre de comptes ou de raison" of the Ruiz family, a sixteenth-century merchant family (339–61). See also the texts studied by Bec; as well as Anselmi et al. 1980. Also relevant are the French and particularly the Provençal *livres de raison*: for example, Bonnet 1995.
39 For a joint study of notarial practices in Spain and Italy (notably in the cities of Seville and Genoa), see the studies gathered in Piergiovanni 1994.
40 On the differences between anthology and *canzonieri*, see Cannata 2000: 93–124. Also of interest is Stemmler 1991. Marotti (1995) differentiates between personal and professional anthologies. The latter are modelled after medieval florilegia and Renaissance commonplace books (18). On early Spanish florilegia, see, for example, Aldama Roy and Muñoz Jiménez (2005),

and the bibliography cited therein. On florilegia and troubadour poetry, see Meneghetti 1991. On relevant phenomena concerning anthologies and miscellanies, see Benedict 1996; Lerer 2003.

41 Dutschke 1998; and the studies collected by Picone and Cazalé-Bérard 1998.
42 For insights into these books in later centuries, see Crane 1993; Havens 2001; and Moss 1996.
43 On relevant practices in a wider context of written culture, see Bouza Álvarez 2003; Castillo Gómez 2006.
44 On the relationship between *libri di famiglia* and literature and autobiography, see Cicchetti and Mordenti 1985–2001, 2: 97–111. On the relationship between memory and poetry, see for example Fedi 1990: esp. 52–80. For the development of autobiography in relation to early modern urban craftsmen narratives see Amelang (1998) and the bibliography cited therein.
45 On the book of memory and the *Vita nuova* see for example Pasquini 2003; also Holmes 2000: 120–44.
46 For example, Petrucci 1995: 145–68.

8. Arranging the Compilation

1 [If the flesh that I receive / incarnates my bones thanks to you / you will do more than God did / incarnating me while alive. / Oh what a mystery you shall perform / my lady, so visible / and noteworthy / that to all you will be able to say / that I was born before / I was incarnated.] On the importance and various aspects of rubrication in Iberian cancioneros/cancioneiros, see the studies collected in Garribba 2008; Gómez-Bravo 2002; Marino 1998, and the bibliography cited therein. For studies on the rubrics of earlier texts, see the insightful analysis on the *Libro de buen amor* by Domínguez 1997; and Lawrance 1997. Dagenais (1994) has stressed the importance of the rubrics in the *Libro de buen amor* in helping to furnish a unified reading of the book and build the Arcipreste as *auctor* (170). See also important insights into paratextual practices in Busby 2002, 1: 127–365, passim.
2 [Stanza written by a gentleman to a lady who sent him some quince flesh (paste) when he was too thin.]
3 [To support gentility / with such weak fortress / I take as a vain thing / because through the barbican / nature will provide it. / Thus will gracefulness remain / with such bad defence, / and if the moat is deep / it will also work as a ladder / on the wrinkled part / if the struggle is intense.] Rubric: [From the Almirante de Castilla to don Juan de Mendoza, upon seeing one day that his beard was shaved, because his face had many wrinkles.]
4 See this use of the term "rubric" by Italian humanists in the vocabulary studied by Rizzo 1973: 59.

5 [Rubrica.ce. feminine gender. Meaning the red color or the line or ruler or edge dyed red in order to mark something or the mark or dye itself. Sapientiae 13. Rubric is also said of the summary of the book or the chapter, which is written in red]. Rubrics in fifteenth-century cancioneros were written at times in red ink and at other times in black, depending on the copyist's choice. For example, TP1 has rubrics in both red and black, reflecting the choice of two different copyists. In MH2 rubrics are all in black ink.
6 "Comiençase la tabla & rubricas del libro dicho Lilio: El qual es partido en siete partes o libros: & cada vn libro tiene sus propios capitulos" [Here begins the table of contents and rubrics of the book called Lilio, which is divided into seven parts or books, and each book has its own chapters] (Gordonio 1991: fol. 2v). "Desdel CCCCII capitulo fasta CCCC e XXII, non cuenta ninguna cosa, ca sson commo rubricas de los capitulos de adelante" [In chapters 402 to 422 not much is told, because they are more like rubrics of the chapters that follow] (Juan Manuel 1983: 648).
7 [Thus now, most noble king and lord, I made some titles for the chapters in which I divided these summaries, even though all poets, following the loftiness and the superior manner of their style, proceed without titles. But I must add them in order to make the work clear for those who will read it in the vernacular.]
8 [As a result, they will become more capacitated to understand with greater clarity the discussions, since it frequently happens that what is well known in a place or time is completely unknown in other places or centuries.]
9 Egan 1983; Meneghetti 1984; Quain 1986. For Provençal vidas as commentary, see Poe 1980.
10 On ordinatio and compilatio, see the seminal study by Parkes 1991a and further considerations by Rouse 1992, as well as Hathaway 1989.
11 On the impact of the mendicant orders, see Nascimento 1995: 246; for the influence of university copying techniques on other books, including liturgical texts, see Marichal 1990.
12 On this topic, as practised by the escribanos in Granada, see the relevant observations by Moreno Trujillo 1995: 94–8.
13 See, for example, the registers studied by Pérez-Bustamante and Rodríguez Adrados 1995: 158.
14 [Anyone who has a tale to tell must include at the beginning all kinds of divisions that are known to him, in such a way that those who listen to it may enjoy it and learn from it, and so that all may be able to better understand the story]; [And these divisions of the parts of the tale into books are made so that those who may read them should not be annoyed by long arguments. For this same reason titles and chapters are used in books, for the purpose of differentiating the various parts, and so that through the titles one may more

easily find the part of the book that one is looking for]. The rubric or docket marks the inclusion of the single document into a book in current compilational practices. For example, in their prescriptive notes on how to do a palaeographic transcription, Romero Tallafigo et al. (1995) explain that the palaeographer needs to write a summary or "extract" of the contents of the document (87).

15 Parkes 1991a and R. Rouse 1975. See also Hasenohr 1990: esp. 274–5. For important insights on tables of contents in fifteenth-century Portugal, see Dionísio 2005.
16 Not only are the leaves to be foliated, but also the alphabet is to be used to divide them into parts, so that the reader knows not only the folio but also the specific place in the folio. The same system was used in the nineteenth century by Migne in the Patrologiae.
17 For this argument, see Meneghetti 1984: 47–8. Meneghetti (1984) postulates the existence of rubrics in the *Liederbücher* (30).
18 See a useful summary study in Avalle 1993: 61–2, 107–11, and passim. See also the cogent studies by Burgwinkle 1990 and 1997.
19 Avalle 1993: 107–11. For the Galician-Portuguese lyric, see Tavani 1969. But see A. Taylor's (1991) study on the concept of minstrel manuscripts.
20 For the intentional design of the chansonniers as books through the use of rubrics, illuminations, indexes, and other elements, see Burgwinkle 1999. See also Bertolucci Pizzorusso 1978, 1989, 1991. On the use of rubrics and other textual strategies in the formation of Guiraut Riquier's book, see the perceptive study by Bossy 1991. See also Holmes 2000: 101–19. On the formation of Petrarch's *canzoniere*, see the classic study by Wilkins 1951.
21 On this interpretation of the function of the vidas and razos, see Jauss 1977.
22 On the categories and other aspects of cancioneiro rubrication, see Gonçalves 1991 and 1994.
23 On rubrication in the *Cantigas de Santa María*, see Schaffer 1991.
24 Rouse (1992) makes the argument that the physical layout of manuscripts was largely dependent on the needs of readers rather than the wish of authors (127).
25 On the growth and textual implications of late-medieval silent reading, see the seminal studies by Saenger, esp. 1997. See also Gómez-Bravo 1999a, 2005 and the bibliography cited therein.
26 For example Zumthor 1993: 363–90. For a useful survey of the characteristics of manuscript culture in the Spanish Golden Age, see Bouza Álvarez 2001.
27 [poetic compositions talk through such concealment that they seem dark and hidden to the unskilled and to those who are well-versed they seem clear and manifest.] See Weiss 1990: 84–106.
28 [for it would not be proper for the knowledgeable and the ignorant to speak similarly [...] And for this reason it was not only necessary to speak

secretively and separately from the common people, but further it was
necessary to cloak and conceal through fictitiousness and diverse genres of
speech and rhetorical figures.] Don Juan Manuel (1994) made a similar,
though positive, remark on the need of the writer to take into account
readers' different capabilities in understanding (*Conde Lucanor* 12).

29 Copyists took the rubrics verbatim from their exemplar or wrote them using
first-hand information (Boffey 1985: 64–7).

30 Villasandino's poems must have circulated in various copies, as is evident
from the book with his works that appear in Queen Isabel's inventories
(listed in Ferrandis 1943: 156). For a useful overview of the manuscript copies
of the *Cancionero de Juan Alfonso de Baena* (1993), see the introductory
study by Dutton and González Cuenca (xix–xxxi). I have studied the
implications of the long textual transmission on versification (in Gómez-
Bravo 1999b).

31 [They say that the said Alfonso Álvarez de Villasandino wrote this poem for
King Enrique, father of the king our lord, when he was under guardianship,
but it cannot be believed to have been written by him since it errs in some
rhymes, notwithstanding that it is very good and stings where it should]; [The
said Alfonso Álvarez de Villasandino wrote this very famous and fine poem
for the love and praise of the said doña Juana de Sosa. But because the said
King Enrique the elder asked him to write it, others are of the opinion that he
wrote it for the queen of Navarre.]

32 [Diegarias, royal treasurer of the Catholic Queen and King, married a son or
nephew of his to a relative of Cardinal don Pedro González de Mendoza. He
invited all of his relatives to Segovia, but forgot or pretended to forget
Rodrigo de Cota the elder, native of this city of Toledo. Aware of it, Cota
celebrated the wedding with this epithalamium. Upon reading it, Queen
Isabel said that he seemed to be robbing his own house.]

33 [Juan Fernández being sick sent a request to Andrés Martín Pineda to lend
him the book of his poems and with the book, he sent these stanzas.]

34 Text in Knust 280–1. The author goes on to explain in further detail the perils
of such visitations.

35 [In addition, in order to keep your modesty, you should not allow anyone to
go into your chamber while you sleep and even less so if you are in your bed,
with the exception of the women in the household, because it would be a
great lack of decency for you to be in bed and talking with a man.]

36 [Juan Álvarez, because he saw her ill in bed and being so much in love with
her and so disturbed, could not and did not dare talk to her]; [Other stanzas
by the author to a lady whom he saw ill in bed.]

37 [Here he calls the rubrics, chapters and divisions arguments, since argument
means that through which the truth of the matter may be known. And for this

reason syllogisms, which reach a conclusion through their premises in order to know the truth of a given question or doubt, are called arguments in logic [...] And Ovid called the titles to the books of the *Aeneid* arguments, since they sum up or even gather the content of each book. It is in this latter meaning that it is referred to here. And the arguments that the said don Enrique placed here at the beginning of the books are not the same as those used by Ovid, because he thought those were too obscure, and therefore wrote them in a simpler manner. And he did similarly with the chapters and divisions.]

38 [And beside these paragraphs, let the words be distinguished by marked capital letters and be decorated in yellow, so that all of them without confusion may come to the attention of the reader, in such a way that no noteworthy doubt may be left unsolved and all that the textual fabric contains may appear clearly and the art of the ordenador may shine with pleasing and scientific rule.]

39 On the tradition of the association of text and weaving, see Chartier 2007: 83–104, esp. 97–8.

40 Huot (1987b) has proposed that scribes used rubrication as a means to provide a reading of the text.

41 Text of the contract in Perea Rodríguez and Madrid Souto 2009. Gómez Manrique (2003) refers to the "copilaçion de mis obras trobadas" in the introduction to his works (97).

42 [may lay their eyes on this reading and may command that whatever by chance and to someone's detriment I could not or knew not how to correct or order in a better way may be corrected and amended. And if the clearer intellect of Your Lordship or that of the other readers may find something to be incorrect or altered ... or any variation in their titles I entreat Your Lordship and beg everyone that they may forgive me and correct that which may not seem correct to them. And whoever should find the mark of another in his works may cross it out and put his own and let it be done similarly by whoever finds his works free of mark ... And please forgive the way in which I collected all these works, because even though I worked diligently, it was not in my hand to have all the works included here from the true originals or from a trustworthy record from the authors who wrote them, it being a nearly impossible task given the passage of time and the distance from the places where these works were made. And because all intellects naturally very much like order and not everyone likes or dislikes the same matters, I introduced order and distinctions in this work by way of parts and subject matters.] Although the use of the term "titles" is ambiguous here, as it may refer to either the titles of the poems or the titles of their authors, though most likely the former, it is clear that it alludes to rubrication. Similarly, although "mark" makes reference to a visual marking, it is obvious here that

it specifically refers to rubrication, as Castillo would be encouraging authors to amend the texts as well as the attribution and contextualization of their poems. His appeal to add an authorial mark to those poems that lack any can hardly be understood as anything but the right of claiming authorship of an anonymously published poem by marking it as one's own on the physical page of the cancionero.

43 On the notion of co-production in Iberian cancioneros/cancioneiros, see Gómez-Bravo 2003.
44 [Title. Also the name or title on a book, almost like *tutulus* from *tueor*, because the author intends to defend the work from being attributed to another.]
45 The explanatory rubrics preceding Santa Fe's poems are only found in the *Cancionero de Palacio* (SA7), where his poem series in honour of King Alfons V (el Magnànim) appear in chronological order. In such cases, rubrics provide the key information to date a particular poem (Tato 1997).
46 [A few verses which I believe were taken from some old ones written by a Jew.] Similarly, the rubric in ID4622 refers to the author in the third person and to the compiler in the first: "Muy scientifico don pablo obispo de burgos ... fizo ... vna obra ... de la qual obra quise aqui" [emphasis added].
47 For MN24 and RC1, see Vozzo Mendia 1995: 181. For the *Cançoner des Masdovelles*, see Beltrán 2003.
48 For LB2, see Whetnall 1997; for Cerverí's compilation, see the in-depth study by Cabré 1999.
49 [That even the printers have left their mark, by putting rubrics or summaries at the beginning of each part, briefly narrating the content of each: something needless according to the practices of old authors.]
50 See the pertinent comments on the formation of fifteenth- and sixteenth-century poetic compilations by Sánchez Mariana 1987.
51 [Antón de Montoro to Juan Poeta on a short poem which Poeta stole from him and then gave to the queen.]
52 [Montoro to the queen on the subject that Juan de Valladolid son of the town crier said that he had made some stanzas that Montoro had made and he was forwarding.]
53 [Love letter without signature or address so that it may not be known for whom it is or whose it is, and it was written by Villalobos for his lady Violante Artal, the fibre of his heart.]

9. The Book of Fragments

1 On Hoccleve and England, see Knapp 2001. For Gringore and other French authors, see Brown 1995.
2 Burgwinkle (1997) has emphasized the economic advantages of such practices.

3 On these self-canonizing strategies, see Kendrick (1992: esp. 848–9) for the creation of a chronological organization that uses rubrics.
4 Pulgar (1982) consciously worked for a new type of historiography more akin to his humanistic interests, as he states in his Letra XXXIII (108). Pérez Priego has reviewed the trends in the historiography and biography of the time in his introduction to *Claros varones* (Pulgar 2007: 18–53).
5 On the manuscripts that contain Gómez Manrique's work, see Beltrán 1998: 57–61. For the purposes of this chapter, I follow the order found in MP3. The fragmentary state of MN24 and its current disorganized state do not allow for a clear analysis. Gómez Manrique had a copy of his own cancionero, as listed in the 1490 post-mortem inventory of his belongings, as well as Santillana's cancionero (in Caunedo del Potro 1991: 109).
6 [And pondering how to fulfil his command I had some of my works searched through the bottom of my chests, where they were as they deserved, and tried to have some more from others, who poorly judging their true merit had them in better places. And thus I began making a compilation of them.]
7 [I have decided to humble myself in order to please you and fulfil your command; in obedience of which I am sending with this my servant this compilation of my works that you have so eagerly requested, and which would better be torn than compiled. However poorly written and decorated it may be, it is even more inexpertly composed and organized, because as you will see, the writing and ornamentation are the work of artisans who are better skilled than the one who composed it.]
8 [But these works, when in the possession of such a great lord as your mercy is, in whose house so many relatives and nobles concur, for whom they must by necessity be published.]
9 In addition to Manrique's and Santillana's individual cancioneros, their poetry circulated together in manuscript dissemination circles and copied together to form a textual unit, as for example in SA1, although there the part dedicated to Gómez Manrique's poetry is less attentively treated, as noted by Marcos Álvarez (2005: 127–8).
10 For further information on this inscription and additional bibliography, see Paz y Meliá's edition of Gómez Manrique's (1991) *Cancionero* 1, xxv–xxxvi, and 2: 318–19; and 2003: 39, 656. See also Palencia Flores 1943: 42, and plate after title page.
11 Gómez-Bravo (2011) has studied the aspects of Álvarez Gato's biopoetic compilation described here.
12 Loureiro (2000) has emphasized the importance of considering the closely intertwined relationship of the rhetorical, political, and ethical components of autobiography (esp. 1–30). I do not use the term "autobiography" in its

contested allusion to a well-defined genre, but rather in its more etymological sense of giving an account of one's life, closer to what has been called "life writing" (Olney 1998: xv), or "giving an account of oneself" (Butler 2005).

13 For an interdisciplinary approach to the various elements of narrative identity, see the studies compiled in McAdams, Josselson, and Lieblich (2006). See also Eakin (1999, 2008), as well as Smith and Watson 2001. On the influence of traumatic experiences on the configuration of autobiography and self-representation, see Gilmore 2001.

14 For example, poem ID3121, which can be dated to 1464 or ca 1465 (n. 59 in the order of appearance in MH2), is copied after ID3114, which is n. 48 in MH2 and has been dated in 1466. See Beltrán's insightful comments on MH2 (1998: 61–5), and his chronology of the compositions in MH2 (62), as well as Beltrán 2006: 221–4.

15 For an overview of the codex, in spite of some inaccuracies, the introductory study to the edition by Artiles Rodríguez is still useful (Álvarez Gato 1928: xxxi–xlix). See also the more recent and accurate Beltrán 1998, and 2006: 221–4; as well as the detailed description by Moreno 2012.

16 [from here onward there is no poetry or writing but that of devotion and good doctrine.]

17 Márquez Villanueva 1960: 28, 161, and passim.

18 The lines stated one of the assertions commonly found in love poetry that elevated women to divine status: "ni dezir que hay otro dios / en la tierra ni en el Cielo" [nor to say there is another God / either on earth or in Heaven], and were changed to "ni dezir que hay entre nos / otra tal en este suelo" [nor to say there is among us / one such other on this ground] (ID3105, vv. 39–40). On this and other occurrences of censorial intervention in 11CG, see the introduction by González Cuenca in Castillo 2004, 1: 79–82.

19 The dispute may be followed through the *Libros de acuerdos* (1932–87: 284–5), starting in January 1484, and becoming the subject of heated debate in the concejo in 1494 (3: 72, 75–7, 90–1), 1495 (3: 143, 190), and 1497 (3: 297–9, 304). He was also an active participant in concejo meetings throughout the decades of the 1480s through at least the late 1490s, as attested to by the *Libros de acuerdos*.

20 On fray Hernando de Talavera and his relationship with Álvarez Gato, see Márquez Villanueva 1960: 33–5, 143, 160–1, 250, 304–6, 392–6.

21 Dutton (1990–1) conjectures that it may be Álvarez Gato's signature (1: 579).

22 [this book is of two parts / made out of mud and gold / one half is all truths / the other all vanities / for which I, miserable, cry / because when I was a young colt / lacking any sense / my body wanted one thing / and my soul now wants another.]

23 For an insightful study of Santillana's attitude toward poetry and age, see Weiss 1990: 166–81.
24 [because this book begins with sinful stanzas on the delights of love, typical of youth, and proceeding speaks on matters of reason and at the end on profitably spiritual and contemplative matters, he who wrote it made this stanza.]
25 A similar idea of progression and circularity is expressed in the epitaph that was placed on Álvarez Gato's tomb (ID3090) (fol. 11r).
26 Such order exists in other compilations. For the organization of clusters of poems along a storyline in Juan del Encina, see the perceptive study by Navarrete 1995.
27 I have studied several aspects of Montoro's poetry in relation to his socio-economic standing in Gómez-Bravo 2010.
28 For an overview of the cancioneros that contain Montoro's poetry, see, for example, the edition by Ciceri and Puértolas (Montoro 1990: 35–9); and additional considerations in Gómez-Bravo 2010.
29 [Another one of his to the said corregidor because he ordered him to play a game of canes.]
30 [Works of Antón de Montoro by another name called the Tailor.]

References

Abbas, Ackbar. 2002. "Cosmopolitan De-scriptions: Shanghai and Hong Kong." In *Cosmopolitanism*, ed. Carol Appadurai Breckenridge, et al., 209–28. Durham, NC: Duke University Press.
Ahumada Batlle, Eulàlia de, ed. 2003. *Epistolaris d'Hipòlita Roís de Liori i d'Estefania de Requesens (segle XVI)*. Valencia: Universitat de València.
Albarracín Navarro, Joaquina. 2002. "Memorial a propósito de los alfaquíes de la Granada mudéjar." In *En el epílogo del Islam andalusí: la Granada del siglo XV*, ed. Celia del Moral Molina, 283–306. Granada: Grupo de Investigación "Ciudades Andaluzas Bajo el Islam," Universidad de Granada.
Aldama Roy, Ana Mª, and Mª José Muñoz Jiménez. 2005. "La cultura literaria a través de los florilegios medievales." In *En la pizarra: los últimos hispanorromanos de la meseta*, ed. Isabel Velázquez Soriano and Manuel Santonja Gómez, 185–99. Burgos: Fundación Instituto Castellano y Leonés de la Lengua.
Alfonso X, King of Castile and Leon. 1986–9. *Cantigas de Santa María*. 3 vols. Ed. Walter Mettmann. Madrid: Castalia.
Alfonso X, King of Castile and Leon. 2001. *General Estoria. Primera parte.* 2 vols. Ed. Pedro Sánchez Prieto-Borja. Madrid: Fundación José Antonio de Castro.
Alfonso X, King of Castile and Leon. 2004. *Siete partidas*. BNM I 766. Ed. Pedro Sánchez Prieto. Alcalá de Henares: Universidad de Alcalá de Henares; Real Academia, CORDE.
Alighieri, Dante. 1960. "Vita Nuova." In *Opere minori*. Ed. Alberto del Monte. Milano: Rizzoli.
Almeida, Manuel Lopes de, et al., eds. 1961. *Monumenta Henricina*. Vol. 3. Coimbra: Comissão Executiva das Comemorações do Quinto Centenário da Morte do Infante D. Henrique.

Álvarez Gato, Juan. 1928. *Obras completas*. Ed. Jenaro Artiles Rodríguez. Madrid: Blass.

Álvarez Llopis, Elisa, Emma Blanco Campos, and José Ángel García de Cortázar y Ruiz de Aguirre, eds. 1994. *Colección diplomática de Santo Toribio de Liébana*. Santander: Fundación Marcelino Botín.

Alves, Ana Maria. 1985. *Iconologia do poder real no período manuelino*. Lisboa: Imprensa Nacional–Casa da Moeda.

Amelang, James S. 1998. *The Flight of Icarus: Artisan Autobiography in Early Modern Europe*. Stanford, CA: Stanford University Press.

Anglade, Joseph, ed. 1926. *Las flors del gay saber. Institut d'estudis catalans, Secció Filològica, Memòries, 1: 2.* Barcelona: Institució Patxot.

Anselmi, Gian-Mario, Fulvio Pezzarossa, and Luisa Avellini. 1980. *La "memoria" dei mercatores: tendenze ideologiche, ricordanze, artigianato in versi nella Firenze del Quattrocento*. Bologna: Pàtron.

Aparici Martí, Joaquín. 2008. "De libros y de representaciones figurativas: cultura material entre los artesanos y comerciantes segorbinos y castellonenses (siglo XV)." *Millars: Espai i historia* 31: 25–48.

Arn, Mary-Jo. 2008. *The Poet's Notebook: The Personal Manuscript of Charles d'Orléans (Paris, BnF, MS fr. 25458)*. Turnhout: Brepols.

Arredondo, María Soledad, Pierre Civil, and Michel Moner, eds. 2009. *Paratextos en la literatura española: (siglos XV–XVIII)*. Madrid: Casa de Velázquez.

Arribas Arranz, Filemón. 1948. "Rollos procesales de papel." *Revista de Archivos, Bibliotecas y Museos* 54: 5–25.

Arribas Arranz, Filemón. 1953. *Documentos de los Reyes Católicos relacionados con Valladolid*. Valladolid: Impr. Sever-Cuesta.

Arribas Arranz, Filemón. 1964a. *Los escribanos públicos en Castilla durante el siglo XV*. Madrid: Junta de Decanos de los Colegios Notariales de España. Offprint of *Estudios Históricos. Centenario de la Ley del notariado* 1: 167–260.

Arribas Arranz, Filemón. 1964b. *Un formulario documental del siglo XV de la Cancillería Real Castellana*. Valladolid: Impr. Sever-Cuesta.

Asenjo González, María. 1986. *Segovia: la ciudad y su tierra a fines del medievo*. Segovia: [s.n.].

Asenjo González, María. 1999. *Espacio y sociedad en la Soria medieval (siglos XIII–XV)*. Soria: Diputación Provincial.

Asenjo González, María. 2006. "La aristocratización política en Castilla y el proceso de participación urbana (1252–1520)." In Nieto Soria: 133–96.

Asenjo González, María, ed. 2009. *Oligarchy and Patronage in Late Medieval Spanish Urban Society*. Turnhout: Brepols.

Asensio, Eugenio. 1967. "La peculiaridad literaria de los conversos." *Anuario de Estudios Medievales* 4: 327–51.
Astarita, Carlos. 2005. *Del feudalismo al capitalismo: cambio social y político en Castilla y Europa Occidental, 1250–1520*. Valencia: Universitat de València; Granada: Universidad de Granada.
Avalle, D'Arco Silvio. 1993, 1961. *I manoscritti della letteratura in lingua d'oc.* Ed. Lino Leonardi. Torino: Einaudi. First publ. as *La letteratura medievale in lingua d'oc nella sua tradizione manoscritta: Problemi di critica testuale.* Torino: Einaudi.
Avalle-Arce, Juan Bautista, ed. 1974. *Las memorias de Gonzalo Fernández de Oviedo*. Vol. 2. Chapel Hill: University of North Carolina Department of Romance Languages.
Avenoza Vera, Gemma. 2007. "El exilio de Lope García de Salazar." In López Castro and Cuesta Torre 1: 265–76.
Aznar Vallejo, Eduardo, and Antonio García-Baquero González, eds. *Andalucía 1492: razones de un protagonismo*. Sevilla: Sociedad Estatal para la Exposición Universal Sevilla 92-Algaida.
Baldissera, Andrea, and Giuseppe Mazzocchi, eds. 2005. *I canzonieri di Lucrezia / Los cancioneros de Lucrecia: atti del convegno internazionale sulle raccolte poetiche iberiche dei secoli XV–XVII: Ferrara, 7–9 ottobre 2002*. Padova: Unipress.
Barrientos, Lope de. 1946. *Refundición de la Crónica del Halconero.* Ed. Juan de Mata Carriazo. Madrid: Espasa-Calpe.
Barrios Aguilera, Manuel, and Mercedes García-Arenal, eds. 2006. *Los plomos del Sacromonte: invención y tesoro*. València: Universitat de València.
Barros, João de. 1988. *Terceira decada da Asia de Joam de Barros: Dos feytos que os Portugueses fizeram no descobrimento e conquista dos mares e terras do Oriente*. Lisboa: Imprensa Nacional-Casa da Moeda.
Bartolini, Alessandra. 1956. "Il canzoniere castigliano di San Martino delle Scale (Palermo)." *Bolletino del Centro di Studi Filologici e Linguistici Siciliani* 4: 147–87.
Basas Fernández, Manuel. 1963. *El consulado de Burgos en el siglo XVI*. Madrid: CSIC, Escuela de Historia Moderna.
Bataillon, Louis J. 1989. "Exemplar, pecia, quaternus." In Weijers: 206–19.
Batlle, Carmen. 1981. "Las bibliotecas de los ciudadanos de Barcelona en el siglo XV." In *Livre et lecture en Espagne et en France sous l'Ancien Régime: colloque de la Casa de Velázquez*, 15–31. Paris: A.D.P.F.
Bäuml, Franz H. 1980. "Varieties and Consequences of Medieval Literacy and Illiteracy." *Speculum* 55 (2): 237–65. http://dx.doi.org/10.2307/2847287.

Bayo, Juan Carlos. 2000. "Problemas de transmisión textual en los *Milagros de Nuestra Señora*." In *Text and Manuscript in Medieval Spain: Papers from the King's College Colloquium*, ed. David Hook, 51–78. London: King's College London, Dept. of Spanish and Spanish-American Studies.

Bec, Christian. 1967. *Les marchands écrivains: affaires et humanisme à Florence (1375–1434)*. Paris: Mouton.

Beceiro Pita, Isabel. 1994. "El escrito, la palabra y el gesto en las tomas de posesión señoriales." *Studia historica. Historia medieval* 12: 53–82.

Beceiro Pita, Isabel. 2000. "La educación: un derecho y un deber del cortesano." In *La enseñanza en la Edad Media: X Semana de Estudios Medievales, Nájera, 1999*, ed. José-Ignacio de la Iglesia Duarte, 175–206. Logroño: Gobierno de la Rioja, Instituto de Estudios Riojanos.

Beceiro Pita, Isabel. 2007. *Libros, lectores y bibliotecas en la España medieval*. Murcia: Nausícaä.

Beceiro Pita, Isabel, and Ricardo Córdoba de la Llave. 1990. *Parentesco, poder y mentalidad: la nobleza castellana: siglos XII–XV*. Madrid: CSIC.

Beltrán, Vicenç. 1995a. "Dos *liederblätter* quizá autógrafos de Juan del Encina y una posible atribución." *Revista de Literatura Medieval* 7: 41–91.

Beltrán, Vicenç. 1995b. "Tipología y génesis de los cancioneros: las grandes compilaciones y los sistemas de clasificación." *Cultura Neolatina* 55: 233–65.

Beltrán, Vicenç. 1996. "Tipología y génesis de los cancioneros: Juan Fernández de Híjar y los cancioneros por adición." *Romance Philology* 50: 1–19.

Beltrán, Vicenç. 1998. "Tipología y génesis de los cancioneros: los cancioneros de autor." *Revista de Filología Española* 78: 49–101.

Beltrán, Vicenç. 1999a. "Tipologia i gènesi dels cançoners: el Cançoner J, Ms. Esp. 225 de la Bibliothèque Nationale de Paris." In Fortuño Llorens and Martínez i Romero 1: 337–52.

Beltrán, Vicenç. 1999b. "Tipología y génesis de los cancioneros: la organización de los materiales." In *Estudios sobre poesía de cancionero*, ed. Vicenç Beltrán et al., 9–54. Noia: Toxosoutos.

Beltrán, Vicenç. 2001. "*La poesía es un arma cargada de futuro*: Poética y propaganda política en el *Cancionero de Baena*." In Serrano Reyes and Fernández Jiménez: 15–52.

Beltrán, Vicenç. 2003. "La disfressa de l'amor cortès: Joan Berenguer de Masdovelles i el seu cançoner." *Cancionero General* 1: 9–28.

Beltrán, Vicenç. 2004. "Anonymity and Opaque Attributions in Late-Medieval Poetry Compilations." *Scriptorium* 58: 26–47.

Beltrán, Vicenç. 2005–6. "Del pliego de poesía (manuscrito) al pliego poético (impreso)." *Incipit* 25–6: 21–56.

Beltrán, Vicenç. 2006. "Del cartapacio al cancionero." In *Convivio. Estudios sobre la poesía de cancionero*, ed. Vicenç Beltrán and Juan Paredes, 193–226. Granada: Universidad de Granada.
Beltrán, Vicenç. 2007. "El plec poètic de Tarragona i el Comte de Prades." In López Castro and Cuesta Torre 1: 311–22.
Beltrán, Vicenç. 2009. "Tipología y génesis de los cancioneros: del *Liederblatt* al cancionero." In *La lirica romanza del Medioevo. Storia, tradizioni, interpretazioni. Atti del VI convegno triennale della Società Italiana di Filologia Romanza*, ed. Furio Brugnolo and Francesca Gambino, 445–72. Padova: Unipress.
Benedict, Barbara M. 1996. *Making the Modern Reader: Cultural Mediation in Early Modern Literary Anthologies*. Princeton, NJ: Princeton University Press.
Benito Ruano, Eloy, ed. 2004. *El Libro del limosnero de Isabel la Católica*. Madrid: Real Academia de la Historia.
Berceo, Gonzalo de. 1984. *La Vida de San Millán de la Cogolla*. Ed. Brian Dutton. London: Tamesis.
Berceo, Gonzalo de. 1992. *Vida de Santo Domingo de Silos; Poema de Santa Oria*. Ed. Aldo Ruffinatto. Madrid: Espasa-Calpe.
Berger, Philippe. 1987. *Libro y lectura en la Valencia del Renacimiento*. 2 vols. Valencia: Alfons el Magnànim.
Bermejo Cabrero, José Luis. 1979. "Los primeros secretarios de los reyes." *Anuario de Historia del Derecho Español* 49: 187–296.
Bermúdez Aznar, Agustín. 1974. *El corregidor en Castilla durante la Baja Edad Media (1348–1474)*. Murcia: Departamento de Historia del Derecho, Universidad de Murcia.
Bernáldez, Andrés. 1962. *Memorias del reinado de los Reyes Católicos*. Ed. Manuel Gómez-Moreno and Juan de Mata Carriazo. Madrid: Real Academia de la Historia.
Bertolucci Pizzorusso, Valeria. 1978. "Il canzoniere di un trovatore: il 'libro' di Guiraut Riquier." *Medioevo Romanzo* 5: 216–59.
Bertolucci Pizzorusso, Valeria. 1989. "Libri e canzonieri d'autore nel Medioevo: prospettive di ricerca." In *Morfologie del testo medievale*. Bologna: Il Mulino. 125–46.
Bertolucci Pizzorusso, Valeria. 1991. "Osservazioni e proposte per la ricerca sui canzonieri individuali." In Tyssens: 273–301.
BETA: http://bancroft.berkeley.edu/philobiblon
[*Biblia de Alba*]. *Biblia (Antiguo testamento). Traducida del hebreo al castellano por rabi Mose Arragel de Guadalfajara (1422–1433?) y publicada por el duque de Berwick y de Alba*. 1920–2. 2 vols. Ed. Antonio Paz y Meliá, et al. Madrid: Imprenta Artistica.

Biblia de Ferrara. 1996. Ed. Moshe Lazar. Madrid: Fundación José Antonio de Castro.

Biblia ladinada: Escorial I.J.3. 1995. 2 vols. Ed. Moshe Lazar. Madison: Hispanic Seminary of Medieval Studies.

Biblia romanceada: Real Academia de la Historia Ms. 87, 15th century. 1994. Ed. Moshe Lazar, Francisco J. Pueyo Mena, and Andrés Enrique-Arias. Madison: Hispanic Seminary of Medieval Studies.

Binayán Carmona, Narciso. 1983. "De la nobleza vieja ... a la nobleza vieja." In *Estudios en homenaje a Don Claudio Sánchez Albornoz en sus 90 años*, 4: 103–38. Buenos Aires: Instituto de Historia de España.

Blanchard, Alain, ed. 1989. *Les Débuts du codex: actes de la journée d'étude.* Turnhout: Brepols.

Blecua, Alberto. 1974–9. "'Perdióse un cuaderno ...': Sobre los *Cancioneros de Baena.*" *Anuario de Estudios Medievales* 9: 229–66.

Bloom, Jonathan. 2001. *Paper before Print: The History and Impact of Paper in the Islamic World.* New Haven, CT: Yale University Press.

Bofarull y Mascaró, Próspero de. 1847–51. *Procesos de las antiguas cortes y parlamentos de Cataluña, Aragón y Valencia.* Vol. 5. Barcelona: Monfort.

Boffey, Julia. 1985. *Manuscripts of English Courtly Love Lyrics in the Later Middle Ages.* Cambridge: D.S. Brewer.

Boffey, Julia. 1994. "Annotation in Some Manuscritps of *Troilus and Criseyde.*" *English Manuscript Studies, 1100–1700* 5: 1–17.

Boffey, Julia, and John J. Thompson. 1989. "Anthologies and Miscellanies: Production and Choice of Texts." In *Book Production and Publishing in Britain, 1375–1475*, ed. Jeremy Griffiths and Derek Albert Pearsall, 279–315. Cambridge/New York: Cambridge University Press.

Bohigas, Pere. 1960–7. *La ilustración y la decoración del libro manuscrito en Cataluña. Contribución al estudio de la historia de la miniatura catalana.* 3 vols. Barcelona: Asociación de Bibliófilos de Barcelona.

Bonachía Hernando, Juan Antonio. 1988. *El señorío de Burgos durante la baja Edad Media (1255–1508).* Valladolid: Universidad de Valladolid.

Bonnet, Marie Rose. 1995. *Livres de raison et de comptes en Provence: fin du XIVe siècle-début du XVIe siècle.* Aix-en-Provence: Publications de l'Université de Provence.

Bono Huertas, José. 1979. *Historia del derecho notarial español. I: La Edad Media.* 2 vols. Madrid: Junta de Decanos de los Colegios Notariales de España.

Bono Huertas, José. 1990. *Breve introducción a la diplomática notarial española.* Sevilla: Consejería de Cultura y Medio Ambiente.

Bono Huertas, José. 1994. "El notariado español en la época colombina." In Piergiovanni: 41–72.

Bono Huertas, José, and Carmen Ungueti-Bono. 1986. *Los protocolos sevillanos de la época del descubrimiento: introducción, catálogo de los protocolos del siglo XV y colección documental.* Sevilla: Junta de Decanos de los Colegios Notariales de España – Colegio Notarial de Sevilla.

Borchert, Till-Holger, et al. 2002. *The Age of Van Eyck: The Mediterranean World and Early Netherlandish Painting, 1430–1530.* New York: Thames and Hudson.

Bornstein, George. 2001. *Material Modernism: The Politics of the Page.* Cambridge/New York: Cambridge University Press.

Bosch, Lynette M.F. 2000. *Art, Liturgy, and Legend in Renaissance Toledo: The Mendoza and the Iglesia Primada.* University Park: Pennsylvania State University Press.

Bossy, Michel-André. 1991. "Cyclical Composition in Guiraut Riquier's Book of Poems." *Speculum* 66 (2): 277–91. http: //dx.doi.org/10.2307/2864145.

Boto Varela, Gerardo. 1995. *La memoria perdida: la Catedral de León, 917–1255.* León: Diputación Provincial de León.

Botta, Patrizia. 2005. "La rubricación cancioneril de las letras de justadores." In *Dejar hablar a los textos: homenaje a Francisco Márquez Villanueva,* ed. Pedro M. Piñero Ramírez, 1: 173–92. Sevilla: Universidad de Sevilla.

Bouchard, Constance B. 2002. "Monastic Cartularies: Organizing Eternity." In *Charters, Cartularies, and Archives: The Preservation and Transmission of Documents in the Medieval West. Proceedings of a Colloquium of the Commission Internationale de Diplomatique (Princeton and New York, 16–18 September 1999),* ed. Adam J. Kosto and Anders Winroth, 22–32. Toronto: Pontifical Institute of Mediaeval Studies.

Bourdieu, Pierre. 1993. *The Field of Cultural Production.* New York: Columbia University Press.

Bourdieu, Pierre. 1996. *The State Nobility: Elite Schools in the Field of Power.* Stanford, CA: Stanford University Press.

Bourgain, Pascale. 1989. "La naissance officielle de l'oeuvre: l'expression métaphorique de la mise au jour." In Weijers: 195–205.

Bourgain, Pascale. 1991. "Les chansonniers lyriques latins." In Tyssens: 61–84.

Bouza Álvarez, Fernando J. 2001. *Corre manuscrito: una historia cultural del Siglo de Oro.* Madrid: Marcial Pons.

Bouza Álvarez, Fernando J. 2003. *Palabra e imagen en la Corte: cultura oral y visual de la nobleza en el Siglo de Oro.* Madrid: Abada Editores.

Branca, Vittore, ed. 1986a. *Mercanti scrittori: ricordi nella Firenze tra Medioevo e Rinascimento.* Milano: Rusconi.

Branca, Vittore. 1986b. "Il tipo boccacciano di rubriche-sommari e il suo riflettersi nella tradizione del Filostrato." In *Book Production and Letters in* the Western European Renaissance: Essays in Honour of Conor Fahey, ed. Anna Laura Lepschy, John Took, and Dennis E. Rhodes, 17–31. London: Modern Humanities Research Association.

Branner, Robert. 1967. "The Saint-Quentin Rotulus." Scriptorium 21: 252–60.

Brea, Mercedes. 1999. "De las *vidas* y *razós* a las rúbricas explicativas." Estudis Romànics 11: 35–49.

Brown, Cynthia Jane. 1995. Poets, Patrons, and Printers: Crisis of Authority in Late Medieval France. Ithaca, NY: Cornell University Press.

Brunetti, Giuseppina. 1993. "Intorno al Liederbuch di Peire Cardenal ed ai 'libri d' autore': alcune riflessioni sulla tradizione della lirica fra XII e XIII secolo." In Actes du XXe Congrès International de Linguistique et Philologie Romanes. Université de Zurich (6–11 avril 1992), 5: 57–71. Tübingen: Francke.

Burgwinkle, William E. 1990. Razos and Troubadour Songs. New York: Garland.

Burgwinkle, William E. 1997. Love for Sale: Materialist Readings of the Troubadour Razo Corpus. New York: Garland.

Burgwinkle, William E. 1999. "The Chansonniers as Books." In The Troubadours: An Introduction, ed. Simon Gaunt and Sarah Kay, 246–62. Cambridge/New York: Cambridge University Press. http://dx.doi.org/10.1017/CBO9780511620508.018.

Busby, Keith. 2002. Codex and Context: Reading Old French Verse Narrative in Manuscript. 2 vols. Amsterdam; New York: Rodopi.

Butler, Judith. 2005. Giving An Account of Oneself. New York: Fordham University Press. http://dx.doi.org/10.5422/fso/9780823225033.001.0001.

Cabañas González, María Dolores. 1982. "La reforma municipal de Fernando de Antequera en Cuenca." Anuario de Estudios Medievales 12: 381–98.

Cabanes Pecourt, Mª Desamparados, and Pilar Pueyo Colomina. 2001. Formulario zaragozano del siglo XV. Zaragoza: El Justicia de Aragón.

Cabré, Miriam. 1999. Cerverí de Girona and His Poetic Traditions. London: Tamesis.

Cabrera Muñoz, Emilio. 1999. "Nobleza y señoríos en Andalucía durante la Baja Edad Media." In La nobleza peninsular: 89–119.

Cabrera Muñoz, Emilio. 2001. "Los conversos de Baena en el siglo XV." In Serrano Reyes and Fernández Jiménez: 85–120.

Cabrera Sánchez, Margarita. 1998. Nobleza, oligarquía y poder en Córdoba al final de la Edad Media. Córdoba: Universidad de Córdoba-CajaSur.

Cabrera Sánchez, Margarita. 2002. "El papel de los universitarios en la Córdoba del siglo XV." In Poder y sociedad en la Baja Edad Media hispánica: estudios en homenaje al profesor Luis Vicente Díaz Martín, ed. Carlos M. Reglero de la Fuente, 1: 333–56. Valladolid: Universidad de Valladolid.

Cabrera Sánchez, Margarita. 2003. "Los caballeros de premia en Córdoba durante el siglo XV." In *Andalucía medieval: actas del III Congreso de Historia de Andalucía, Córdoba, 2001*, 6: 99–122. Córdoba: Obra Social y Cultural CajaSur.
Cañas Gálvez, Francisco de Paula. 2007. *El itinerario de la corte de Juan II de Castilla (1418–1454)*. Madrid: Sílex.
Cañas Gálvez, Francisco de Paula. 2008. "Los burócratas como grupo de poder: su influencia y participación en la vida urbana y en las luchas de bandos (Castilla, primera mitad del siglo XV)." In *El contrato político en la Corona de Castilla: cultura y sociedad políticas entre los siglos X al XVI*, ed. François Foronda and Ana Isabel Carrasco Machado, 391–412. Madrid: Dykinson.
Cancioneiro Geral de Garcia de Resende. 1990–8. 5 vols. Ed. Aida Fernanda Dias. Lisboa: Imprensa Nacional – Casa da Moeda.
Cancionero de Juan Alfonso de Baena. 1993. Ed. Brian Dutton and Joaquín González Cuenca. Madrid: Visor.
Cancionero de Juan del Encina. 1989. Facsim. Madrid: Real Academia Española.
Canellas, Vidal de. 1989. *Vidal mayor*. 2 vols. Ed. Antonio Ubieto Arteta, et al. Huesca: Excma. Diputación Provincial, Instituto de Estudios Altoaragoneses.
Canellas López, Ángel, and Josep Trenchs i Odena. 1988. *Folia Stuttgartensia*. Zaragoza: Cátedra "Zurita," Institución Fernando el Católico.
Cannata, Nadia. 2000. *Il canzoniere a stampa (1470–1530): tradizione e fortuna di un genere fra storia del libro e letteratura*. Roma: Bagatto libri.
Cantera Burgos, Francisco. 1952. *Alvar García de Santa María; historia de la Judería de Burgos y de sus conversos más egregios*. Madrid: Instituto Arias Montano.
Carceller Cerviño, María del Pilar. 2000. "El ascenso político de Miguel Lucas de Iranzo. Ennoblecimiento y caballería al servicio de la monarquía." *Boletín del Instituto de Estudios Giennenses* 176: 11–30.
Carlé, María del Carmen. 1981. "Caminos del ascenso en la Castilla bajo medieval." *Cuadernos de Historia de España* 65–6: 207–76.
Carlé, María del Carmen. 1993. *Una sociedad del siglo XV: los castellanos en sus testamentos*. Buenos Aires: Universidad Católica Argentina, Instituto de Historia de España.
Carpio Dueñas, Juan Bautista. 2000. *La tierra de Córdoba: el dominio jurisdiccional de la ciudad durante la Baja Edad Media*. Córdoba: Universidad de Córdoba – Obra Social y Cultural CajaSur.
Carrasco Manchado, Ana Isabel. 2006. *Isabel I de Castilla y la sombra de la ilegitimidad: propaganda y representación en el conflicto sucesorio (1474–1482)*. Madrid: Sílex.

Carrasco Pérez, Juan. 2000. "Los libros de Cuentas de la Tesorería de Tiebas, según el inventario de 1328." *Príncipe de Viana* 61: 673–94.
Carreres Valls, Ricardo. 1936. *El llibre a Catalunya, 1338–1590*. Barcelona: Altés.
Carrete Parrondo, Carlos. 1986. *Proceso inquisitorial contra los Arias Dávila segovianos: un enfrentamiento social entre judíos y conversos*. Salamanca: Universidad Pontificia de Salamanca; Granada: Universidad de Granada.
Carrillo de Huete, Pedro. 1946. *Crónica del halconero de Juan II*. Ed. Juan de Mata Carriazo. Madrid: Espasa-Calpe.
Cartagena, Alfonso de. 1969. *La rethorica de M. Tullio Cicerone*. Ed. Rosalba Mascagna. Napoli: Liguori.
Cartagena, Teresa de. 1967. *Arboleda de los enfermos y Admiraçión operum Dei*. Ed. Lewis Joseph Hutton. Madrid: Fundación Conde de Cartagena.
Casas Rigall, Juan, ed. 2007. *Libro de Alexandre*. Madrid: Castalia.
Castillo, Hernando del. 2004. *Cancionero general*. 5 vols. Ed. Joaquín González Cuenca. Madrid: Castalia.
Castillo Gómez, Antonio. 2006. *Entre la pluma y la pared: una historia social de la escritura en los Siglos de Oro*. Madrid: Akal.
Cátedra, Pedro M. 1994. *Sermón, sociedad y literatura en la Edad Media: San Vicente Ferrer en Castilla (1411–1412): estudio bibliográfico, literario y edición de los textos inéditos*. Valladolid: Junta de Castilla y León.
Cátedra, Pedro M. 2009. "La literatura funcionarial en tiempos de los Reyes Católicos." In *Siempre soy quien ser solía: estudios de literatura española medieval en homenaje a Carmen Parrilla*, ed. Antonio Chas Aguión and Cleofé Tato García, 57–82. A Coruña: Universidade da Coruña.
Caunedo del Potro, Betsabé. 1991. "Un inventario de bienes de Gómez Manrique." In *Estudios de Historia Medieval. Homenaje a Luis Suárez*, ed. Vicente Ángel Álvarez Palenzuela, Miguel Ángel Ladero Quesada, and Julio Valdeón Baruque, 95–114. Valladolid: Universidad de Valladolid.
Caunedo del Potro, Betsabé. 2006. "La formación y educación del mercader." In *El comercio en la Edad Media*, ed. José Ignacio de la Iglesia Duarte, 417–54. Logroño: Instituto de Estudios Riojanos.
Ceballos-Escalera y Gila, Alfonso, marqués de la Floresta. 1993. *Heraldos y reyes de armas en la corte de España*. Madrid: Prensa y Ediciones Iberoamericanas S.A.
Celso, Hugo de. 2000, 1553. *Repertorio vniversal de todas las leyes destos reynos de Castilla: abreuiadas y reduzidas en forma de repertorio decisiuo*. Facsim. Ed. Javier Alvarado Planas. Madrid: Centro de Estudios Políticos y Constitucionales y Boletín Oficial del Estado.
Cepeda, Isabel Vilares. 1987. "Os Livros da Rainha D. Leonor, segundo o códice 11352 da Biblioteca Nacional, Lisboa." *Revista da Biblioteca Nacional*, ser. 2 2: 51–81.

Cervantes Saavedra, Miguel de. 1994. *Rinconete y Cortadillo*. Ed. Florencio Sevilla Arroyo and Antonio Rey Hazas. Alcalá de Henares: Centro de Estudios Cervantinos; Real Academia, CORDE.

Cervantes Saavedra, Miguel de. 1998. *Don Quijote de la Mancha*. Ed. Francisco Rico. Barcelona: Instituto Cervantes-Crítica; Real Academia, CORDE.

Chacón Gómez-Monedero, Francisco Antonio. 2005. "El primer registro de Simón Fernández de Moya, escribano público de Cuenca (1423)." *Espacio, Tiempo y Forma, Serie III, Historia Medieval* 18: 71–128.

Challet, Vicent, ed. 2007. *La sociedad política a fines del siglo XV en los reinos ibéricos y en Europa: ¿élites, pueblo, súbditos? = La société politique à la fin du XVe siècle dans les royaumes ibériques et en Europe: élites, peuple, sujets?: actes du colloque franco-espagnol de Paris, 26–9 mai 2004*. Valladolid: Universidad de Valladolid; Paris: Publications de la Sorbonne.

Chartier, Roger. 2007. *Inscription and Erasure: Literature and Written Culture from the Eleventh to the Eighteenth Century*. Philadelphia: University of Pennsylvania Press.

Chartier, Roger, and Hans-Jürgen Lüsebrink, eds. 1996. *Colportage et lecture populaire: imprimés de large circulation en Europe XVIe–XIXe siècles: actes du colloque des 21–24 avril 1991, Wolfenbüttel*. Paris: Institut Mémoires de l'édition contemporaine / Maison des sciences de l'homme.

Chaves, Álvaro Lopes de. 1983. *Livro de apontamentos (1438–1489)*. Ed. Anastásia Mestrinho Salgado and Abílio José Salgado. Lisboa: Imprensa Nacional–Casa da Moeda.

Checa Cremades, Fernando, et al. 2004. *Isabel la Católica. La magnificencia de un reinado. Quinto centenario de Isabel la Católica, 1504–2004*. Madrid: Sociedad Estatal de Conmemoraciones Culturales; Valladolid: Junta de Castilla y León.

Cicchetti, Angelo, and Raul Mordenti. 1985–2001. *I libri di famiglia in Italia*. 2 vols. Roma: Edizioni di Storia e Letteratura.

Clanchy, M.T. 1981. "'Tenacious Letters': Archives and Memory in the Middle Ages." *Archivaria* 11: 115–25.

Clanchy, M.T. 1993. *From Memory to Written Record, England 1066–1307*. Oxford/Cambridge, MA: Blackwell.

Clemencín, Diego. 1821. *Elogio de la reina católica Doña Isabel, al que siguen varias ilustraciones sobre su reinado*. Madrid: Sancha

Clemens, Raymond, and Timothy Graham. 2007. *Introduction to Manuscript Studies*. Ithaca, NY: Cornell University Press.

Cluzel, Irénée. 1957–8. "Princes et troubadours de la maison royale de Barcelone-Aragon." *Boletín de la Real Academia de Buenas Letras de Barcelona* 27: 321–73.

Collantes de Terán Delorme, Francisco. 1972–80. *Inventario de los papeles del Mayordomazgo del siglo XV.* 2 vols. Sevilla: Ayuntamiento, Delegación de Cultura.

Collantes de Terán Sánchez, Antonio. 1980. "El mundo urbano." In *Historia de Andalucía, 3: Andalucía del medievo a la modernidad (1350–1504)*, ed. Manuel González Jiménez and José Enrique López de Coca Castañer, 189–316. Madrid: Cupsa; Barcelona: Planeta.

Collantes de Terán Sánchez, Antonio. 1992a. "Los mercaderes." In Aznar Vallejo and García-Baquero González: 185–211.

Collantes de Terán Sánchez, Antonio. 1992b. "Una sociedad abierta." In Aznar Vallejo and García-Baquero González: 243–63.

Colón, Cristóbal. 1992. *Textos y documentos completos.* Ed. Consuelo Varela and Juan Gil. Madrid: Alianza.

Cómez Ramos, Rafael. 1987. "El retrato de Alfonso X, el Sabio, en la primera Cantiga de Santa María." *Studies on the Cantigas de Santa Maria: Art, Music, and Poetry: Proceedings of the International Symposium on the Cantigas de Santa Maria of Alfonso X, el Sabio (1221–1284) in Commemoration of Its 700th Anniversary Year–1981 (New York, 19–21 November)*, ed. Israel J. Katz, et al., 35–52. Madison, WI: Hispanic Seminary of Medieval Studies.

Conde y Delgado de Molina, Rafael. 1998. "Archivos y archiveros en la Edad Media Peninsular." In *Historia de los archivos y de la archivística en España*, ed. Juan José Generelo Lanaspa, and Ángeles Moreno López, 13–28. Valladolid: Universidad de Valladolid.

Contreras, Jaime. 1995. "Los primeros años de la Inquisición: guerra civil, monarquía, mesianismo y herejía." In *El Tratado de Tordesillas* 2: 681–703.

Córdoba, Fernando de. 2002. *Suma de la flor de cirugía.* Ed. José Ignacio Pérez Pascual. Noia: Toxoutos.

Córdoba de la Llave, Ricardo, and José Luis del Pino García. 1988. "Los servicios sustitutivos en la guerra de Granada: el caso de Córdoba (1460–1492)." In *Relaciones exteriores del Reino de Granada: IV Coloquio de Historia Medieval Andaluza*, ed. Cristina Segura Graíño, 185–210. Almería: Instituto de Estudios Almerienses.

Corral García, Esteban. 1987. *El escribano de Concejo en la Corona de Castilla (siglos XI al XVII).* Burgos: Ayuntamiento.

Cortés Alonso, Vicenta. 1986. *La escritura y lo escrito: paleografía y diplomática de España y América en los siglos XVI y XVII.* Madrid: Instituto de Cooperación Iberoamericana.

Cortes de los antiguos reinos de León y de Castilla. 1861–1903. 5 vols. Madrid: Rivadeneyra.

Costa, Marithelma. 2001. *Bufón de palacio y comerciante de ciudad: la obra del poeta cordobés Antón de Montoro*. Córdoba: Diputación Provincial de Córdoba.
Covarrubias, Sebastián de. 1998. *Tesoro de la lengua castellana o española*. Facsim. Ed. Martín de Riquer. Barcelona: Alta Fulla.
Crane, Mary Thomas. 1993. *Framing Authority: Sayings, Self, and Society in Sixteenth-Century England*. Princeton, NJ: Princeton University Press.
Crespo Rico, Miguel Ángel, José Ramón Cruz Mundet, and José Manuel Gómez Lago, eds. 1992. *Colección documental del Archivo Municipal de Mondragón*. Vol. 2. Donostia: Eusko Ikaskuntza.
Crónica anónima de Enrique IV de Castilla, 1454–1474: Crónica castellana. 1991. 2 vols. Ed. María del Pilar Sánchez-Parra García. Madrid: Ediciones de la Torre.
Crónica de don Álvaro de Luna, Condestable de Castilla, Maestre de Santiago. 1940. Ed. Juan de Mata Carriazo. Madrid: Espasa-Calpe.
Cruselles Gómez, José María. 1998. *Els notaris de la ciutat de València: activitat profesional i comportament social a la primera meitat del segle XV*. Barcelona: Fundació Noguera.
Cuesta Gutiérrez, Luisa. 1947. *Formulario notarial castellano del siglo XV*. Madrid: Ministerio de Justicia.
Cummins, John G. 1973. "Pero Guillén de Segovia y el ms. 4114." *Hispanic Review* 41 (1): 6–32. http://dx.doi.org/10.2307/471871.
Cuñat Ciscar, Virginia M. 1998. *Documentación medieval de la Villa de Laredo, 1200–1500*. Santander: Fundación Marcelino Botín.
Dacosta, Arsenio. 2007. *El "Libro del linaje de los señores de Ayala" y otros textos genealógicos: materiales para el estudio de la conciencia del linaje en la Baja Edad Media*. Bilbao: Universidad del País Vasco = Euskal Herriko Unibertsitatea.
Dagenais, John. 1994. *The Ethics of Reading in Manuscript Culture: Glossing the Libro de Buen Amor*. Princeton, NJ: Princeton University Press.
Daly, Lloyd W. 1973. "Rotuli: Liturgy Rolls and Formal Documents." *Greek, Roman and Byzantine Studies* 14: 332–8.
D'Ancona, Paolo, and Erhard Aeschlimann. 1949. *Dictionnaire des miniaturistes du Moyen Age et de la Renaissance dans les différentes contrées de l'Europe*. Milan: U. Hoepli.
D'Andrea, Antonio. 1980. "La struttura della *Vita Nuova*: le divisioni delle rime." *Yearbook of Italian Studies* 4: 13–40.
Delclaux, Federico. 1973. *Imágenes de la Virgen en los códices medievales de España*. Madrid: Dirección General de Bellas Artes, Ministerio de Educación y Ciencia, Patronato Nacional de Museos.

Deleuze, Gilles, and Félix Guattari. 1987. *A Thousand Plateaus: Capitalism and Schizophrenia*. Minneapolis: University of Minnesota Press.

Delicado, Francisco. 1994. *Retrato de la Lozana andaluza*. Ed. Claude Allaigre. Madrid: Cátedra.

De Looze, Laurence. 1991. "Signing off in the Middle Ages: Medieval Textuality and Strategies of Authorial Self-Naming." In *Vox intexta: Orality and Textuality in the Middle Ages*, ed. Alger Nicolaus Doane and Carol Braun Pasternack, 162–78. Madison: University of Wisconsin Press.

Del Treppo, Mario. 1972. *I mercanti catalani e l'espansione della corona d'Aragona nel secolo XV*. Napoli: L'Arte tipografica.

Del Treppo, Mario. 1976. *Els mercaders catalans i l'expansió de la corona catalano-aragonesa al segle XV*. Barcelona: Curial.

Derrida, Jacques. 1988. *The Ear of the Other: Otobiography, Transference, Translation: Texts and Discussions with Jacques Derrida*. Ed. Christine MacDonald. Lincoln: University of Nebraska Press.

Deschamps, Eustache. 1994. *L'art de dictier*. Ed. Deborah M. Sinnreich-Levi. East Lansing, MI: Colleagues Press.

Diago Hernando, Máximo. 1998–9. "El contador Fernán Alonso de Robles: nuevos datos para su biografía." *Cuadernos de Historia de España* 75: 117–34.

Diago Hernando, Máximo. 2001. *Le comunidades di Castiglia (1520–1521). Una rivolta urbana contro la monarchia degli Asburgo*. Milano: Edizione Unicopli.

Díaz de Games, Gutierre. 1997. *El Victorial*. Ed. Rafael Beltrán Llavador. Salamanca: Universidad de Salamanca.

Díaz de Montalvo, Alfonso. 1995. *Text and concordance of the "Ordenanzas reales"*: I–1338. Biblioteca Nacional, Madrid, Huete, Álvaro de Castro, 1484. Ed. Ivy A. Corfis. Madison: Hispanic Seminary of Medieval Studies; Real Academia, CORDE.

Di Camillo, Ottavio. 1996. "Las teorías de la nobleza en el pensamiento ético de Mosén Diego de Valera." In Menéndez Collera and Roncero López: 223–37.

Díez Borque, José María. 1983. "Manuscrito y marginalidad poética en el XVII hispano." *Hispanic Review* 51 (4): 371–92. http: //dx.doi.org/10.2307/472874.

Díez Garretas, María Jesús. 1983. *La obra literaria de Fernando de la Torre*. Valladolid: Universidad de Valladolid.

Díez Garretas, María Jesús, and María Wenceslada de Diego Lobejón. 2000. *Un cancionero para Alvar García de Santamaría: "Diversas virtudes y vicios" de Fernán Pérez de Guzmán*. Tordesillas: Instituto de Estudios de Iberoamérica y Portugal.

DiFranco, Ralph, José J. Labrador, and C. Ángel Zorita, eds. 1989. *Cartapacio de Francisco Morán de la Estrella*. Madrid: Patrimonio Nacional.

Dinshaw, Carolyn. 1989. *Chaucer's Sexual Poetics*. Madison: University of Wisconsin Press.
Dionísio, João. 2005. "Tables of Contents in Portuguese Late Medieval Manuscripts." In *The Book as Artefact, Text and Border*, ed. Anne Mette Hansen, 89–109. Amsterdam-New York: Rodopi.
La documentación notarial y la historia: actas del II Coloquio de Metodología Histórica Aplicada. 1984. 2 vols. Madrid: Junta de Decanos de los Colegios Notariales de España; Santiago de Compostela: Universidad de Santiago.
Dolbeau, François. 1989. "Noms de livres." In Weijers: 79–99.
Domingo Palacio, Timoteo, ed. 1907. *Documentos del Archivo General de la Villa de Madrid*. Vol. 3. Madrid: Imp. y Lit. Municipal.
Domínguez, César. 1997. "*Ordinatio* y rubricación en la tradición manuscrita: El *Libro de buen amor* y las *cánticas de serrana* en el Ms. S." *Revista de Poética Medieval* 1: 71–112.
Domínguez Bordona, Jesús. 1929. *Exposición de códices miniados españoles; catálogo*. Madrid: Sociedad Española Amigos del Arte.
Domínguez Bordona, Jesús. 1969. *Spanish Illumination*. New York: Hacker Art Books.
Dufour, Jean. 2009. *Les rouleaux des morts*. Turnhout: Brepols.
Durán Cañameras, Félix. 1955. "Notas para la Historia del Notariado Catalán." *Estudios históricos y documentos de los Archivos de Protocolos* 3: 71–207.
Dutschke, Dennis. 1998. "Il libro miscellaneo: problemi di metodo tra Boccaccio e Petrarca." In Picone and Cazalé-Bérard: 95–112.
Dutton, Brian, ed. 1990–1. *El cancionero del siglo XV, c. 1360–1520*. 7 vols. Biblioteca Española del Siglo XV. Salamanca: Universidad de Salamanca.
Eakin, Paul John. 1999. *How Our Lives Become Stories: Making Selves*. Ithaca, NY: Cornell University Press.
Eakin, Paul John. 2008. *Living Autobiographically: How We Create Identity in Narrative*. Ithaca, NY: Cornell University Press.
Edwards, John. 1982. *Christian Córdoba: The City and Its Region in the Late Middle Ages*. Cambridge/New York: Cambridge University Press.
Edwards, John. 2000. *The Spain of the Catholic Monarchs, 1474–1520*. Malden, MA: Blackwell.
Edwards, John. 2004. *Ferdinand and Isabella*. New York: Pearson-Longman.
Egan, Margarita. 1983. "Commentary, *Vita Poetae*, and *Vida*: Latin and Old Provençal 'Lives of Poets.'" *Romance Philology* 37: 36–48.
Eiximenis, Francesc. 1993. *Libro de las doñas* (Escorial, Real Biblioteca; Manuscrito H.III.20, ca 1448). Transcr. Gracia Lozano López. *ADMYTE: Archivo digital de manuscritos y textos españoles*, CD-ROM. Vol. 1. Madrid: Micronet.

Elia, Paola, ed. 2002. *El "Pequeño Cancionero" (Ms. 3788 BNM). Notas críticas y edición*. Noia: Toxoutos.
Elia, Paola. 2003. "La relación del *Pequeño Cancionero* (Ms. 3788 BNM) con tres ejemplares de cancioneros canónicos." In Serrano Reyes 1: 543–54.
Elias, Norbert, et al. 2000. *The Civilizing Process: Sociogenetic and Psychogenetic Investigations*. Oxford; Malden, MA: Blackwell.
Encina, Juan del. 1991. *Teatro completo*. Ed. Miguel Ángel Pérez Priego. Madrid: Cátedra.
Enríquez del Castillo, Diego. 1994. *Crónica de Enrique IV*. Ed. Aureliano Sánchez Martín. Valladolid: Universidad de Valladolid.
Enríquez Fernández, Javier, Concepción Hidalgo de Cisneros Amestoy, Araceli Lorente Ruigómez, and Adela Martínez Lahidalga. 1993. *Libro de visitas del corregidor, 1508–1521; y Libro de fábrica de Santa María, 1498–1517, de la villa de Lequeitio*. Donostia: Eusko Ikaskuntza.
Escalona Monge, Julio, Pilar Azcárate Aguilar-Amat, and Miguel Larrañaga Zulueta. 2002. "De la crítica diplomática a la ideología política. Los diplomas fundacionales de San Pedro de Arlanza y la construcción de una identidad para la Castilla medieval." In Sáez 2002a: 159–206.
Escartí, Vicent Josep. 1998. *Memòria privada: literatura memorialística valenciana del segle XV al XVIII*. València: Edicions 3 i 4; Eliseu Climent.
Escobar Camacho, José Manuel. 1989. *Córdoba en la Baja Edad Media: evolución urbana de la ciudad*. Córdoba: Caja Provincial de Ahorros de Córdoba.
Escudero, José Antonio. 1976. *Los secretarios de Estado y del despacho (1474–1724)*. 4 vols. Madrid: Instituto de Estudios Administrativos.
Esperabé Arteaga, Enrique. 1914–17. *Historia pragmática é interna de la Universidad de Salamanca*. 2 vols. Salamanca: Francisco Núñez Izquierdo.
Estepa Díez, Carlos, et al. 2002. *La Catedral de León: mil años de historia*. León: Edilesa.
Exposición antológica del tesoro documental, bibliográfico y arqueológico de España. 1959. Madrid: [s.n.].
Ezell, Margaret J. 1999. *Social Authorship and the Advent of Print*. Baltimore: Johns Hopkins University Press.
Febvre, Lucie, and Henri-Jean Martin. 1976. *The Coming of the Book: The Impact of Printing, 1450–1800*. London: N.L.B.
Fedi, Roberto. 1990. *La memoria della poesia: canzonieri, lirici e libri di rime nel Rinascimento*. Roma: Salerno.
Fernández Conde, Francisco Javier. 1971. *El libro de los testamentos de la Catedral de Oviedo*. Roma: Iglesia Nacional Española.

Fernández Conde, Francisco Javier. 1982. *Historia de la Iglesia en España*. Vol. 2. Part 1–2. *La Iglesia en la España de los siglos VIII al XIV*. Madrid: Biblioteca de Autores Cristianos.

Fernández de Heredia, Juan. 1955. *Obras*. Ed. Rafael Ferreres. Madrid, Espasa-Calpe.

Fernández de Heredia, Juan. 1995. *Gran Crónica de España*, I. Ms. 10133 BNM. Ed. R. af Geijerstam. Madison: Hispanic Seminary of Medieval Studies; Real Academia, CORDE.

Fernández de Oviedo, Gonzalo. 1870. *Libro de la camara real del Prínçipe Don Juan e offiçios de su casa e seruiçio ordinario*. Madrid: [s.n.].

Fernández de Santaella, Rodrigo. 1992. *Vocabulario eclesiástico*. Ed. Gracia Lozano. Madison: Hispanic Seminary of Medieval Studies; Real Academia, CORDE.

Fernández Gallardo, Luis. 2004. "Sobre la crónica real en el siglo xv. Un nuevo manuscrito de la 'Refundición del Halconero.'" *En la España Medieval* 27: 285–316.

Fernández Gallardo, Luis. 2006. "La biografía como memoria estamental. Identidades y conflictos." In Nieto Soria: 423–88.

Fernández Gómez, Marcos, Pilar Ostos Salcedo, and María Luisa Pardo Rodríguez. 1993. *El libro de privilegios de la Ciudad de Sevilla: estudio introductorio y transcripción*. Sevilla: Ayuntamiento de Sevilla-Universidad de Sevilla.

Fernández González, Etelvina. 2001. "Del santo *Mandilyon* a la Verónica: sobre la vera icona de Cristo en la edad media." In *Imágenes y promotores en el arte medieval: miscelánea en homenaje a Joaquín Yarza Luaces*, ed. Mª Luisa Melero Moneo, et al., 353–71. Bellaterra: Universitat Autònoma de Barcelona.

Ferrandis, José, ed. 1943. *Inventarios reales (Juan II a Juana la Loca)*. Madrid: CSIC, Instituto Diego Velázquez.

Ferrari, Anna. 1998. *Filologia classica e filologia romanza: esperienze ecdotiche a confronto: atti del convegno, Roma, 25–27 maggio 1995*. Spoleto: Centro italiano di studi sull'alto Medioevo.

Ferreira, Manuel Pedro. 1986. *O som de Martin Codax: sobre a dimensão musical da lírica galego-portuguesa, séculos XII–XIV*. Lisboa: UNISYS-Imprensa Nacional-Casa da Moeda.

Ferrer i Mallol, María-Teresa. 1974. "La redacció de l'instrument notarial a Catalunya: cèdules, manuals, llibres i cartes." *Estudios históricos y documentos de los Archivos de Protocolos* 4: 29–191.

Ferrer i Mallol, María-Teresa. 1980. "Cartes i billets privats en els manuals del notari barceloní Narcis Guerau Gili (segle XV)." In *Estudis de llengua i*

literatura catalanes oferts a R. Aramon i Serra en el seu setantè aniversari, ed. Manuel Jorba, 2: 197–217. Barcelona: Curial.
Flandrin, Jean-Louis, and Carole Lambert. 1998. *Fêtes gourmandes au Moyen Age*. Paris: Imprimerie nationale.
Flores, Juan de. 1981. *Triunfo de amor*. Ed. Antonio Gargano. Pisa: Giardini.
Floriano, Antonio Cristino. 1950. *El libro registro de Corias*. 2 vols. Oviedo: Instituto de Estudios Asturianos del Patronato José Mª Quadrado (C.S.I.C.).
Folger, Robert. 2003. *Generaciones y semblanzas: Memory and Genealogy in Medieval Iberian Historiography*. Tübingen: G. Narr.
Font Rius, José María. 1988. *Constitucions de Catalunya: incunable de 1495*. Facsim. Barcelona: Generalitat de Catalunya, Departament de Justícia.
Forradellas, Joaquín. 1986. *Cartapacio Poético del Colegio de Cuenca*. Salamanca: Diputación de Salamanca.
Fortuño Llorens, Santiago, and Tomàs Martínez i Romero, eds. 1999. *Actes del VII Congrés de l'Associació Hispànica de Literatura Medieval (Castelló de la Plana, 22–6 de setembre de 1997)*. 3 vols. Castelló de la Plana: Universitat Jaume I.
Fossier, François. 1980–1. "Chroniques universelles en forme de rouleau à la fin du Moyen Age." *Bulletin de la Société nationale des antiquaries de France* 163–83.
Foucault, Michel. 1986. "What Is an Author?" In *Critical Theory since 1965*, ed. Hazard Adams and Leroy Searle, 138–48. Tallahassee: University Press of Florida.
Franco Mata, María Ángela. 1976. *Escultura gótica en León*. León: Institución Fray Bernardino de Sahagún de la Excma. Diputación Provincial.
Franco Mata, María Ángela. 1998. *Escultura gótica en León y provincia: 1230–1530*. León: Diputación de León, Instituto Leonés de Cultura.
Fuchs, Barbara. 2003. *Passing for Spain: Cervantes and the Fictions of Identity*. Urbana: University of Illinois Press.
Fuentes Guerra, Rafael. 1955. *Juan de Mena, poeta insigne y cordobés modesto*. Córdoba: Artística.
Galíndez de Carvajal, Lorenzo. 1875–8. *Anales breves del reinado de los Reyes Católicos don Fernando y doña Isabel*. In *Crónicas de los reyes de Castilla, desde don Alfonso el Sabio, hasta los católicos don Fernando y doña Isabel*, ed. Cayetano Rosell, 3: 533–65. Madrid: Rivadeneyra.
Gallego Barnés, Andrés. 1982. *Juan Lorenzo Palmireno, 1524–1579: un humanista aragonés en el Studi General de Valencia*. Zaragoza: Institución Fernando el Católico.
Gamble, Harry Y. 1995. *Books and Readers in the Early Church: A History of Early Christian Texts*. New Haven, CT: Yale University Press.

García, Honorio. 1949. "El *Libre del Repartiment* y la practica notarial de su tiempo." *Boletín de la Sociedad Castellonense de Cultura* 15: 493–9.
Garcia, Michel. 1979. "La Colección de Martínez de Burgos (siglo XV)." *Hommage des Hispanistes français à Noël Salomon*, 335–49. Barcelona: Laia.
García de Cortázar, José Ángel. 1994. "Los marcos de relación social: el predominio de la aldea y la ciudad." In *La época del gótico en la cultura española: c. 1220–c. 1480*, ed. José Ángel García de Cortázar, 83–132. Madrid: Espasa Calpe.
García de Salazar, Lope. 1967. *Las bienandanzas e fortunas*. 4 vols. Ed. Ángel Rodríguez Herrero. Bilbao: Diputación de Vizcaya.
García de Valdeavellano, Luis. 1991, 1969. *Orígenes de la burguesía en la España medieval*. Madrid: Espasa Calpe.
García Leal, Alfonso. 2000. *El registro de Corias*. Oviedo: Real Instituto de Estudios Asturianos.
García López, Aurelio. 1995. "La Correspondencia del Conde de Tendilla." *Wad-Al-Hayara* 22: 65–122.
García Marín, José María. 1974. *El oficio público en Castilla durante la baja Edad Media*. Sevilla: Universidad de Sevilla.
García Valle, Adela. 1999. *El notariado hispánico medieval: consideraciones histórico-diplomáticas y filológicas*. València: Universitat de València.
García y García, Antonio. 1963. "Los manuscritos jurídicos medievales de la Hispanic Society of America." *Revista española de derecho canónico* 18: 501–60.
Garribba, Aviva, ed. 2008. *De rúbricas ibéricas*. Roma: Aracne.
Garrido Atienza, Miguel. 1992, 1910. *Las capitulaciones para la entrega de Granada*. Facsim. Intr. José Enrique López de Coca Castañer. Granada: Universidad de Granada.
Garriga, Carlos. l99l. "Control y disciplina de los oficiales públicos en Castilla: la 'visita' del Ordenamiento de Toledo (1480)." *Anuario de Historia del Derecho Español* 60: 215–390.
Geary, Patrick J. 1994. *Phantoms of Remembrance: Memory and Oblivion at the End of the First Millenium*. Princeton, NJ: Princeton University Press.
Genette, Gérard. 1997. *Paratexts: Thresholds of Interpretation*. Cambridge/New York: Cambridge University Press. http://dx.doi.org/10.1017/CBO9780511549373
Gerbet, Marie-Claude. 1989. *La nobleza en la Corona de Castilla. Sus estructuras sociales en Extremadura (1454–1516)*. Cáceres: Institución Cultural "El Brocense."
Gerbet, Marie-Claude. 1997. *Las noblezas españolas en la Edad Media. Siglos XI–XV*. Madrid: Alianza.

Gerli, E. Michael. 1995. "Antón de Montoro and the Wages of Eloquence: Poverty, Patronage, and Poetry in 15th-Century Castile." *Romance Philology* 48: 265–76.

Gerli, E. Michael. 1996. "Performing Nobility: Mosén Diego de Valera and the Poetics of *Converso* Identity." *La Corónica* 25 (1): 19–36.

Gibert y Sánchez de la Vega, Rafael. 1949. *El Concejo de Madrid. Su organización en los siglos XII a XV*. Madrid: [s.n.].

Giddens, Anthony. 1990. *The Consequences of Modernity*. Stanford, CA: Stanford University Press.

Gil, Juan. 1989. "El archivo colombino de La Cartuja. El inventario de 1544 y un problema anejo." In *Historia de La Cartuja de Sevilla: de ribera del Guadalquivir a recinto de la Exposición Universal*, 147–60. Madrid: Turner; Sevilla: Sociedad Estatal para la Exposición Universal de 1992.

Gilmore, Leigh. 2001. *The Limits of Autobiography: Trauma and Testimony*. Ithaca, NY: Cornell University Press.

Gimeno Blay, Francisco M. 1985. *La escritura gótica en el País Valenciano después de la conquista del siglo XIII*. Valencia: Universidad de Valencia.

Gimeno Blay, Francisco M. 2006. *Escribir, reinar: la experiencia gráfico-textual de Pedro IV el Ceremonioso (1336–1387)*. Madrid: Abada Editores.

Gimeno Blay, Francisco, and María Teresa Palasí Fas. 1986. "Del negocio y del amor: el diario del mercader Pere Seriol." *Saitabi* 36: 37–55.

Gomes, Rita Costa. 1998. "A curialização da nobreza." In *O Tempo de Vasco da Gama*, ed. Diogo Ramada Curto, 179–87. Lisboa: DIFEL; Difusão Editorial.

Gómez-Bravo, Ana M. 1999a. "Cantar decires y decir canciones: género y lectura de la poesía cuatrocentista castellana." *Bulletin of Hispanic Studies* 76: 169–87.

Gómez-Bravo, Ana M. 1999b. "La poesía cuatrocentista castellana: tipificación y problemas textuales en el *Cancionero de Baena*." *Romance Philology* 52: 39–57.

Gómez-Bravo, Ana M. 1999c. "Retórica y poética en la evolución de los géneros poéticos cuatrocentistas." *Rhetorica: A Journal of the History of Rhetoric* 17 (2): 137–75. http://dx.doi.org/10.1525/rh.1999.17.2.137.

Gómez-Bravo, Ana M. 2002. "Práctica poética y cultura manuscrita en el *Cancioneiro geral* de Resende." In *Iberia cantat: estudios sobre poesía hispánica medieval*, ed. Juan Casas Rigall and Eva Mª Díaz Martínez, 1–24. Santiago de Compostela: Universidade de Santiago de Compostela.

Gómez-Bravo, Ana M. 2003. "'*A huma senhora que lhe disse*': Sobre la naturaleza social de la autoría y la noción de texto en el *Cancioneiro geral* de Resende y la lírica cancioneril ibérica." *La Corónica* 32 (1): 43–64. http://dx.doi.org/10.1353/cor.2003.0034.

Gómez-Bravo, Ana M. 2004. "Memorias y archivos. Modelos de producción textual y antologías poéticas del siglo XV." *Cancionero General* 2: 53–87.

Gómez-Bravo, Ana M. 2005. "El espacio de la escritura: sobre la localización de la actividad cultural en la España y el Portugal del cuatrocientos." In *Actas del IX Congreso Internacional de la Asociación Hispánica de Literatura Medieval* (A Coruña, 18–22 de septiembre de 2001), ed. Carmen Parrilla García and Mercedes Pampín, 2: 353–73. Noia: Toxosoutos.

Gómez-Bravo, Ana M. 2010. "Ser social y poética material en la obra de Antón de Montoro, mediano converso." *Hispanic Review* 78 (2): 145–67. http://dx.doi.org/10.1353/hir.0.0105.

Gómez-Bravo, Ana M. 2011. "Vida en fragmentos: el *libro* de Juan Álvarez Gato y la memoria autobiográfica." *Romance Quarterly* 58 (3): 231–48. http://dx.doi.org/10.1080/08831157.2011.576980.

Gómez-Bravo, Ana M. 2013. "Situation and Textual Mediation: Toward a Material Poetics of the Fifteenth-Century Lyric." *La Corónica* 41 (2): 35–60.

Gómez Mampaso, Mª Valentina. 1980. "Profesiones de los judaizantes españoles en tiempos de los Reyes Católicos, según los legajos del Archivo Histórico Nacional de Madrid." In *La Inquisicion española: nueva visión, nuevos horizontes*, ed. Joaquín Pérez Villanueva, 671–87. Madrid: Siglo Veintiuno de España.

Gómez-Menor, José. 1970. *Cristianos nuevos y mercaderes de Toledo*. Toledo: Librería Gómez-Menor.

Gómez Moreno, Ángel. 1985. "Dos decires de recuesta y algunas notas sobre poemas sueltos en el siglo XV." *Revista de Filología Española* 65: 109–14.

Gómez Moreno, Ángel. 2001. "Judíos y conversos en la prosa castellana medieval (con un excurso sobre el círculo cultural del marqués de Santillana)." In *Judíos en la literatura española*, ed. Iacob M. Hassán and Ricardo Izquierdo Benito, 57–86. Cuenca: Universidad de Castilla-La Mancha.

Gómez Pérez, José. 1969–70. "Dos canciones juglarescas del siglo XV." *Anuario de filología (Venezuela)* 7–9: 19–30.

Gómez Redondo, Fernando. 2001. "El Victorial de Gutierre Díaz de Games." In Martin: 191–210.

Gonçalves, Elsa. 1991. "Sur la lyrique galégo-portugaise: Phénoménologie de la constitution des chansonniers ordonnés par genres." In Tyssens: 447–64.

Gonçalves, Elsa. 1994. "O sistema das rubricas atributivas e explicativas nos cancioneiros trovadorescos galego-portugueses." In *Actas do XIX congreso internacional de lingüística e filoloxía románicas: Sección IX Filoloxía medieval e renacentista (Santiago de Compostela 1989)*, ed. Ramón Lorenzo, 7: 979–90. A Coruña: Fundación Pedro Barrié de la Maza.

González Alonso, Benjamín. 1970. *El corregidor castellano (1348–1808)*. Madrid: Instituto de Estudios Administrativos.
González Alonso, Benjamín. 1981. *Sobre el Estado y la administración de la Corona de Castilla en el Antiguo Régimen: las comunidades de Castilla y otros estudios*. Madrid: Siglo Veintiuno de España.
González Alonso, Benjamín. 2001. "La reforma del gobierno de los concejos en el reinado de Isabel." In *Isabel La Católica y la política: ponencias presentadas al I Simposio sobre el reinado de Isabel La Católica, celebrado en las ciudades de Valladolid y México en el otoño de 2000*, ed. Julio Valdeón Baruque, 293–313. Valladolid: Ámbito.
González de Caldas, Victoria. 2000. *¿Judíos o cristianos? El Proceso de fe "Sancta Inquisitio."* Sevilla: Universidad de Sevilla.
González Rolán, Tomás, Antonio Moreno Hernández, and Pilar Saquero Suárez-Somonte. 2000. *Humanismo y teoría de la traducción en España e Italia en la primera mitad del siglo XV: edición y estudio de la Controversia Alphonsiana (Alfonso de Cartagena vs. L. Bruni y P. Candido Decembrio)*. Madrid: Clásicas.
González Sánchez, Carlos Alberto. 2007. *Homo viator, homo scribens: cultura gráfica, información y gobierno en la expansión atlántica, siglos XV–XVII*. Madrid: Marcial Pons Historia.
Goody, Jack. 1986. *The Logic of Writing and the Organization of Society*. Cambridge/New York: Cambridge University Press. http://dx.doi.org/10.1017/CBO9780511621598
Gordonio, Bernardo de. 1991. *Lilio de medicina: un manual básico de medicina medieval*. Ed. John Cull and Brian Dutton. Madison: Hispanic Seminary of Medieval Studies.
Gracia Boix, Rafael. 1982. *Colección de documentos para la historia de la Inquisición de Córdoba*. Córdoba: Monte de Piedad y Caja de Ahorros.
Gracián, Antonio. 1962. *Diurnal de Antonio Gracián, Secretario de Felipe II*. Ed. P. Gregorio de Andrés, O. S. A. Vol. 5 of *Documentos para la historia del Monasterio de San Lorenzo el Real de el Escorial*. Madrid: [s.n.].
Green, Richard Firth. 1980. *Poets and Princepleasers: Literature and the English Court in the Late Middle Ages*. Toronto; Buffalo: University of Toronto Press.
Greenblatt, Stephen. 1980. *Renaissance Self-Fashioning: From More to Shakespeare*. Chicago: University of Chicago Press.
Greetham, David C. 1994. *Textual Scholarship: An Introduction*. New York: Garland.
Greetham, David C. 1999. *Theories of the Text*. Oxford / New York: Oxford University Press. http://dx.doi.org/10.1093/acprof:oso/9780198119937.001.0001

Gudiol, Josep, and Santiago Alcolea i Blanch. 1986. *Pintura gótica catalana*. Barcelona: Polígrafa.
Gudiol i Cunill, Josep. 1955. *Els primitius: la miniatura catalana*. Barcelona: Llibreria Canuda.
Guerrero Navarrete, Yolanda. 1986. *Organización y gobierno en Burgos durante el reinado de Enrique IV de Castilla, 1453–1476*. Madrid: Universidad Autónoma.
Gutiérrez García-Muñoz, Almudena E. 2002. "Originales y copias: la conservación en el Archivo de la Catedral de Sigüenza (siglo XII)." In Sáez 2002a: 133–42.
Guyotjeannin, Olivier. 1989. "Le vocabulaire de la diplomatique en latin medieval." In Weijers: 120–34.
Hanna, Ralph. 1986. "Booklets in Medieval Manuscripts: Further Considerations." *Studies in Bibliography* 39: 100–11.
Hasenohr, Geneviève. 1990. "Les Systèmes de repérage textuel." In Martin and Vezin: 273–88.
Hathaway, Neil. 1989. "*Compilatio*: From Plagiarism to Compiling." *Viator* 20: 19–44.
Havens, Earle. 2001. *"Of Common Places, or Memorial Books": A Seventeenth-Century Manuscript from the James Marshall and Marie-Louise Osborn Collection*. New Haven, CT: Beinecke Rare Book and Manuscript Library.
Hennessy, Rosemary. 1993. *Materialist Feminism and the Politics of Discourse*. New York: Routledge.
Hernández Gassó, Héctor. 2007. "El funcionariado letrado y su dimensión literaria en la corte de los Reyes Católicos: el caso de Alonso Ramírez de Villaescusa." In López Castro and Cuesta Torre 1: 685–97.
Heusch, Carlos. 2000. *La caballería castellana en la baja edad media: textos y contextos*. With Jesús D. Rodríguez Velasco. Montpellier: Université de Montpellier III.
Hiatt, Alfred. 2004. *The Making of Medieval Forgeries: False Documents in Fifteenth-Century England*. London: British Library and University of Toronto Press.
Holmes, Olivia. 2000. *Assembling the Lyric Self: Authorship from Troubadour Song to Italian Poetry Book*. Minneapolis: University of Minnesota Press.
Horozco, Sebastián de. 1981. *Relaciones históricas toledanas*. Ed. Jack Weiner. Toledo: I.P.I.E.T.
Howsam, Leslie. 2006. *Old Books and New Histories: An Orientation to Studies in Book and Print Culture*. Toronto: University of Toronto Press.
Huot, Sylvia. 1987a. *From Song to Book: The Poetics of Writing in Old French Lyric and Lyrical Narrative Poetry*. Ithaca, NY: Cornell University Press.

Huot, Sylvia. 1987b. "The Scribe as Editor: Rubrication as Critical Apparatus in Two Manuscripts of the *Roman de la Rose*." *L'Esprit Créateur* 27: 67–78.

Iglesias, J. Antoni. 2002. "*Scrit de mà del dit deffunct*: copias autógrafas en una librería profesional, la del jurista barcelonés Lluís Destorrent, +1466 (con una breve nota sobre codicología y paleografía descriptiva)." In Sáez 2002a: 273–88.

Infantes de Miguel, Víctor. 1993. "En busca del lector perdido: la recepción de la poesía culta (1543–1600)." *Edad de Oro* 12: 141–8.

Infantes de Miguel, Víctor, and Juan Carlos Conde. 1997. "Antes de partir: un poema taurino antijudaico en el Toledo medieval (¿1489?)." In Macpherson and Penny: 91–108.

Iradiel Murugarren, Francisco Paulino. 1999. "Ciudades, comercio y economía artesana." In *La historia medieval en España: un balance historiográfico (1968–1998) / XXV Semana de Estudios Medievales, Estella, 14 a 18 de julio de 1998*, 603–58. Pamplona: Gobierno de Navarra, Departamento de Educación y Cultura.

Jager, Eric. 2000. *The Book of the Heart*. Chicago: University of Chicago Press.

Jara Fuente, José Antonio. 2001. "La ciudad y la otra caballería: realidad político-social e imaginario de los caballeros ('villanos')." In Martin: 27–44.

Jauss, Hans Robert. 1977. "Littérature médiévale et expérience esthétique (actualité des Questions de littérature de R. Guiette)." *Poétique* 31: 322–36.

Jeay, Claude. 2001. "La naissance de la signature dans les cours royale at princières de France (XIVe–XVe siècle)." In *Auctor et auctoritas: invention et conformisme dans l'écriture médiévale: actes du colloque tenu à l'Université de Versailles-Saint-Quentin-en-Yvelines, 14–16 juin 1999*, ed. Michel Zimmermann, 457–75. Paris: École des chartes.

Jiménez Monteserín, Miguel. 1980. *Introducción a la Inquisición española*. Madrid: Editora Nacional.

Johnston, Mark D. 1996. "Poetry and Courtliness in Baena's Prologue." *La Corónica* 25 (1): 93–105.

Jones, R.O. 1962. "Isabel la Católica y el amor cortés." *Revista de Literatura* 21: 55–64.

Kasten, Lloyd A., ed. 1957. *Poridat de las poridades*. Madrid: S. Aguirre.

Keller, John Esten, and Annette Grant Cash. 1998. *Daily Life Depicted in the Cantigas de Santa Maria*. Lexington: University Press of Kentucky.

Keller, John E., and Richard P. Kinkade. 1983. "Iconography and Literature: Alfonso Himself in Cantiga 209." *Hispania* 66 (3): 348–52. http://dx.doi.org/10.2307/342307.

Kelly, Thomas Forrest. 1996. *The Exultet in Southern Italy*. New York: Oxford University Press.

Kendrick, Laura. 1992. "The Monument and the Margin." *South Atlantic Quarterly* 91: 835–64.
Kennedy, William J. 1984. "Petrarchan Audiences and Print Technology." *Journal of Medieval and Renaissance Studies* 14: 1–20.
Knapp, Ethan. 2001. *The Bureaucratic Muse: Thomas Hoccleve and the Literature of Late Medieval England*. University Park: Pennsylvania State University Press.
Knust, Hermann, ed. 1878. *Dos obras didácticas y dos leyendas sacadas de manuscritos de la Biblioteca del Escorial*. Madrid: Sociedad de Bibliófilos españoles. 251–93.
Kren, Thomas, ed. 1983. *Renaissance Painting in Manuscripts: Treasures from the British Library*. New York: Hudson Hills Press.
LaCapra, Dominick. 1982. "Rethinking Intellectual History and Reading Texts." In *Modern European Intellectual History: Reappraisals and New Perspectives*, ed. Dominick LaCapra and Steven L. Kaplan, 47–85. Ithaca, NY: Cornell University Press.
Lacarra, María Jesús. 2007. "Género y recepción de las Memorias de Leonor López de Córdoba (1361/1362–1430)." In López Castro and Cuesta Torre 2: 731–41.
Lacarra de Miguel, José María. 1927. "Fuero de Estella." *Anuario de Historia del Derecho Español* 4: 404–51.
Lacarra Ducay, María Carmen. 1989. "Las miniaturas del Vidal Mayor: estudio histórico-artístico". In Vidal de Canellas, *Vidal mayor* 1: 113–66.
Lacarra Ducay, María Carmen. 1993. *Catedral y Museo Diocesano de Jaca*. Zaragoza: IberCaja.
Ladero Quesada, Miguel Ángel. 1973. *La Hacienda Real de Castilla en el siglo XV*. La Laguna: Universidad de La Laguna.
Ladero Quesada, Miguel Ángel. 1986. "Corona y ciudades en la Castilla del siglo XV." *En la España medieval* 8: 551–74.
Ladero Quesada, Miguel Ángel. 1990. ""Aristócratas y marginales: aspectos de la sociedad castellana en la Celestina." *Espacio, Tiempo y Forma, Serie III, Historia Medieval* 3: 95–120.
Ladero Quesada, Miguel Ángel. 1992a. *Andalucía en torno a 1492: estructuras, valores, sucesos*. Madrid: MAPFRE.
Ladero Quesada, Miguel Ángel. 1992b. "Los judeoconversos en la Castilla del siglo XV." *Historia 16* 194: 39–52.
Ladero Quesada, Miguel Ángel. 1996–7. "La consolidación de la nobleza en la Baja Edad Media." In *Nobleza y sociedad en la España Moderna*, ed. María del Carmen Iglesias, 1: 19–45. Oviedo: Nobel.

Ladero Quesada, Miguel Ángel. 1999a. *La España de los Reyes Católicos*. Madrid: Alianza.

Ladero Quesada, Miguel Ángel. 1999b. *Legislación hacendística de la Corona de Castilla en la Baja Edad Media*. Madrid: Real Academia de la Historia.

Ladero Quesada, Miguel Ángel. 2005. *Hernando de Zafra, secretario de los Reyes Católicos*. Madrid: Dykinson.

Lagares Díez, Xoán Carlos. 1998. "A edición das rúbricas explicativas dos cancioneiros medievais galego-portugueses: implicacións históricas e literarias." *Edición y anotación de textos: actas del I Congreso de Jóvenes Filólogos, A Coruña, 25–28 de septiembre de 1996*, ed. Carmen Parrilla García, 1: 351–9. A Coruña: Universidade da Coruña.

Lagares Díez, Xoán Carlos. 2000. *E por esto fez este cantar: sobre as rubricas explicativas dos cancioneiros profanos galego-portugueses*. Santiago de Compostela: Laiovento.

Lalinde Abadía, Jesús. 1970. *Los medios personales de gestión del poder público en la historia española*. Madrid: Instituto de Estudios Admistrativos.

Lalou, Elisabeth, ed. 1992. *Les Tablettes à écrire de l'Antiquité à l'époque moderne*. Turnhout: Brepols.

Landry, Donna, and Gerald MacLean. 1993. *Materialist Feminisms*. Cambridge, MA: Blackwell.

Lapeyre, Henri. 1955. *Une famile de marchands: les Ruiz. Contribution à l'étude du commerce entre la France et l'Espagne au temps de Philippe II*. Paris: A. Colin.

Lara Garrido, José. 2005. "Inéditos de Pietro Bembo en un manuscrito de Ferrara." In Baldissera and Mazzocchi: 503–7.

La Sorsa, Saverio. 1947. *Pasquinate, cartelli, satire e motteggi popolari*. Napoli: Libreria scientifica editrice.

Lawrance, Jeremy N.H. 1981. "Juan Alfonso de Baena's Versified Reading List: A Note on the Aspirations and the Reality of Fifteenth-Century Castilian Culture." *Journal of Hispanic Philology* 5: 101–22.

Lawrance, Jeremy N.H. 1984. "Nueva luz sobre la biblioteca del Conde de Haro: Inventario de 1455." *El Crotalón. Anuario de Filología Española* 1: 1073–111.

Lawrance, Jeremy N.H. 1985. "The Spread of Lay Literacy in Late Medieval Castile." *Bulletin of Hispanic Studies* 62 (1): 79–94. http://dx.doi.org/10.1080/1475382852000362079.

Lawrance, Jeremy N.H. 1997. "The Rubrics in MS S of the *Libro de buen amor*." In Macpherson and Penny: 223–52.

Layna Serrano, Francisco. 1942. *Historia de Guadalajara y sus Mendozas en los siglos XV y XVI*. 4 vols. Madrid: Aldus.

Lejeune, Philippe. 1982. "The Autobiographical Contract." In *French Literary Theory Today: A Reader*, ed. Tzetvan Todorov, 192–222. Cambridge/New York: Cambridge University Press; Paris: Editions de la Maison des sciences de l'homme.
León, Fray Luis de. 1991. *La perfecta casada*. In *Obras completas castellanas*. Vol. 1. Ed. P. Félix García. Madrid: Biblioteca de Autores Cristianos.
León Tello, Pilar. 1979. *Judíos de Toledo*. 2 vols. Madrid: CSIC, Instituto B. Arias Montano.
Lerer, Seth. 2003. "Medieval English Literature and the Idea of the Anthology." *PMLA* 118 (5): 1251–67. http://dx.doi.org/10.1632/003081203X68018.
Levinas, Emmanuel. 1994. *Outside the Subject*. Stanford, CA: Stanford University Press.
Levinas, Emmanuel. 1998. *Otherwise than Being or Beyond Essence*. Pittsburgh, PA: Duquesne University Press.
El Libro de los privilegios concedidos a los mercaderes genoveses establecidos en Sevilla (siglos XIII–XVI). 1992. Madrid: Tabapress.
Libros de acuerdos del Concejo madrileño 1464–1600. 1932–87. 5 vols. Ed. Agustín Millares Carlo, Jenaro Artiles Rodríguez, Agustín Gómez Iglesias, Carmen Rubio Pardos, et al. Madrid: Artes Gráficas Municipales.
Linehan, Peter A., and Patrick N. R. Zutshi. 2007. "*Fiat A*: The Earliest Known Roll of Petitions Signed by the Pope (1307)." *English Historical Review* 122 (498): 998–1015. http://dx.doi.org/10.1093/ehr/cem215.
Livermore, Harold V. 2000. "The Formation of the *Cancioneiros*." In *Essays on Iberian History and Literature, from the Roman Empire to the Renaissance*, 107–47. Aldershot, Hampshire, UK; Brookfield, VT: Ashgate.
López Castro, Armando, and María Luzdivina Cuesta Torre, eds. 2007. *Actas del XI Congreso Internacional de la Asociación Hispánica de Literatura Medieval (Universidad de León, 20 al 24 de septiembre de 2005)*. 2 vols. León: Universidad de León.
López de Coca Castañer, José E. and Ángel Galán Sánchez, ed. 1991. *Las ciudades andaluzas (siglos XIII–XVI): actas del IV Coloquio Internacional de Historia Medieval de Andalucía*. Málaga: Universidad de Málaga.
López de Toro, M. José. 1964. *Miniatures espagnoles et flamandes dans les collections d'Espagne*. Bruxelles: Bibliothèque Albert 1er.
López Gómez, Óscar. 2004. "Claves del sistema de pacificación ciudadana desarrollado por los Reyes Católicos en Toledo (1475–1485)." *En la España Medieval* 27: 165–93.
López Martínez, Nicolás. 1954. *Los judaizantes castellanos y la Inquisición en tiempo de Isabel la Católica*. Burgos: Aldecoa.

López Nieto, Juan C. 1999. "Nueve cartas, con autógrafos, de Gómez Manrique al Ayuntamiento de Toledo." *Voz y letra: Revista de literatura* 10: 37–80.

López Pita, Paulina, Carmelo Luis López, and Gregorio del Ser Quijano. 2002. *Documentación medieval de la casa de Velada: Instituto Valencia de Don Juan.* 2 vols. Ávila: Institución "Gran Duque de Alba"-Obra Cultural de la Caja de Ahorros de Ávila.

López Villalba, José Miguel. 1997. *Las actas de sesiones del concejo medieval de Guadalajara.* Madrid: UNED.

Lorenzo Cadarso, Pedro Luis. 1994. "Esplendor y decadencia de las oligarquías conversas de Cuenca y Guadalajara (XV–XVI)." *Hispania: Revista española de historia* 54: 53–94.

Lorenzo Gradín, Pilar. 2000. "Trovadores, cronología y razós en los cancioneros gallego-portugueses." *Actas del VIII Congreso Internacional de la Asociación Hispánica de Literatura Medieval: Santander, 22–26 de septiembre de 1999, Palacio de la Magdalena, Universidad Internacional Menéndez Pelayo,* ed. Margarita Freixas and Silvia Iriso, 2: 1105–25. Santander: Asociación Hispánica de Literatura Medieval.

Losa Contreras, Carmen. 1999. *El Concejo de Madrid en el tránsito de la Edad Media a la Edad Moderna.* Madrid: Dykinson.

Loureiro, Ángel G. 2000. *The Ethics of Autobiography: Replacing the Subject in Modern Spain.* Nashville, TN: Vanderbilt University Press.

Love, Harold. 1987. "Scribal Publication in Seventeenth-Century England." *Transactions of the Cambridge Bibliographical Society* 9: 130–54.

Lucas Álvarez, Manuel. 1998. *Tumbo A de la catedral de Santiago.* Sada, A Coruña: Ediciós do Castro.

Lucena, Juan de. 1892a. *Epístola exhortatoria a las letras.* In Paz y Meliá: 209–17.

Lucena, Juan de. 1892b. *Libro de vida beata.* In Paz y Meliá: 104–205.

Lucía Megías, José Manuel. 1997. *Actas del VI Congreso Internacional de la Asociación Hispánica de Literatura Medieval (Alcalá de Henares, 12–16 de septiembre de 1995).* 2 vols. Alcalá de Henares: Universidad de Alcalá de Henares.

Lunenfeld, Marvin. 1987. *Keepers of the City: The Corregidores of Isabella I of Castile (1474–1504).* Cambridge/New York: Cambridge University Press. http://dx.doi.org/10.1017/CBO9780511898051

Macpherson, Ian, and Ralph J. Penny, eds. 1997. *The Medieval Mind: Hispanic Studies in Honour of Alan Deyermond.* London: Tamesis.

Mackay, Angus. 1986. "The Lesser Nobility in the Kingdom of Castile." In *Gentry and Lesser Nobility in Late Medieval Europe,* ed. Michael Jones, 159–80. New York: St Martin's Press.

Mackay, Angus. 1991. "La conflictividad social urbana." In López de Coca Castañer and Galán Sánchez: 509–24.
Madurell i Marimón, Josep María. 1958. "La cofradía de la Santa Trinidad de los conversos de Barcelona." *Sefarad* 18: 60–82.
Madurell i Marimón, Josep María. 1974. *Manuscrits en català anteriors a la impremta (1321–1474): contribució al seu estudi*. Barcelona: Associació Nacional de Bibliotecaris, Arxivers i Arqueòlegs.
Madurell i Marimón, Josep María, and Arcadi Garcia i Sanz. 1973. *Comandas comerciales barcelonesas de la baja Edad Media*. Barcelona: Colegio Notarial.
Madurell i Marimón, Josep María, and Jordi Rubió i Balaguer. 1955. *Documentos para la historia de la imprenta y librería en Barcelona, 1474–1553*. Barcelona: Gremios de Editores, de Libreros y de Maestros Impresores.
Mandingorra Llavata, María Luz. 2002. "La configuración de la identidad privada: diarios y libros de memorias en la Baja Edad Media." In *La conquista del alfabeto: escritura y clases populares*, ed. Antonio Castillo Gómez, 131–52. Gijón: Trea.
Mann, Michael. 1986. *The Sources of Social Power*. 2 vols. Cambridge/New York: Cambridge University Press. http://dx.doi.org/10.1017/CBO9780511570896
Manrique, Gómez. 1991, 1885. *Cancionero*. 2 vols. Ed. Antonio Paz y Meliá. Palencia: Diputación Provincial de Palencia, Departamento de Cultura.
Manrique, Gómez. 2003. *Cancionero*. Ed. Francisco Vidal González. Madrid: Cátedra.
Manuel, Juan. 1983. *Crónica abreviada*. In *Obras completas*. Vol. 2. Ed. José Manuel Blecua. Madrid: Gredos.
Manuel, Juan. 1994. *El conde Lucanor*. Ed. Guillermo Serés. Barcelona: Crítica.
Maravall, José Antonio. 1986. *Estado moderno y mentalidad social (Siglos XV a XVII)*. Vol. 2. Madrid: Alianza.
Marcos Álvarez, Francisco de B. 2005. "El códice 1865 de la Universidad de Salamanca (SA1): observaciones sobre su composición e hipótesis sobre su origen." In Baldissera and Mazzocchi: 125–44.
Marcuello, Pedro. 1987. *Cancionero*. Ed. José Manuel Blecua. Zaragoza: Institución Fernando el Católico.
Marichal, Robert. 1990. "Les manuscrits universitaires." In Martin and Vezin: 211–17.
Marín Martínez, Tomás, et al. 1997. *Paleografía y diplomática*. 2 vols. Madrid: UNED.
Marín Padilla, Encarnación. 1998. *Maestre Pedro de la Cabra: médico converso aragonés del siglo XV, autor de unas coplas de arte menor*. Madrid: Encarnación Marín Padilla.

Marino, Nancy F. 1998. "A Life of Their Own: Reading the Rubrics of the *Cancionero de Baena*." *Romance Notes* 38: 311–9.
Marino, Nancy F. 2006. "Fernando de la Torre's 'Juego de Naipes,' a Game of Love." *La Corónica* 35 (1): 209–47. http://dx.doi.org/10.1353/cor.2006.0014.
Marotti, Arthur F. 1995. *Manuscript, Print, and the English Renaissance Lyric*. Ithaca, NY: Cornell University Press.
Márquez Villanueva, Francisco. 1957. "Conversos y cargos concejiles en el siglo XV." *Revista de Archivos, Bibliotecas y Museos* 63: 503–40.
Márquez Villanueva, Francisco. 1960. *Investigaciones sobre Juan Álvarez Gato: contribución al conocimiento de la literatura castellana del siglo XV*. Madrid: RAE.
Martí, Sadurní. 1997. "El Cançoner del Marquès de Barberà (S1 / BM1) Descripció codicològica (1)." *Boletín Bibliográfico de la Asociación Hispánica de Literatura Medieval* 11: 463–502.
Martí, Sadurní. 1998. "Fonts i problemes del Cançoner del Marquès de Barberà (S^1/BM1)." In *Atti del XXI Congresso Internazionale di Linguistica e Filologia Romanza, Centro di studi filologici e linguistici siciliani, Università di Palermo 18–24 settembre 1995*, ed. Giovanni Ruffino, 6: 311–24. Tübingen: Niemeyer.
Martin, Georges, ed. 2001. *La chevalerie en Castille à la fin du Moyen Âge: aspects sociaux, idéologiques et imaginaires*. Paris: Ellipses.
Martin, Henri-Jean, and Jean Vezin, eds. 1990. *Mise en page et mise en texte du livre manuscrit*. Paris: Éditions du Cercle de la Librairie-Promodis.
Martín Gaite, Carmen. 2000. *Nubosidad variable*. Barcelona: Anagrama.
Martín Gaite, Carmen. 2002. *Cuadernos de todo*. Ed. María Vittoria Calvi. Barcelona: Random House Mondadori.
Martín Postigo, María de la Soterraña. 1959. *La Cancillería castellana de los Reyes Católicos*. Valladolid: Universidad de Valladolid.
Martínez de Toledo, Alfonso. 1987. *Arcipreste de Talavera o Corbacho*. Ed. Michael Gerli. Madrid: Cátedra.
Martínez Ferrando, J. Ernesto. 1961. "Aportación de datos acerca del Archivo Real de Barcelona y de sus archiveros durante los reinados de Juan II y Fernando el Católico." In *Fernando el Católico y la cultura de su tiempo*, Jordi Rubió i Balaguer et al., 77–109. Zaragoza: Institución Fernando el Católico.
Martorell, Francesc, ed. 1926. *Epistolari del segle XV. Recull de cartes privades*. Barcelona: Barcino.
Martos, Josep Lluís. 1999. "El *Còdex de Cambridge* del Trinity College, R. 14. 17 (X2): descripció i estudi." In Fortuño Llorens and Martínez Romero 2: 443–60.
Martos, Josep Lluís. 2003. "Los espacios en blanco y la estructura del Cançoner del Marqués de Barberà." In *Proceedings of the Twelfth Colloquium*, ed. Alan

Deyermond and Jane Whetnall, 57–65. London: Department of Hispanic Studies, Queen Mary and Westfield College, University of London.

Martos, Josep Lluís. 2005. "La restauración de las obras de Ausis March: los cancioneros impresos del siglo XVI." In Baldissera and Mazzocchi: 409–26.

Martos, Josep Lluís. 2011. *Del impreso al manuscrito en los cancioneros*. Alcalá de Henares: Centro de Estudios Cervantinos.

Martz, Linda. 2003. *A Network of Converso Families in Early Modern Toledo: Assimilating a Minority*. Ann Arbor: University of Michigan Press.

Mata Carriazo, Juan de, ed. 1953. "Los anales de Garci Sánchez, jurado de Sevilla." *Anales de la Universidad Hispalense* 15: 3–63.

McAdams, Dan P., Ruthellen Josselson, and Amia Lieblich, eds. 2006. *Identity and Story: Creating Self in Narrative*. Washington, DC: American Psychological Association. http://dx.doi.org/10.1037/11414-000

McGann, Jerome J. 1991. *The Textual Condition*. Princeton, NJ: Princeton University Press.

McKenzie, D.F. 1999. *Bibliography and the Sociology of Texts*. Cambridge/New York: Cambridge University Press. http://dx.doi.org/10.1017/CBO9780511483226

Medina de Mendoza, Francisco. 1853. "Vida del cardenal d. Pedro Gonzalez de Mendoza." *Memorial histórico español* 22: 147–310.

Melis, Federigo. 1962. *Aspetti della vita economica medievale: studi nell'Archivo Datini di Prato*. Siena: Monte dei Paschi di Siena; Firenze: L.S. Olschki (distr.).

Memorias de don Enrique IV de Castilla. 1835–1913. Madrid: Fortanet.

Mena, Juan de. 1986. *Antología de su obra poética*. Ed. José María Azáceta. Barcelona: Plaza y Janés.

Mena, Juan de. 1989. *Obras completas*. Ed. Miguel Ángel Pérez Priego . Barcelona: Planeta.

Mendoza, Fray Íñigo de. 1968. *Cancionero*. Ed. Julio Rodríguez-Puértolas. Madrid: Espasa-Calpe.

Meneghetti, Maria Luisa. 1984. *Il pubblico dei trovatori: ricezione e riuso dei testi lirici cortesi fino al XIV secolo*. Modena: Mucchi.

Meneghetti, Maria Luisa. 1991. "Les florileges dans la tradition lyrique des troubadours." In Tyssens: 43–56.

Menéndez Collera, Ana, and Victoriano Roncero López, eds. 1996. *Nunca fue pena mayor: estudios de literatura española en homenaje a Brian Dutton*. Cuenca: Universidad de Castilla–La Mancha.

Menéndez Pidal, Ramón. 1914. "*Elena y María* (disputa del clérigo y el caballero). Poesía leonesa inédita del siglo XIII." *Revista de Filología Española* 1: 52–96.

Milán, Luis. 1874. *Libro intitulado El cortesano, Libro de motes de damas y caballeros*. Madrid: Aribau y Cª.

Milán, Luis. 1982. *Libro de motes de damas y caballeros*. Facsim. Valencia: Librerías París-Valencia.

Millares Carlo, Agustín. 1927. *Índice y extractos del Libro horadado del Concejo madrileño (siglos XV–XVI)*. Madrid: Impr. Municipal.

Millares Carlo, Agustín. 1983. *Tratado de paleografía española*. 3 vols. With José Manuel Ruiz Asencio. Madrid: Espasa Calpe.

Minnis, Alastair J. 1992. "Authors in Love: The Exegesis of Late-Medieval Love-Poets." In *The Uses of Manuscripts in Literary Studies: Essays in Memory of Judson Boyce Allen*, ed. Charlotte C. Morse, Penelope Reed Doob, and Marjorie Curry Woods, 161–91. Kalamazoo: Western Michigan University.

Miquel Rosell, Francisco, ed. 1945. *Liber feudorum maior: Cartulario real que se conserva en el Archivo de la Corona de Aragon*. 2 vols. Barcelona: CSIC.

Miralles, Melcior. 1992. *Dietari del capellà d'Alfons V el Magnànim*. Ed. María Desamparados Cabanes Pecourt. Zaragoza: Anúbar Ediciones.

Miralles, Melcior. 1999. *Dietari del capellà d'Alfons el Magnànim: ficcio*. Ed. José Vicente Gómez Bayarri and Josep Giner i Ferrando. Valencia: Oronella Serveis Editorials Valencians.

Mitre Fernández, Emilio. 1968. *Evolución de la nobleza en Castilla bajo Enrique III (1396–1406)*. Valladolid: Universidad de Valladolid.

Monsalvo Antón, José María. 1995. "Historia de los poderes medievales, del Derecho a la Antropología (el ejemplo castellano: monarquía, concejos y señoríos en los siglos XII–XV)." In *Historia a debate: actas del Congreso Internacional "A historia a debate," celebrado el 7–11 de julio de 1993 en Santiago de Compostela*, ed. Carlos Barros, 4: 81–150. Santiago de Compostela: Debate.

Monsalvo Antón, José María. 2003. "Gobierno municipal, poderes urbanos y toma de decisiones en los concejos castellanos bajomedievales (consideraciones a partir de concejos salmantinos y abulenses)." In *Las sociedades urbanas*: 409–88.

Montero Tejada, Rosa María. 1996. *Nobleza y sociedad en Castilla: el linaje Manrique (siglos XIV–XVI)*. Madrid: Caja de Madrid.

Montero Tejada, Rosa María. 1999. "Monarquía y gobierno concejil: continos reales en las ciudades castellanas". In *V Reunión Científica Asociación Española de Historia Moderna*. Vol. 2: *La administración municipal de la Edad Moderna*, ed. José Manuel de Bernardo Ares, 577–89. Cádiz: Universidad de Cádiz; Asociación de Historia Moderna.

Montero Tejada, Rosa María, and María José García Vera. 1992. "La alta nobleza en la Cancillería real castellana del siglo XV." *Espacio, Tiempo y Forma, Serie III, Historia Medieval* 5: 163–210.

Montero Vallejo, Manuel. 1992, 1987. *El Madrid medieval*. Madrid: El Avapiés.
Montoro, Antón de. 1990. *Cancionero*. Ed. Marcella Ciceri and Julio Rodríguez-Puértolas. Salamanca: Universidad de Salamanca.
Montoya Martínez, Jesús, and Ana Domínguez Rodríguez, eds. 1999. *El Scriptorium alfonsí: de los Libros de Astrología a las Cantigas de Santa María*. Madrid: Editorial Complutense.
Morales García-Goyena, Luis. 1906–7. *Documentos históricos de Málaga, recogidos directamente de los originales*. 2 vols. Granada: López Guevara.
Morandini, Antonietta, Guglielmo De Angelis d'Ossat, and Mario Tesi. 1986. *Biblioteca Medicea Laurenziana*. Firenze: Nardini.
Moreno, Manuel. "Descripción codicológica: MH2." In *An Electronic Corpus of 15th Century Castilian Cancionero Manuscripts*. http://cancionerovirtual.liv.ac.uk (accessed 15 May 2013).
Moreno Hernández, Carlos. 1989. *Obra poética*. Pero Guillén de Segovia. Madrid: Fundación Universitaria Española.
Moreno Trujillo, Mª Amparo. 1995. "Diplomática Notarial en Granada en los inicios de la Modernidad (1505–1520)." In Ostos Salcedo and Pardo Rodríguez: 75–125.
Moreno Trujillo, Mª Amparo. 2002. "Registro oficial, registro personal: la dualidad de la correspondencia del Conde de Tendilla." In *La correspondencia en la historia: modelos y prácticas de la escritura epistolar: actas del VI Congreso Internacional de Historia de la Cultura Escrita*, ed. Carlos Sáez and Antonio Castillo Gómez, 1: 205–30. Madrid: Calambur.
Moreno Trujillo, Mª Amparo, María José Osorio Pérez, and Juan María de la Obra Sierra, eds. 2007. *Escribir y gobernar: el último registro de correspondencia del conde de Tendilla (1513–1515)*. Intro. Mª Amparo Moreno Trujillo. Granada: Universidad de Granada. Book and CD-ROM.
Morison, Samuel Eliot. 1942. *Admiral of the Ocean Sea, a Life of Christopher Columbus*. Boston: Little, Brown. http://dx.doi.org/10.2307/210008
Moss, Ann. 1996. *Printed Commonplace-Books and the Structuring of Renaissance Thought*. Oxford: Clarendon Press. http://dx.doi.org/10.1093/acprof:oso/9780198159087.001.0001
Mota Placencia, Carlos. 1990. "La obra poética de Alfonso Álvarez de Villasandino." Unpubl. PhD diss. Universitat Autònoma de Barcelona. Microfiche.
Moxó y Ortiz de Villajos, Salvador de. 1969a. "Los Cuadernos de alcabalas. Orígenes de la legislación tributaria castellana." *Anuario de Historia del Derecho Español* 39: 317–450.
Moxó y Ortiz de Villajos, Salvador de. 1969b. "De la nobleza vieja a la nobleza nueva. La transformación nobiliaria castellana en la baja Edad Media." *Cuadernos de Historia: anexos de la Revista "Hispania"* (Madrid) 3: 1–210.

Moxó y Ortiz de Villajos, Salvador de. 1975. "La promoción política y social de los 'letrados' en la Corte de Alfonso XI." *Hispania: Revista española de historia* 129: 5–30.

Nader, Helen. 1979. *The Mendoza Family in the Spanish Renaissance, 1350–1550*. New Brunswick, NJ: Rutgers University Press.

Narbona Vizcaíno, Rafael. 2003. "Vida pública y conflictividad urbana en los reinos hispánicos (siglos XIV–XV)." In *Las sociedades urbanas*: 541–89.

Narkiss, Bezalel. 1982. *Hebrew Illuminated Manuscripts in the British Isles: A Catalogue Raisonné*. 2 vols. Jerusalem-Oxford: Oxford University Press.

Nascimento, Aires A. 1993. "As livrarias dos Príncipes de Avis." *Biblos* 69: 265–87.

Nascimento, Aires A. 1995. "O livro de teologia: génese de uma estrutura e estruturação de uma ciência." *Didaskalia* 25: 235–55.

Navarrete, Ignacio. 1995. "The Order of the Poems in Encina's 1496 *Cancionero*." *Bulletin of Hispanic Studies* 72 (2): 147–63. http://dx.doi.org/10.1080/1475382952000372147.

Navarro Espinach, Germán. 2006. "Los protagonistas del comercio: oficios e identidades sociales en la España bajomedieval." In *El comercio en la Edad Media: XVI semana de estudios medievales, Nájera y Tricio (1–5 de agosto 2005)*, ed. José Ignacio de la Iglesia Duarte, 147–87. Logroño: Instituto de Estudios Riojanos.

Nebrija, Antonio de. 1979. *Diccionario Latino-Español (Salamanca, 1492)*. Facsim. Ed. Germán Colón and Amadeu-J. Soberanas. Barcelona: Puvill.

Nebrija, Antonio de. 1981. *Introductiones Latinae*. Facsim. Salamanca: Universidad de Salamanca.

Nichols, Stephen G., and Siegfried Wenzel, eds. 1996. *The Whole Book: Cultural Perspectives on the Medieval Miscellany*. Ann Arbor: University of Michigan Press.

Nieto Cumplido, Manuel. 1973. *Córdoba en el siglo XV*. Córdoba: Diputación Provincial.

Nieto Soria, José Manuel, ed. 2006. *La monarquía como conflicto en la corona castellano-leonesa (c. 1230–1504)*. Madrid: Sílex.

La nobleza peninsular en la Edad Media: VI Congreso de Estudios Medievales. 1999. Ávila: Fundación Sánchez-Albornoz.

Nordström, Carl-Otto. 1967. *The Duke of Alba's Castilian Bible: A Study of the Rabbinical Features of the Miniatures*. Uppsala: Almqvist and Wiksells.

O'Callaghan, Joseph F. 1990. *A History of Medieval Spain*. Ithaca, NY: Cornell University Press.

Oliveira, António Resende de. 1994. *Depois do espectáculo trovadoresco: a estrutura dos cancioneiros peninsulares e as recolhas dos séculos XIII e XIV*. Lisboa: Edições Colibri.

Olney, James. 1998. *Memory and Narrative: The Weave of Life-Writing*. Chicago: University of Chicago Press.
Ordenações Manuelinas. 1984. Vol. 2. Coimbra, 1797. Reprint. Lisboa: Fundação Calouste Gulbekian.
Orejudo, Antonio, ed. 1993. *Cartas de batalla*. Barcelona: PPU.
Orlandelli, Gianfranco. 1965. "Genesi dell'ars notariae nel secolo XIII." *Studi Medievali Ser. 3* 6: 329–66.
Ortúñez de Calahorra, Diego. 1975. *Espejo de príncipes y cavalleros*. Vol. 6. Ed. Daniel Eisenberg. Madrid: Espasa-Calpe.
Ostos Salcedo, Pilar. 1994. "Diplomática notarial en la época colombina: fases de redacción y forma documental." In Piergiovanni: 189–212.
Ostos Salcedo, Pilar. 1998. "Aranceles notariales de Córdoba (1482–1495)." *Historia, Instituciones, Documentos* 25: 503–24.
Ostos Salcedo, Pilar. 2005. *Notariado, documentos notariales y Pedro González de Hoces, veinticuatro de Córdoba*. Sevilla: Universidad de Sevilla; Córdoba: Universidad de Córdoba.
Ostos Salcedo, Pilar, and María Luisa Pardo Rodríguez, eds. 1995. *El notariado andaluz en el tránsito de la Edad Media a la Edad Moderna: I Jornadas sobre el Notariado en Andalucía, del 23 al 25 de febrero de 1994*. Sevilla: Ilustre Colegio Notarial de Sevilla.
Pacheco Sampedro, Rogelio. 1997. "El signum manuum en el cartulario del monasterio de San Juan de Caaveiro (s. IX–XIII)." *Signo. Revista de historia de la cultura escrita* 4: 27–37.
Paden, William D. 2004–5. "Roll versus Codex: The Testimony of Roll Cartularies." *Rivista di studi testuali* 6–7: 153–90.
Paden, William D. 2011. "Lyrics on Rolls." In *"Li premerains vers": Essays in Honor of Keith Busby*, ed. Catherine M. Jones and Logan E. Whalen, 325–40. Amsterdam-New York: Rodopi.
Palencia, Alfonso de. 1967. *Universal vocabulario en latín y en romance*. Facsim. Madrid: Comisión Permanente de la Asociación de Academias de la Lengua Española.
Palencia, Alfonso de. 1998. *Gesta Hispaniensia: ex annalibus suorum dierum collecta*. Vol. 1. Ed. Robert Brian Tate and Jeremy Lawrance. Madrid: Real Academia de la Historia.
Palencia, Alfonso de. 1998, 1909. *Guerra de Granada*. Facsim. Trans. Antonio Paz y Meliá. Intr. Rafael Gerardo Peinado Santaella. Granada: Universidad de Granada.
Palencia Flores, Clemente. 1943. *El poeta Gómez Manrique, corregidor de Toledo; discurso leído el día 28 de marzo de 1943, en la recepción pública de la Real Academia de Bellas Artes y Ciencias Históricas de Toledo*. Toledo: Editorial Católica Toledana.

Palenzuela Domínguez, Natalia. 2003. *Los mercaderes burgaleses en Sevilla a fines de la Edad Media.* Sevilla: Universidad de Sevilla.

Pardo Rodríguez, María Luisa. 1992. "Notariado y monarquía: los escribanos públicos de la ciudad de Sevilla en el reinado de los Reyes Católicos." *Historia, Instituciones, Documentos* 19: 317–26.

Pardo Rodríguez, María Luisa. 1994. "Notariado y cultura en la época colombina." In Piergiovanni: 147–86.

Pardo Rodríguez, María Luisa. 2000. "La escribanía mayor del Concejo de Sevilla en la Edad Media." In *La diplomatique urbaine en Europe au Moyen Age: actes du congrès de la Commission internationale de diplomatique, Gand, 25–9 août 1998,* ed. Walter Prevenier and Thérèse de Hemptinne, 357–81. Leuven: Garant.

Pardo Rodríguez, María Luisa. 2002. *Señores y escribanos: el notariado andaluz entre los siglos XIV y XVI.* Sevilla: Universidad de Sevilla.

Parkes, Malcolm B. 1991a. "The Influence of the Concepts of *Ordinatio* and *Compilatio* on the Development of the Book." In Parkes 1991b: 35–70.

Parkes, Malcolm B. 1991b. *Scribes, Scripts and Readers: Studies in the Communication, Presentation and Dissemination of Medieval Texts.* London/Rio Grande, OH: Hambledon.

Parkes, Malcolm B., and A.I. Doyle. 1991. "The Production of Copies of the *Canterbury Tales* and the *Confessio Amantis* in the Early Fifteenth Century." In Parkes 1991b: 201–48.

Parkhurst, Charles P. 1941. "The Madonna of the Writing Christ Child." *The Art Bulletin* 23: 292–306.

Pascual Martínez, Lope. 1981. "Estudios de diplomática castellana. El documento privado y público en la Baja Edad Media: los escribanos." *Miscelánea Medieval Murciana* 8: 119–90.

Pascual Martínez, Lope. 1983. "El notariado en la Baja Edad Media: escribas y documentos (Cataluña, Valencia y Mallorca)." *Miscelánea Medieval Murciana* 10: 197–220.

Pasquini, E. 2003. "La 'vita nova' di Dante: autobiografia come 'memoria selettiva.'" In *In quella parte del libro de la mia memoria: verità e finzioni dell' "io" autobiografico,* ed. Francesco Bruni, 57–67. Venezia: Marsilio; Fondazione Giorgio Cini.

Paz y Meliá, Antonio. 1892. *Opúsculos literarios de los siglos XIV a XVI.* Madrid: Sociedad de Bibliófilos Españoles.

Penna, Mario, ed. 1959. *Prosistas castellanos del siglo XV.* Biblioteca de Autores Españoles, vol. 116. Madrid: Atlas.

Perea Rodríguez, Óscar. 2003. "El Cancionero de Baena como fuente historiográfica de la Baja Edad Media castellana: el ejemplo de Ruy López Dávalos." In Serrano Reyes 1: 293–333.

Perea Rodríguez, Óscar. 2005. "Enrique IV de Castilla en la poesía de cancionero: algún *afán* ignorado entre las *mil congoxas* conocidas." *Cancionero General* 3: 33–71.
Perea Rodríguez, Óscar. 2008. "Minorías en la España de los Trastámara (II): judíos y conversos." *eHumanista: Journal of Iberian Studies* 10: 353–468.
Perea Rodríguez, Óscar, and Raquel Madrid Souto. 2009. "Una efeméride lírico-mercantil: quinto centenario de la firma del contrato para la primera edición del *Cancionero general* (1509–2009)." *Cancionero General* 7: 71–93.
Pérez, Martín. 2002. *Libro de las confesiones: una radiografía de la sociedad medieval española*. Ed. Antonio García y García, Bernardo Alonso Rodríguez, and Francisco Cantelar Rodríguez. Madrid: Biblioteca de Autores Cristianos.
Pérez-Bustamante, Rogelio. 1974. "El juramento de los oficiales del reino de Castilla, 1252–1474." *Moneda y Crédito* 129: 211–27.
Pérez-Bustamante, Rogelio. 1985. *El registro notarial de Dueñas*. Palencia: Diputación Provincial de Palencia; Madrid: Fundación Matritense del Notariado.
Pérez-Bustamante, Rogelio, and Antonio Rodríguez Adrados. 1995. *Los registros notariales de Madrid, 1441–1445*. Madrid: Fundación Matritense del Notariado.
Pérez-Bustamante, Rogelio, and José Manuel Calderón Ortega. 1983. *El marqués de Santillana: biografía y documentación*. Santillana del Mar: Taurus.
Pérez Chozas, Ángel, Agustín Millares Carlo, and Eulogio Varela Hervías. 1932. *Documentos del Archivo General de la Villa de Madrid*. Madrid: Artes gráficas.
Pérez de Guzmán, Fernán. 1965. *Generaciones y semblanzas*. Ed. Robert Brian Tate. London: Tamesis.
Pérez de Salinas, Ochoa. 1980. *Libro mayor del "Banquero de Corte" de los Reyes Católicos, Ochoa Pérez de Salinas (1498–1500)*. Ed. Amando Represa Rodríguez. Intr. Felipe Ruiz Martín. Bilbao: Banco de Bilbao.
Pérez Priego, Miguel Ángel. 1992. "Historia y literatura en torno al príncipe D. Juan: la Representación sobre el poder del amor de Juan del Encina." In *Historias y ficciones: coloquio sobre la literatura del siglo XV. Actas del Coloquio Internacional organizado por el Departament de Filologia Espanyola de la Universitat de València, celebrado en Valencia los días 29, 30 y 31 de octubre de 1990*, ed. Rafael Beltrán Llavador, José Luis Canet Vallés, and Josep Lluís Sirera, 337–49. València: Universitat de València.
Pérez Priego, Miguel Ángel. 1997. "Noticia sobre Alonso Ramírez de Villaescusa, su Espejo de corregidores y el Directorio de príncipes." In Lucía Megías 2: 1169–78.
Pérez Priego, Miguel Ángel, ed. 1999. *Poesía lírica*. Íñigo López de Mendoza marqués de Santillana. Madrid: Cátedra.

Pérez Priego, Miguel Ángel. 2005. "Los versos españoles a Lucrezia Borgia y sus damas." In Baldissera and Mazzocchi: 313–24.
Pescador, Carmela. 1961–4. "La caballería popular en León y Castilla". *Cuadernos de Historia de España* 33–4: 101–238; 35–6: 56–201; 37–8: 88–198; 39–40: 169–260.
Pescador del Hoyo, María del Carmen. 1972–3. "Aportaciones al estudio de Juan Álvarez Gato." *Anuario de Estudios Medievales* 8: 305–47.
Petrucci, Armando, ed. 1963. *Il protocollo notarile di Coluccio Salutati (1372–1373)*. Milano: Giuffrè.
Petrucci, Armando. 1992. "Dalla minuta al manoscritto d'autore." In *Lo spazio letterario del medioevo. 1: Il medioevo latino.* Vol. 1: *La produzione del testo*, ed. Guglielmo Cavallo, Claudio Leonardi, and Enrico Menestò, 353–72. Roma: Salerno.
Petrucci, Armando. 1995. *Writers and Readers in Medieval Italy: Studies in the History of Written Culture*. New Haven, CT: Yale University Press.
Petrucci, Armando. 1998. *Writing the Dead: Death and Writing Strategies in the Western Tradition*. Stanford, CA: Stanford University Press.
Pezzarossa, Fulvio. 1979. "La memorialista fiorentina tra Medioevo e Rinascimento." *Lettere Italiane* 31: 96–138.
Phillips, William D. 1978. "State Service in Fifteenth-Century Castile. A Statistical Study of the Royal Appointees." *Societas: A Review of Social History* 8: 115–36.
Phillips, William D. 1986. "University Graduates in Castilian Royal Service." In *Estudios en homenaje a Don Claudio Sánchez Albornoz en sus 90 años*, 4: 475–90. Buenos Aires: Instituto de Historia de España.
Philobiblon. http://sunsite.berkeley.edu/Philobiblon
Picone, Michelangelo, and Claude Cazalé-Bérard, eds. 1998. *Gli zibaldoni di Boccaccio: memoria, scrittura, riscrittura: atti del seminario internazionale di Firenze-Certaldo, 26–8 aprile 1996*. Firenze: F. Cesati.
Piergiovanni, Vito, ed. 1994. *Tra Siviglia e Genova: notaio, documento e commercio nell'età colombiana: atti del Convegno internazionale di studi storici per le celebrazioni colombiane, Genova, 12–14 marzo 1992*. Milano: A. Giuffrè.
Pike, Ruth. 1972. *Aristocrats and Traders: Sevillian Society in the Sixteenth Century*. Ithaca, NY: Cornell University Press.
Pino Rebolledo, Fernando. 1972. *Diplomática Municipal. Reino de Castilla 1474–1520*. Valladolid: Universidad de Valladolid.
Pino Rebolledo, Fernando. 1991. *Tipología de los documentos municipales (siglos XII–XVII)*. Valladolid: Universidad de Valladolid - Asociación para la Defensa y Conservación de los Archivos.

Planas, Josefina. 1998. *El esplendor del gótico catalán: la miniatura a comienzos del siglo XV*. Lleida: Universitat de Lleida.
Plaza Cuervo, M. Teresa. 1995. "Notas para una edición crítica del *Cancionero de Gallardo o San Román*." In *Medievo y Literatura. Actas del V Congreso de la Asociación Hispánica de Literatura Medieval* (Granada, 27 septiembre–1 octubre 1993), ed. Juan Paredes Núñez, 4: 75–84. Granada: Universidad de Granada.
Poe, Elizabeth Wilson. 1980. "Old Provençal Vidas as Literary Commentary." *Romance Philology* 33: 510–18.
Poe, Elizabeth Wilson. 2000. *Compilatio: Lyric Texts and Prose Commentaries in Troubadour Manuscript H (Vat. Lat. 3207)*. Lexington, KY: French Forum.
Polo Martín, Regina. 1999. *El régimen municipal de la Corona de Castilla durante el reinado de los Reyes Católicos (organización, funcionamiento y ámbito de actuación)*. Madrid: Colex.
Ponce Cárdenas, Jesús. 2005. "Sobre el poeta burlesco García de Astorga: una hipótesis cronológica y dos notas léxicas." In Baldissera and Mazzocchi: 165–77.
Porras Arboledas, Pedro Andrés. 2003. "Una actuación de la Inquisición cordobesa. Las penitencias pecuniarias de 1533–1538." In Serrano Reyes 2: 375–418.
Portugal, Francisco de. 1943. *Arte de galantaria*. Ed. Joaquim Ferreira. Pôrto: Domingos Barreira.
Powers, James F. 1988. *A Society Organized for War: The Iberian Municipal Militias In the Central Middle Ages, 1000–1284*. Berkeley: University of California Press.
Prieto Cantero, Amalia. 1970. "Cartas autógrafas de los Reyes Católicos (1474–1502)." In *Isabel la Católica en la opinión de españoles y extranjeros: siglos XV al XX*, ed. Vicente Rodríguez Valencia, 3: 59–147. Valladolid: Instituto "Isabel la Católica" de Historia Eclesiástica.
Prieto Escanciano, Eduardo, ed. 1997. *Testamentos de los Reyes de León (Libro de las Estampas)*. 2 vols. León: Universidad de León.
Puig, Isidre. 2005. *Jaume Ferrer II: pintor de la Paeria de Lleida*. Lleida: Pagès-Institut Municipal d'Acció Cultural de Lleida.
Pulgar, [Fernando de]. 1943. *Crónica de los Reyes Católicos*. 2 vols. Ed. Juan de Mata Carriazo. Madrid: Espasa-Calpe.
Pulgar, [Fernando de]. 1982. *Letras*. Ed. Paola Elia. Pisa: Giardini.
Pulgar, [Fernando de]. 2007. *Claros varones de Castilla*. Ed. Miguel Ángel Pérez Priego. Madrid: Cátedra.
Quain, Edwin A. 1986. *The Medieval Accessus ad Auctores*. New York: Fordham University Press.

Quand le livre était manuscrit: présentation de l'Institut de recherche et d'histoire des textes. 1992. Paris-Orléans: C.N.R.S.-I.R.H.T.

Quintanilla Raso, María Concepción. 1999. "La renovación nobiliaria en la Castilla bajomedieval: entre el debate y la propuesta." In *La nobleza peninsular*: 255–96.

Quintanilla Raso, María Concepción, ed. 2006. *Títulos, grandes del reino y grandeza en la sociedad política. Fundamentos en la Castilla medieval.* Madrid: Sílex.

Rábade Obradó, María del Pilar. 1992. "Las lugartenencias de escribanías como conflicto: un ejemplo de la época de los Reyes Católicos." *Espacio, Tiempo y Forma, Serie III, Historia Medieval* 5: 211–28.

Rábade Obradó, María del Pilar. 1993. *Una élite de poder en la corte de los Reyes Católicos: los Judeoconversos.* Madrid: Sigilo.

Rábade Obradó, María del Pilar. 1994. "Un letrado en el Madrid del siglo XV: el escribano Alfonso González." In *Las diferentes historias de letrados y analfabetos: actas del congreso celebrado en Pastrana, 1 a 3 de julio, 1993*, ed. Carlos Sáez and Joaquín Gómez-Pantoja, 125–33. Alcalá de Henares: Universidad de Alcalá de Henares.

Rábade Obradó, María del Pilar. 1995a. "El acceso al oficio notarial en el siglo XV: la toma de posesión de Juan González de Madrid." *Anales del Instituto de Estudios Madrileños* 35: 361–88.

Rábade Obradó, María del Pilar. 1995b. "El origen de los archivos del Santo Oficio: una aproximación al valor histórico de las fuentes inquisitoriales." In *El Tratado de Tordesillas* 2: 751–9.

Rábade Obradó, María del Pilar. 1996. "Conversos, inquisición y criptojudaísmo en el Madrid de los Reyes Católicos." *Anales del Instituto de Estudios Madrileños* 36: 249–68.

Rábade Obradó, María del Pilar. 2002. "Los judeoconversos en tiempos de Isabel la Católica." In *Sociedad y economía en tiempos de Isabel La Católica: ponencias presentadas al II Simposio sobre el Reinado de Isabel la Católica, celebrado en las ciudades de Valladolid y Buenos Aires en el otoño de 2001*, ed. Julio Valdeón Baruque, 201–28. Valladolid: Ámbito.

Rábade Obradó, María del Pilar. 2008. "El entorno judeoconverso de la Casa y Corte de Isabel la Católica." In *Las relaciones discretas entre las monarquías hispana y portuguesa: las casas de las reinas (siglos XV-XIX)*, ed. J. Martínez Millán, and Mª P. Marçal Lourenço, 2: 887–917. Madrid: Polifemo.

Rada y Delgado, Juan de Dios de la. 1874. "Sepulcro de Don Juan II en la Cartuja de Miraflores de Burgos." *Museo español de antigüedades* 3: 293–324.

Ramírez de Arellano, Rafael. 1900a. "Antón de Montoro y su testamento." *Revista de Archivos, Bibliotecas y Museos* 4: 484–9.

Ramírez de Arellano, Rafael. 1900b. "Ilustraciones a la biografía de Antón de Montoro. El motín de 1473 y las ordenanzas de los aljabibes." *Revista de Archivos, Bibliotecas y Museos* 4: 723–35.

Raulston, Stephen Boykin. 1993. "*Cartas de batalla*: Literature and Law in Fifteenth-Century Spain." Unpubl. PhD diss., University of California, Berkeley.

Real Academia Española. Banco de datos (CORDE) [on line]. *Corpus diacrónico del español*. http://www.rae.es (accessed 15 May 2013).

Reed, R. 1972. *Ancient Skins, Parchments and Leathers*. London/New York: Seminar Press.

Resende, Garcia de. 1994. *Chronica de el-rei D. João II*. In *Livro das obras de Garcia de Resende*. Ed. Evelina Verdelho. Lisboa: Fundação Calouste Gulbenkian.

Los Reyes bibliófilos: junio-septiembre 1986, Biblioteca Nacional, Madrid. 1986. Madrid: Dirección General del Libro y Bibliotecas.

Reynolds, Catherine. 2003. "Illuminators and the Painter's Guilds." In *Illuminating the Renaissance: The Triumph of Flemish Manuscript Painting in Europe*, ed. Thomas Kren and Scot McKendrick, 14–33. Los Angeles, CA: J. Paul Getty Museum.

Ricoeur, Paul. 1992. *Oneself as Another*. Chicago: University of Chicago Press.

Riera i Viader, Sebastià, and Manuel Rovira i Solà. 2004. "Estudio histórico y codicológico." In *Llibre verd de Barcelona*, ed. Jaume Sobrequés i Callicó, Sebastià Riera i Viader, Manuel Rovira i Solà, and Tomàs de Montagut i Estragués, 167–214. Barcelona: Base – Ajuntament de Barcelona.

Riquer, Martín de. 1963. *Lletres de batalla; cartells de deseiximents i capítols de passos d'armes*. Vol. 1. Barcelona: Barcino.

Riquer, Martín de. 1995. *Vidas y retratos de trovadores*. Barcelona: Galaxia Gutenberg – Círculo de Lectores.

Riu-Barrera, Eduard, Albert Torra, and Alfred Pastor. 1999. *La Capilla de Santa Agueda del Palacio Real Mayor de Barcelona: historia y restauraciones*. Barcelona: Generalitat de Catalunya.

Rivera Garretas, María-Milagros. 2007. "Los testamentos de Juana de Mendoza, camarera mayor de Isabel la Católica, y de su marido el poeta Gómez Manrique, corregidor de Toledo (1493 y 1490)." *Anuario de Estudios Medievales* 37 (1): 139–80. http://dx.doi.org/10.3989/aem.2007.v37.i1.36.

Rizzo, Silvia. 1973. *Il lessico filologico degli umanisti*. Roma: Edizioni di storia e letteratura.

Robbins, R.H. 1939. "The 'Arma Christi' Rolls." *Modern Language Review* 34 (3): 415–21. http://dx.doi.org/10.2307/3717702.

Roberts, Colin H., and T.C. Skeat. 1983. *The Birth of the Codex*. London/New York: Published for the British Academy by the Oxford University Press.
Robinson, P.R. 1980. "The 'Booklet': A Self-Contained Unit in Composite Manuscripts." *Codicologica* 3: 46–69.
Rodríguez Adrados, Antonio. 1988. "La Pragmática de Alcalá, entre Las Partidas y la Ley del Notariado." In *Homenaje a Juan Berchmans Vallet de Goytisolo*, 7: 517–813. Madrid: Junta de Decanos de los Colegios Notariales de España.
Rodríguez-Moñino, Antonio. 1968. *Poesía y cancioneros (siglo XVI): discurso leído ante la Real Academia Española el día 20 de octubre de 1968 en su recepción pública*. Madrid: Real Academia Española.
Rodríguez Velasco, Jesús D. 1996. *El debate sobre la caballería en el siglo XV: la tratadística caballeresca castellana en su marco europeo*. Valladolid: Consejería de Educación y Cultura.
Rodríguez Velasco, Jesús D. 2010. *Order and Chivalry: Knighthood and Citizenship in Late Medieval Castile*. Philadelphia: University of Pennsylvania Press.
Rojas, Fernando de. 2000. *La Celestina; tragicomedia de Calisto y Melibea*. Ed. Dorothy S. Severin. Intr. Stephen Gilman. Madrid: Alianza.
Román, Comendador. 1990. *Coplas de la Pasión con la Resurrección*. Ed. Giuseppe Mazzocchi. Firenze: La Nuova Italia.
Romero Tallafigo, Manuel, Laureano Rodríguez Liañez, and Antonio Sánchez González. 1995. *Arte de leer escrituras antiguas: paleografía de lectura*. Huelva: Universidad de Huelva.
Root, Robert K. 1913. "Publication before Printing." *PMLA* 28 (3): 417–31. http://dx.doi.org/10.2307/457029.
Rosenwein, Barbara H. 1989. *To Be the Neighbor of Saint Peter: The Social Meaning of Cluny's Property, 909–1049*. Ithaca, NY: Cornell University Press.
Ross, Michael, and Roger Buehler. 2004. "Identity through Time: Constructing Personal Pasts and Futures." In *Self and Social Identity*, ed. Marilynn B. Brewer and Miles Hewstone, 25–51. Malden, MA: Blackwell.
Rouse, Mary A., and Richard H. Rouse. 1989. "The Vocabulary of Wax Tablets." In Weijers: 220–30.
Rouse, Mary A., and Richard H. Rouse. 1991. "Roll and Codex: The Transmission of the Works of Reinmar von Zweter." In *Authentic Witnesses: Approaches to Medieval Texts and Manuscripts*. Notre Dame, IN: University of Notre Dame Press. 13–29. Repr. from *Münchener Beiträge zur Mediävistik und Renaissance-Forschung* 32 (1982): 107–23.
Rouse, Richard. 1975. "La diffusion en Occident au XIIIe siècle des outils de travail facilitant l'accès aux textes autoritatifs." *Revue des Études Islamiques* 44: 115–47.

Rouse, Richard. 1992. "*Ordinatio* and *Compilatio* Revisited." In *Ad Litteram: Authoritative Texts and Their Medieval Readers*, ed. Mark D. Jordan and Kent Emery, Jr, 113–34. Notre Dame, IN: University of Notre Dame Press.

Rubio García, Luis. 1983. *Documentos sobre el Marqués de Santillana*. Murcia: Universidad de Murcia.

Rubió i Lluch, Antoni, and Albert Balcells. 2000, 1908. *Documents per a la història de la cultura catalana medieval*. 2 vols. Barcelona: Institut d'Estudis Catalans.

Rucquoi, Adeline. 1995. "Las oligarquías urbanas y las primeras burguesías en Castilla." In *El Tratado de Tordesillas* 1: 345–69.

Rucquoi, Adeline. 1997. *Valladolid en la Edad Media*. Vol. 2. Valladolid: Consejería de Educación y Cultura.

Rucquoi, Adeline. 2000. *Historia medieval de la Península Ibérica*. México: El Colegio de Michoacán.

Rufo Ysern, Paulina. 1991. "Extensión del régimen de corregidores en Andalucía en los primeros años del reinado de los Reyes Católicos." In López de Coca Castañer and Galán Sánchez: 55–75.

Ruiz, Juan, Arcipreste de Hita. 1992. *Libro de buen amor*. Ed. Alberto Blecua. Madrid: Cátedra.

Ruiz, Teófilo. 1994. *Crisis and Continuity: Land and Town in Late Medieval Castile*. Philadelphia: University of Pennsylvania Press.

Ruiz García, Elisa. 1999. "El poder de la escritura y la escritura del poder." In *Orígenes de la Monarquía Hispánica: propaganda y legitimación (ca 1400–1520)*, ed. José Manuel Nieto Soria, 275–313. Madrid: Dykinson.

Ruiz García, Elisa. 2006. "*Rex scribens*: discursos de la conflictividad en Castilla (1230–1350)." In Nieto Soria: 359–422.

Rumeu de Armas, Antonio. 1974. *Itinerario de los Reyes Catolicos 1474–1516*. Madrid: CSIC, Instituto Jerónimo Zurita.

Rumeu de Armas, Antonio. 1989. *Libro copiador de Cristóbal Colón: correspondencia inédita con los Reyes católicos sobre los viajes a América*. 2 vols. Madrid: Ministerio de Cultura – Testimonio.

Saenger, Paul. 1975. "Colard Mansion and the Evolution of the Printed Book." *Library Quarterly* 45 (4): 405–18. http://dx.doi.org/10.1086/620426.

Saenger, Paul. 1989. "Books of Hours and the Reading Habits of the Later Middle Ages." In *The Culture of Print: Power and the Uses of Print in Early Modern Europe*, ed. Roger Chartier, 141–73. Princeton, NJ: Princeton University Press.

Saenger, Paul. 1996. "The Impact of the Early Printed Page on the History of Reading." *Bulletin de Bibliophile* 2: 237–301.

Saenger, Paul. 1997. *Space between Words: The Origins of Silent Reading*. Stanford, CA: Stanford University Press.

Sáez, Carlos, ed. 2002a. *Libros y documentos en la Alta Edad Media. Los libros de derecho. Los archivos familiares.* Vol. 2: *Actas del VI Congreso Internacional de Historia de la Cultura Escrita.* Madrid: Calambur.

Sáez, Carlos. 2002b. "La sociedad visual: signos diplomáticos en la Corona de Aragón." In Sáez 2002a: 207–26.

Salinas, Pedro, and Jorge Guillén. 1992. *Correspondencia (1923–1951).* Ed. Andrés Soria Olmedo. Barcelona: Tusquets.

Sánchez León, Pablo. 1998. *Absolutismo y comunidad: los orígenes sociales de la guerra de los comuneros de Castilla.* Madrid: Siglo Veintiuno de España.

Sánchez Mariana, Manuel. 1987. "Los manuscritos poéticos del Siglo de Oro." *Edad de Oro* 6: 201–13.

Sánchez Mariana, Manuel. 1988. "La ejecución de los códices en Castilla en la segunda mitad del siglo XV." In *El libro antiguo español. Actas del primer coloquio internacional (Madrid, 18 al 20 de diciembre de 1986)*, ed. María Luisa López-Vidriero and Pedro M. Cátedra, 317–44. Salamanca: Universidad de Salamanca, Biblioteca Nacional de Madrid and Sociedad Española de Historia del Libro.

Sánchez Mariana, Manuel. 1994. *Introducción al libro manuscrito.* Madrid: Arco Libros.

Sánchez Mariana, Manuel. 1996. "El libro en la Baja Edad Media. Reino de Castilla. " In *Historia ilustrada del libro español.* Vol. 1: *Los manuscritos*, ed. Hipólito Escolar Sobrino, 165–221. Madrid: Fundación Germán Sánchez Ruipérez.

Sánchez Prieto, Ana Belén. 2001. *La casa de Mendoza hasta el tercer Duque del Infantado, 1350–1531. El ejercicio y alcance del poder señorial en la Castilla bajomedieval.* Madrid: Palafox y Pezuela.

Sánchez Saus, Rafael. 1995. "Garcí Sánchez de Arauz, Jurado de Sevilla y autor de los 'Anales.'" *Archivo hispalense: Revista histórica, literaria y artística* 78: 163–70.

Sánchez Saus, Rafael. 1998. "Sevillian Medieval Nobility: Creation, Development and Character." *Journal of Medieval History* 24 (4): 367–80. http://dx.doi.org/10.1016/S0304-4181(98)00012-8.

Sánchez Saus, Rafael. 2004. "Los patriciados urbanos." *Medievalismo: Boletín de la Sociedad Española de Estudios Medievales* 14: 143–56.

Sanchis y Sivera, José. 1929. "Pintores medievales en Valencia." *Archivo de Arte Valenciano* 15: 3–65.

Sanchis y Sivera, José. 1930. "Bibliología valenciana medieval." *Anales del Centro de Cultura Valenciana* 5: 33–56.

Santa Cruz, Alonso de. 1951. *Crónica de los Reyes Católicos.* 2 vols. Ed. Juan de Mata Carriazo. Sevilla: Escuela de Estudios Hispano-Americanos.

Santillana, Íñigo López de Mendoza, marqués de. 1988. *Obras completas.* Ed. Ángel Gómez Moreno and Maximilian P.A.M. Kerkhof. Barcelona: Planeta.
Sanz Fuentes, María Josefa. 1990. "Cancillería y cultura en la Castilla de los siglos XIV y XV." In *Cancelleria e cultura nel Medio Evo. Comunicazioni presentate nelle Giornate di studio della Commissione (Stoccarda, 29–30 agosto 1985). XVI Congresso Internazionale di Scienze Storiche,* ed. Germano Gualdo, 187–99. Città del Vaticano: Archivio Segreto Vaticano.
Sanz Hermida, Jacobo. 1996. "Entretenimiento femenino en la corte de Isabel de Castilla: el *Juego trobado* de Gerónimo de Pinar." In Menéndez Collera and Roncero López: 605–14.
Schaffer, Martha E. 1991. "Epigraphs as a Clue to the Conceptualization and Organization of the *Cantigas de Santa María*." *La Corónica* 19 (2): 57–88.
Ser Quijano, Gregorio del. 1998. *Documentación medieval en archivos municipales abulenses: Aldeavieja, Avellaneda, Bonilla de la Sierra, Burgohondo, Hoyos del Espino, Madrigal de las Altas Torres, Navarredonda de Gredos, Riofrío, Santa Cruz de Pinares y El Tiemblo.* Ávila: Ediciones de la Institución Gran Duque de Alba - Ediciones de la Obra Cultural de Caja de Ahorros.
Serrano Larráyoz, Fernando, and Margarita Velasco Garro. 2010. "Inventario de los hostales reales (siglos XIV–XV): fuentes contables del Archivo Real y General de Navarra." *Príncipe de Viana* 71 (250): 375–500.
Serrano Reyes, Jesús L., ed. 2003. *Cancioneros en Baena. Actas del II Congreso Internacional "Cancionero de Baena." In memoriam Manuel Alvar.* Vol. 1. Baena: Ayuntamiento de Baena.
Serrano Reyes, Jesús, and Juan Fernández Jiménez, eds. 2001. *Juan Alfonso de Baena y su cancionero. Actas del I Congreso Internacional sobre el Cancionero de Baena, Baena, del 16 al 20 de febrero de 1999.* Baena: Ayuntamiento de Baena – Diputación de Córdoba.
Serrano y Sanz, Manuel. 1930. "El archivo colombino de la Cartuja de las Cuevas. Estudio histórico y bibliográfico" *Boletín de la Real Academia de la Historia* 97: 145–256, 534–637.
Serverat, Vincent. 2001. "Royauté et chevalerie: sur un débat *cancioneril* méconnu (MN 24, 105–108)." In Martin: 101–28.
Sesma Muñoz, José Ángel. 1987. *El establecimiento de la inquisición en Aragón (1484–1486): documentos para su estudio.* Zaragoza: Institución Fernando el Católico.
Severin, Dorothy Sherman, ed. 1990. *El Cancionero de Oñate-Castañeda.* Madison, WI: Hispanic Seminary of Medieval Studies.
Severin, Dorothy Sherman, ed., with Fiona Maguire. 2000. *Two Spanish Songbooks: The "Cancionero Capitular de la Colombina" (SV2) and the*

"Cancionero de Egerton" (LB3). Liverpool: Liverpool University Press – Seville: Institución Colombina.

Sherman, William H. 2008. *Used Books: Marking Readers in Renaissance England*. Philadelphia: University of Pennsylvania Press.

Sierra Macarrón, Leonor. 2002. "El aumento de la producción escrita en los tumbos del monasterio de Sobrado de los Monjes (siglos IX–XIII)." In Sáez 2002a: 119–31.

Silva y Verástegui, Soledad de. 1988. *La miniatura medieval en Navarra*. Pamplona: Gobierno de Navarra.

Silva y Verástegui, Soledad de. 1989. "El tema de la Crucifixión en los manuscritos jurídicos medievales." *Cuadernos de arte e iconografía* 2: 159–65.

Sirat, Colette. 2002. *Hebrew Manuscripts of the Middle Ages*. Cambridge/New York: Cambridge University Press.

Sirat, Colette. 2006. *Writing as Handwork: A History of Handwriting in Mediterranean and Western Culture*. Turnhout: Brepols.

Smith, Sidonie. 1987. *A Poetics of Women's Autobiography: Marginality and the Fictions of Self-Representation*. Bloomington: Indiana University Press.

Smith, Sidonie, and Julia Watson. 2001. *Reading Autobiography: A Guide for Interpreting Life Narratives*. Minneapolis: University of Minnesota Press.

Las sociedades urbanas en la España Medieval: XXIX Semana de Estudios Medievales, Estella, 15 a 19 de julio de 2002. 2003. Pamplona: Departamento de Educación y Cultura.

Solana Villamor, María Concepción. 1962. *Cargos de la casa y corte de los Reyes Catolicos*. Valladolid: Universidad de Valladolid.

Spiegel, Gabrielle M. 1997. *The Past as Text: The Theory and Practice of Medieval Historiography*. Baltimore: Johns Hopkins University Press.

Stallybrass, Peter. 2002. "Books and Scrolls: Navigating the Bible." In *Books and Readers in Early Modern England: Material Studies*, ed. Jennifer Lotte Andersen and Elizabeth Sauer, 42–79. Philadelphia: University of Pennsylvania Press.

Steinberg, Justin. 2007. *Accounting for Dante: Urban Readers and Writers in Late Medieval Italy*. Notre Dame, IN: University of Notre Dame Press.

Steiner, Emily. 2003. *Documentary Culture and the Making of Medieval English Literature*. Cambridge/New York: Cambridge University Press.

Stemmler, Theo. 1991. "Miscellany or Anthology? The Structure of Medieval Manuscripts: Ms Harley 2253, for Example." *Zeitschrift für Anglistik und Americanistik* 39: 231–7.

Storey, H. Wayne. 1993. *Transcription and Visual Poetics in the Early Italian Lyric*. New York: Garland.

Street, Florence. 1953. "La vida de Juan de Mena." *Bulletin Hispanique* 55 (2): 149–73. http://dx.doi.org/10.3406/hispa.1953.3354.
Sturm-Maddox, Sara. 1980. "Transformations of Courtly Love Poetry: *Vita Nuova* and *Canzoniere*." In *The Expansion and Transformations of Courtly Literature*, ed. Nathaniel B. Smith and Joseph T. Snow, 128–40. Athens: University of Georgia Press.
Talavera, Fray Hernando de. 1930. *Instrucciónde fray Fernando de Talavera para el régimen interior de su palacio*. Ed. Jesús Domínguez Bordona. Madrid: Tip. de Archivos.
Tato, Cleofé. 1997. "Cronología de una serie poética en elogio de Alfonso V incluida en el *Cancionero de Palacio* (SA7)." In *"Quien hubiese tal ventura:" Medieval Hispanic Studies in Honour of Alan Deyermond*, ed. Andrew M. Beresford, 299–308. London: Department of Hispanic Studies, Queen Mary and Westfield College, University of London.
Tato, Cleofé. 2003. "El Cancionero de Palacio (SA7), ms. 2653 de la Biblioteca Universitaria de Salamanca (I)." In Serrano Reyes 1: 495–523.
Tato, Cleofé. 2005a. "Huellas de un cancionero individual en el *Cancionero de Palacio* (SA7)." In *Los cancioneros españoles: materiales y métodos*, ed. Manuel Moreno and Dorothy Sherman Severin, 59–89. London: Department of Hispanic Studies, Queen Mary, University of London.
Tato, Cleofé. 2005b. "Sobre los cancioneros de autor: el caso de Pedro de Santa Fe." In Baldissera and Mazzocchi: 105–23.
Tavani, Giuseppe. 1969. *Poesia del Duecento nella Penisola Iberica: problemi della lirica galego-portoghese*. Roma: Ateneo.
Taylor, Andrew. 1991. "The Myth of the Minstrel Manuscript." *Speculum* 66 (1): 43–73. http://dx.doi.org/10.2307/2863947.
Taylor, Andrew. 2002. *Textual Situations: Three Medieval Manuscripts and Their Readers*. Philadelphia: University of Pennsylvania Press.
Taylor, Jane H. M. 2007. *The Making of Poetry: Late-Medieval French Poetic Anthologies*. Turnhout: Brepols.
Tendilla, Íñigo López de Mendoza, conde de. 1973–4. *Correspondencia del Conde de Tendilla*. Ed. Emilio Meneses García. Madrid: Real Academia de la Historia.
Tendilla, Íñigo López de Mendoza, conde de. 1996. *Epistolario del Conde de Tendilla (1504–1506)*. 2 vols. Ed. José Szmolka Clares. Transcrip. Mª Amparo Moreno Trujillo, and Mª José Osorio Pérez. Granada: Universidad de Granada.
Thornton, Dora. 1997. *The Scholar in His Study: Ownership and Experience in Renaissance Italy*. New Haven, CT: Yale University Press.
Toro Miranda, Rosa María de. 2001. *Colección diplomática de Santa Catalina de Monte Corbán, 1299–1587*. 2 vols. Santander: Fundación Marcelino Botín.

Torras i Cortina, Miquel, ed. 2003. *El manual del notari Pere Pau Solanelles de l'escrivania pública d'Igualada, 1475–1479*. 2 vols. Lleida: Pagès.
Torre, Alfonso de la. 1991. *Visión deleytable*. 2 vols. Ed. Jorge García López. Salamanca: Universidad de Salamanca.
Torre y del Cerro, Antonio de la. 1950. *Documentos sobre relaciones internacionales de los Reyes Católicos*. Vol. 2. Barcelona: CSIC.
Torre y del Cerro, Antonio de la. 1954. *La casa de Isabel la Católica*. Madrid: CSIC.
Torre y del Cerro, Antonio de la. 1968. *Testamentaría de Isabel la Católica*. Valladolid: [s. n.].
Torre y del Cerro, Antonio de la, and Engracia Alsina de la Torre, eds. 1955–6. *Cuentas de Gonzalo de Baeza, tesorero de Isabel la Católica*. 2 vols. Madrid: CSIC.
Torres Fontes, Juan. 1953. *Itinerario de Enrique IV de Castilla*. Murcia: CSIC, Seminario de Historia de la Universidad de Murcia.
Torres Fontes, Juan. 1966. "La incorporación a la caballería de los judíos murcianos en el siglo XV." *Murgetana* 27: 5–14.
Torres Sanz, David. 1982. *La administración central castellana en la Baja Edad Media*. Valladolid: Universidad de Valladolid.
El Tratado de Tordesillas y su época. 1995. 3 vols. Valladolid: Sociedad V Centenario del Tratado de Tordesillas.
Triggiano, Tonia Bernardi. 1999. "Piety among Women of Central Italy (1300–1600): A Critical Edition and Study of Battista da Montefeltro-Malatesta's Poem in Praise of Saint Jerome." Unpubl. PhD diss., University of Wisconsin – Madison.
El Tumbo de los Reyes Católicos del Concejo de Sevilla. 1968. 5 vols. Ed. Ramón Carande and Juan de Mata Carriazo. Sevilla: Fondo para el Fomento de la Investigación en la Universidad.
Turner, Eric G. 1978. *The Terms Recto and Verso: The Anatomy of the Papyrus Roll*. Bruxelles: Fondation égyptologique Reine Élisabeth.
Tyssens, Madeleine, ed. 1991. *Lyrique romane médiévale: La Tradition des chansonniers: Actes du Colloque de Liège, 1989*. Liège: Faculté de philosophie et lettres de l'Université de Liège.
Ubieto Arteta, Antonio. 1991. *Orígenes de los reinos de Castilla y Aragón*. Zaragoza: Universidad de Zaragoza.
Val Valdivieso, María Isabel del. 1995. "Dinámica social en las ciudades castellanas en torno a 1494." In *El Tratado de Tordesillas* 1: 113–30.
Val Valdivieso, María Isabel del. 1996. "Aspiraciones y actitudes socio-políticas. Una aproximación a la sociedad urbana de la Castilla bajomedieval." In *La ciudad medieval. Aspectos de la vida urbana en la Castilla bajomedieval*,

ed. Juan Antonio Bonachía Hernando, 213–54. Valladolid: Universidad de Valladolid.
Val Valdivieso, María Isabel del. 2001. "Élites urbanas en la Castilla del siglo XV (Oligarquía y común)." In *Elites e redes clientelares na Idade Média: problemas metodológicos: actas do colóquio*, ed. Filipe Themudo Barata, 71–89. Lisboa: Colibri; Evora: CIDEHUS-Universidade de Évora.
Val Valdivieso, María Isabel del. 2005. "Conflictividad social en la Castilla del siglo XV." *Acta historica et archaeologica mediaevalia* 26: 1033–50.
Val Valdivieso, María Isabel del. 2006. "La identidad urbana al final de la Edad Media." *Anales de historia medieval de la Europa atlántica: AMEA* 1: 5–28.
Val Valdivieso, María Isabel del. 2007. "Élites populares urbanas en la época de Isabel I de Castilla." In Challet: 33–48.
Valdeón Baruque, Julio. 1990. "Las oligarquías urbanas." In *Concejos y ciudades en la Edad Media hispánica: II Congreso de Estudios Medievales*. Ávila: Fundación Sánchez-Albornoz. 507–21.
Valdeón Baruque, Julio. 1995. "La conflictividad social en Castilla." In *El Tratado de Tordesillas* 1: 315–24.
Valdeón Baruque, Julio. 2007. "La nobleza y las ciudades en tiempos de Isabel I." In Challet: 21–31.
Valera, Diego de. 1927. *Crónica de los Reyes Católicos*. Ed. Juan de Mata Carriazo. Madrid: Junta para la ampliación de estudios, Centro de Estudios Históricos.
Vallejo, Juan de. 1913. *Memorial de la vida de fray Francisco Jiménez de Cisneros*. Ed. Antonio de la Torre y del Cerro. Madrid: Impr. Bailly-Bailliere.
Valls i Subirà, Oriol. 1978–82. *The History of Paper in Spain*. 3 vols. Madrid: Empresa Nacional de Celulosas.
Vaquero Serrano, María del Carmen. 2005. *Fernán Alvarez de Toledo, secretario de los Reyes Católicos: genealogía de la toledana familia Zapata*. Toledo: Mª del Carmen Vaquero.
Varela, Consuelo, ed. 1987. *Documentos colombinos en la Casa de Alba*. 2 vols. Sevilla: Diputación Provincial de Sevilla; Madrid: Testimonio.
Varela, Consuelo. 1992. *Cristóbal Colón: retrato de un hombre*. Madrid: Alianza.
Vega Carpio, Lope de. 1988. *La Dorotea*. Ed. Edwin S. Morby. Madrid: Castalia.
Velázquez Soriano, Isabel, and Manuel Santonja Gómez, eds. 2005. *En la pizarra: los últimos hispanorromanos de la meseta: exposición*. Burgos: Instituto Castellano y Leonés de la Lengua.
Vendrell de Millás, Francisca. 1958a. "Las poesías inéditas de Juan de Dueñas." *Revista de Archivos, Bibliotecas y Museos* 64: 149–240.
Vendrell de Millás, Francisca. 1958b. "La posición del poeta Juan de Dueñas respecto a los judíos españoles de su época." *Sefarad* 18: 108–13.

Vendrell de Millás, Francisca. 1968. "Retrato irónico de un funcionario converso." *Sefarad* 28: 40–4.

Vicente, Gil. 1983. *Copilaçam de todalas obras.* 2 vols. Ed. Maria Leonor Carvalhão Buescu. Lisboa: Imprensa Nacional-Casa da Moeda.

Villalba Dávalos, Amparo. 1964. *La miniatura valenciana en los siglos XIV y XV.* Valencia: Diputación Provincial.

Villena, Enrique de. 1994. *Obras completas.* Vol. 2. Ed. Pedro M. Cátedra. Madrid: Turner.

Viñas Mey, Carmelo, and Ramón Paz Remolar, eds. 1949. *Relaciones histórico-geográfico-estadísticas de los pueblos de España hechas por iniciativa de Felipe II.* Vol. 1. *Provincia de Madrid.* Madrid: CSIC.

Viñayo González, Antonio, and Etelvina Fernández González. 1985. *Abecedario-bestiario de los códices de Santo Martino.* León: Isidoriana.

Vives, Juan Luis. 1971. *Vives: On Education; A Translation of the "De tradendis disciplinis" of Juan Luis Vives.* Intr. Foster Watson. Totowa, NJ: Rowman and Littlefield.

Vizcaíno Villanueva, María A. 1991–2. "La iglesia de San Salvador en el antiguo Madrid (I)." *Anales de historia del arte* 3: 143–58.

Vozzo Mendia, Lia. 1995. "La lirica spagnola alla corte napoletana di Alfonso d'Aragona: note su alcune tradizioni testuali." *Revista de Literatura Medieval* 7: 173–86.

Walkowitz, Judith R., Myra Jehlen, and Bell Chevigny. 1989. "Patrolling the Borders: Feminist Historiography and the New Historicism." *Radical History Review* 43: 23–43.

Wallensköld, Axel. 1917. "Le ms. Londres, Bibliothèque de Lambeth Palace, misc. rolls 1435." *Mémoires de la Société Néophilologique de Helsingfors* 6: 3–40.

Weijers, Olga, ed. 1989. *Vocabulaire du livre et de l'écriture au Moyen Age: actes de la table ronde, Paris, 24–26 septembre 1987.* Turnhout: Brepols.

Weiss, Julian. 1990. *The Poet's Art: Literary Theory in Castile c 1400–60.* Oxford: The Society for Mediaeval Languages and Literature.

Weissberger, Barbara F. 2004. *Isabel Rules: Constructing Queenship, Wielding Power.* Minneapolis: University of Minnesota Press.

Weissberger, Barbara F. 2005. "Patronage and Politics in the Court of the Catholic Monarchs: The *Cancionero de Pedro Marcuello*." *Studies in Iconography* 26: 175–204.

Whetnall, Jane. 1997. "Unmasking the Devout Lover: Hugo de Urriés in the *Cancionero* de Herberay." *Bulletin of Hispanic Studies* 74: 275–98.

Wilkins, Ernest Hatch. 1951. *The Making of the "Canzoniere" and Other Petrarchan Studies.* Roma: Edizioni di Storia e Letteratura.

Williams, Raymond. 1995. *The Sociology of Culture*. Chicago: University of Chicago Press.
Williams, Sarah Jane. 1969. "An Author's Role in Fourteenth-Century Book Production: Guillaume de Machaut's 'Livre ou je met toutes mes choses.'" *Romania* 90: 433–54.
Winroth, Anders. 2000. *The Making of Gratian's "Decretum."* Cambridge: University of Cambridge; New York: Cambridge University Press.
Wright, Erik Olin. 2009. "Understanding Class. Towards an Integrated Analytical Approach." *New Left Review* 60: 101–16.
Yarza Luaces, Joaquín. 1993. *Los Reyes Católicos: paisaje artístico de una monarquía*. Madrid: Nerea.
Yarza Luaces, Joaquín. 2004. "La ilustración." In *Llibre verd de Barcelona*. 2 vols. Jaume Sobrequés i Callicó, Sebastià Riera i Viader, Manuel Rovira i Solà, Tomàs de Montagut i Estragués, and Joaquín Yarza Luaces. Barcelona: Base – Ajuntament de Barcelona. 257–318.
Yarza Luaces, Joaquín. 2005. *Isabel La Católica: promotora artística*. León: Edilesa.
Zabalza Duque, Manuel. 1998. *Colección Diplomática de los Condes de Castilla: edición y comentario de los documentos de los Condes Fernán González, García Fernández, Sancho García y García Sánchez*. Valladolid: Consejería de Educación y Cultura.
Zabbia, Marino. 1998. "La memoria domestica nella cronachistica notarile del Trecento." *Quellen und Forschungen aus Italienischen Archiven und Bibliotheken* 78: 123–40.
Zarco del Valle, M.R. 1870. *Documentos inéditos para la Historia de las Bellas Artes en España*. Madrid: Viuda de Calero.
Zumthor, Paul. 1993. *La mesure du monde: représentation de l'espace au Moyen Age*. Paris: Seuil.

Index

08AM: Fray Ambrosio Montesino, *Cancionero*, 194
11CG: *Cancionero general* de Hernando del Castillo, 1511, 67, 113, 130, 178, 179, 181, 183, 200, 206, 209, 227n, 267n
13UC: *Cancionero* de Urrea, 179
14CG: *Cancionero general* de Hernando del Castillo, 1514, 66, 67, 113, 178, 179, 209
16RE: *Cancioneiro geral* de Garcia de Resende, 65, 68, 74, 79, 149, 229n
19OB: *Cancionero de obras de burlas*, 209
96JE: *Cancionero* de Juan del Encina, 62, 171

account books, 60, 116, 125, 131, 132, 133, 144, 154, 155, 156, 161, 163, 248n, 258n, 259n
Afonso, Prince, 84
Afonso, Rodrigo, 84
Afonso V, King of Portugal, 142, 189, 193, 227n
Agraz, Juan, 155
Aguilar, don Alonso de, 86, 212
Aguilar, don Pedro de, 208, 210, 212

albalá(es), 101, 123, 125, 128, 129, 130, 131, 137, 170, 209, 212, 248n, 250n
Alexandre, 62
Alfons II, King of Aragon, 103
Alfons V, King of Aragon, 53, 106, 158, 229n, 265n
Alfonso, Prince, 80, 192, 193
Alfonso X, King of Castile and Leon, 21, 23, 40, 41, 62, 72, 79, 82, 85, 109–11, 114, 117, 135, 136, 137, 138, 149, 170, 174, 178, 188, 243n, 245n, 254n, 257n
Alfonso XI, King of Castile and Leon, 10, 17, 22
Álvarez de Toledo, Alfonso, 25
Álvarez de Toledo, Fernán, 25, 224n, 228n
Álvarez de Villasandino, Alfonso, 31, 46, 68, 93, 129, 130, 146, 176, 177, 181, 229n, 263n
Álvarez Gato, Juan, 10, 26, 48, 49, 53–4, 63, 64, 66, 69, 89, 93, 140, 146, 178–9, 184, 185, 189, 194, 195, 196–205, 213, 224n, 230n, 231n, 234n, 263n, 266–8nn
anales, 152, 158, 192

Arias Dávila, Diego, 50, 130, 193–4, 207, 244n, 249n
Arias Dávila, family, 53, 133, 139, 177, 244n, 254n, 263n
Arragel, Mosé, 41, 81, 96–97
ars notaria, 135
Astorga, García de, 44–5
Avellaneda, Beatriz de, Condesa de Castro, 130
Ávila, Alfonso de, 43, 228n
Ávila, Gómez de, Corregidor de Córdoba, 210, 212
Avís, Juan, 212

Baena, Juan Alfonso de, 15, 16, 26, 37, 38, 47, 48, 55, 56, 57, 63, 68, 96, 98, 100, 101, 129, 130, 140, 146, 149, 171, 176, 177, 180, 181, 183, 190, 195, 203, 226n, 229n, 232n, 234n, 249n, 254n
Baeza, Gonzalo de, 60, 131, 156, 244n
Barberino, Francesco da, 188
Barrientos, Lope de, 25, 235n
Barros, João de, 79
Beatriz, Princess, 84
Bembo, Pietro, 121, 122
Berceo, Gonzalo de, 115, 135, 174, 227n
Bernáldez, Andrés, 158
Biblia de Alba, or de Arragel (Alba Bible), 41, 78, 80–1, 96–7, 237n, 240n
Bible (Biblia), 78, 97–8, 240n
billete, 69
blank book(s), 9, 107, 115, 117, 119, 126, 127, 133, 136–40, 141, 147, 150, 153, 161, 252n, 255n, 259n
BM1: Cancionero catalán, 141, 253n
Boccaccio, Giovanni, 162, 168, 180
Bonastre, Bernat de, 71

book of memory, 148, 162, 260n
booklet(s), 6, 7, 9, 35, 70, 92, 99, 104, 105, 106, 109, 111, 114–18, 119, 120–2, 124, 128, 147, 150, 152, 153, 155, 161, 162, 164, 172, 174, 186, 239n, 242n, 244n, 245n, 246n, 259n
Borgia, Lucrezia, 121–2
borrador, 127, 132, 248n
Boscán, Juan, 90
bound book, 35, 99, 107, 114, 115, 116, 117, 118, 121, 122, 124, 132, 133, 134, 137, 138, 139, 140, 141, 146, 244n, 252n, 255n
BU1: *Jardinet d'orats*, 121
BU2: Works by Mena and others, 162
Burgos, Diego de, 38, 39, 227n
Burgos, Gonzalo de, 223n

caballería, 16, 18, 19, 20, 21, 28, 31, 48, 217, 231n
caballero(s), 12, 16, 17, 18, 19, 21, 22, 25, 26, 27, 28, 31, 33, 34, 46, 47, 48, 53, 54, 56, 66, 69, 89, 105, 155, 165, 191, 210, 212, 221n, 223n, 227n, 229n, 230n, 231n, 234n, 238n, 258n
Cabra, Pedro de la, 141
cabreo, 99
caderno(s), 114, 115
Caldes, Ramon de, 103, 104, 135, 168
Cancionero de Baena, 37, 57, 66, 120, 121, 149, 176, 177, 181, 183, 220n, 229n, 232n, 263n
Cançoner des Masdovelles, 184, 265n
Cançoner J, 140
Canellas, Vidal de, Bishop of Huesca, 104, 135, 242n, 251n
Cantigas de Santa María, 82, 109–11, 174, 178, 243n, 262n
Carbonell, Pere Miquel, 99, 140, 158
Cárcel de amor, 14, 64, 92

Carmina burana, 120, 172
Carrillo de Acuña, Alfonso, Archbishop of Toledo, 30, 46, 48, 120, 192, 194, 207, 211
Carrillo de Huete, Pedro, 25, 73, 235n
carta(s), 38, 43, 66, 67, 73, 74, 77, 88, 90, 93, 95, 96, 100, 101, 109, 116, 124, 125, 126, 129, 130, 136, 138, 142, 143, 144, 154, 169, 170, 185, 197, 212, 226n, 238n, 242n, 247n, 249n, 253n
Cartagena, 66
Cartagena, Alfonso de, 76, 153–4, 167, 182, 190
Cartagena, Teresa de, 181, 189
cartapacio(s), 87, 118–20, 246n
Cartapacio de Morán de la Estrella (MP1), 141
Cartapacio poético del Colegio Mayor de Cuenca, 121
cartapàs, 119, 246n
cartel(es), 74, 86, 238n
cartes, 99
cartiellas, 174
cartilla(s), 82, 101, 130, 171, 226n, 242n
cartipas, 119, 246n
cartulary(-ies), 79, 80, 99, 103–4, 134, 136, 148, 169, 236n, 242n, 250n, 251n
Castigos y Dotrinas que vn sabio daua a sus hijas, 178
Castilla, Juana de, 73
Castillo, Hernando del, 181, 182, 183, 206, 265n
Catalina, Queen of Castile, 225n
Catholic monarchs, 13, 23, 53, 60, 102, 112, 138, 158, 224n, 263n. See also Isabel I, Queen of Castile; Fernando II, King of Aragon
cédula, 19, 84, 124, 132, 137, 171, 226n, 242n, 247n

Celestina, 14, 49, 144, 184, 221n, 223n, 225n
Celso, Hugo de, 18, 138, 241n
Cervantes Saavedra, Miguel de, 47, 162
Cerverí de Girona, 184, 265n
CH1: *Cancionero de Pedro Marcuello*, 111–12
Chaucer, Geoffrey, 168
Chaves, Álvaro Lopes de, 84, 142
Cicero, 168
Cisneros, Francisco Jiménez de, Cardinal, 85, 250n
CO1: Cancionero de Coimbra, Works by Pedro Torrella and others, 162
Còdex de Cambridge, 121
coern(s), 114, 115, 119
Colocci-Brancuti (*Cancioneiro da Biblioteca Nacional*), 121
Colonne, Guido delle, 158
Columbus, Christopher (Cristóbal Colón), 68–9, 94, 95, 100, 132, 142, 225n, 236n, 249n, 250n, 253n
commonplace books, 162, 259n
compilador, 180, 181
compilatio, 109, 168, 179, 181, 261n
compilation(s), 9, 10, 38, 64, 66, 68, 70, 79, 90, 91, 92, 93, 96, 100, 101, 102, 103, 104, 116, 120, 121, 131, 135, 136, 140, 141, 144, 146, 147, 148, 160, 163, 164–86, 187–213, 227n, 245n, 254n, 262n, 265n, 266n, 268n
compiler(s), 9, 38, 68, 76, 135, 147, 149, 164, 168, 173, 176, 179, 180, 181, 182, 183, 184, 185, 186, 195, 202, 251n, 265n
componedor, 180, 181, 183, 190
común (commoners), 12, 27
Condestable de Portugal, 190

converso(s), 8, 12, 13, 16, 21, 26, 27, 29, 30, 35, 37, 41, 42–5, 46, 48, 49, 50, 51, 52, 53, 54, 55, 56, 57, 76, 86, 116, 130, 141, 154, 155, 157, 158, 167, 177, 189, 197, 198, 200, 203, 205, 210, 211, 222n, 223n, 224n, 228n, 230n, 232n, 247n, 259n
copilaçion, 89, 91, 170, 189, 190, 258n, 264n
Córdoba, Fernando de, 81
Correa, António, 79
Cota, Rodrigo de, 48, 55, 177, 263n
Covarrubias, Sebastián de, 100, 119, 238n, 258n
cuadernicos, 246n
cuadernillo(s), 115, 118, 244n, 246n
cuadernitos, 119
cuaderno(s), 6, 7, 9, 35, 75, 78, 86, 100, 101, 102, 105, 114–18, 119, 120, 122, 125, 126, 133, 139, 141, 146, 155, 160, 169, 226n, 239n, 242n, 244n, 245n, 246n, 247n, 248n

d'Abella, Berenguer, 71
Dante, 145, 163, 168, 188
Decembrio, Pier Candido, 76, 167, 183, 190
Delicado, Francisco, 55
Deschamps, Eustache, 178, 188
diary, 158, 161, 184
Díaz de Games, Gutierre, 18, 158
Díaz de Montalvo, Alfonso, 18, 35, 94, 102, 138, 170, 221n, 236n, 241n
Díaz de Toledo, Fernán, 30, 224n, 245n
Díaz de Toledo, Pero, 48, 193
Díaz de Vivar, Rodrigo (the Cid), 148
dietari, 158, 192
dietario(s), 199
diornal, 248n
Disputa de Elena y María, 118

diurnal, 127
drawing books, 153, 257n
Duarte I, King of Portugal, 114, 236n
Dueñas, Juan de, 45–6

Eiximenis, Francesc, 167
EM9: composite manuscript, two fragmentary cancioneros in one, 162
Encina, Juan del, 62, 91, 122, 155, 160, 171, 179, 185, 194, 268n
Enrique II, King of Castile, 10
Enrique III, King of Castile, 21, 45, 225n, 232n
Enrique IV, King of Castile, 10, 11, 14, 25, 28, 29, 34, 50, 53, 54, 78, 83, 116, 130, 156, 193, 203, 204, 212, 228n, 235n, 244n, 245n
Enríquez, don Enrique, 15, 157
Enríquez de Cabrera, Fadrique, Almirante de Castilla, 61, 67, 72, 165, 233n, 235n, 260n
Enríquez del Castillo, Diego, 159
envoltorio(s), 9, 92, 94–6, 97, 98, 101, 122, 139, 140, 143, 239n, 252n, 253n
escribanía(s), 29, 34, 35, 36, 37, 40, 125, 181
escribano(s), 8, 13, 15, 17, 18, 21, 22, 23, 24, 25, 26, 28, 29, 30, 33–46, 53, 55, 56, 58, 66, 74, 75, 77, 78, 83, 85, 86, 87, 88, 96, 100, 101, 102, 117, 118, 125, 126, 132, 137, 138, 139, 140, 149, 150, 156, 159, 161, 169, 170, 176, 180, 81, 192, 199, 205, 220n, 222n, 223n, 224n, 225n, 226n, 227n, 229n, 232n, 236n, 237n, 238n, 241n, 242n, 246n, 247n, 248n, 252n, 261n
escritura(s), 15, 59, 63, 66, 72, 89, 94, 100, 101, 137, 150, 152, 156, 181, 190, 226n, 257n

fair copy, 86, 89, 90, 123, 124, 125, 131, 134–6, 137, 138, 139, 140, 141, 142, 144, 146, 147, 148, 157, 158, 181, 205
Fernández de Heredia, Juan, 150
Fernández de Heredia, Juan (poet), 178, 263n
Fernández de Oviedo, Gonzalo, 94, 127, 132, 156, 233n
Fernández de Santaella, Rodrigo, 116, 166, 183, 226n, 241n
Fernando II, King of Aragon, the Catholic, 11, 12, 13, 14, 15, 17, 23, 24, 25, 28, 29, 35, 36, 45, 51, 52, 53, 54, 65, 83, 84, 85, 95, 102, 112, 127, 137, 138, 139, 140, 150, 151, 153, 157, 170, 192, 193, 204, 211, 212, 230n, 240n, 241n, 247n, 256n
Fernando de Antequera, 73
Ferrer, Vicente, 13, 40, 41, 52, 53, 82
Ferrera, Antón de, 226n
Flores, Juan de, 246n
florilegia, 162, 259n, 260n
Flors del gay saber, 178
foja(s), 123, 151, 241n, 245n, 248n
fuero(s), 40, 78, 100, 109, 126, 135, 237n, 241n

Galíndez de Carvajal, Lorenzo, 151, 153
García de Salazar, Lope, 39, 74, 161, 189, 235n
García de Santa María, Alvar, 120, 247n
Gómez de Sandoval, Diego, 129, 130
González, Martín, 37, 225n, 226n
González de Hoces, Pedro, 30, 118, 181, 225n
González de Mendoza, Pedro, Cardinal, 73, 227n, 263n
Gordonio, Bernardo de, 167, 261n

Gorricio, Gaspar de, 69, 142, 253n
Gower, John, 168, 188
Gringore, Pierre, 188, 265n
Guevara, 66, 208
Guillén, Jorge, 3
Guillén de Segovia, Pero, 30, 48
Guiraut Riquier, 169, 170, 188, 262n
Guittone d'Arezzo, 188

Harana, Rodrigo de, 68
HH1: *Cancionero de Oñate-Castañeda*, 206, 221n, 227n
hidalgo(s), 12, 25, 29, 33, 43, 45, 46, 48, 54, 228n, 229n, 231n
Hoccleve, Thomas, 188, 265n
hoja(s), 100, 109, 126, 151, 241n, 248n
Horozco, Sebastián de, 44, 51, 52, 160

index(es), 103, 128, 133, 135, 262n
Inquisition, 13, 40, 51, 52, 54, 57, 72, 74, 95, 97, 98, 118, 127, 132, 133, 139, 140, 150, 159, 160, 193, 200, 211, 228n, 230n, 231n, 235n, 244n, 252n, 254n
Isabel I, Queen of Castile, the Catholic, 10, 11, 12, 13, 14, 17, 23, 24, 25, 26, 28, 35, 36, 45, 51, 52, 53, 54, 60, 61, 64, 65, 73, 75, 80, 82, 83, 84, 88, 93, 100, 102, 106, 112, 114, 115, 116, 118, 125, 127, 131, 132, 133, 137, 138, 139, 140, 142, 150, 151, 152, 153, 157, 159, 160, 170, 177, 178, 185, 191, 192, 193, 199, 203, 204, 206, 207, 208, 210, 211, 212, 222n, 230n, 232n, 240n, 241n, 242n, 244n, 246n, 247n, 255n, 256n, 257n, 263n, 265n
Isabel of Aragon, Queen of Portugal, 115
Isidore of Seville, 168

Jaume I, King of Aragon, 104
Jaume II, King of Aragon, 111
Jerome, 99, 107, 108
Jiménez de Urrea, Pedro Manuel, 91, 179
Joan, Prince, 71
João II, King of Portugal, 83, 84, 85, 142, 150
João III, King of Portugal, 136
Juan, Prince, 118, 127, 132, 156, 200, 247n, 248n
Juan II, King of Castile, 10, 18, 21, 26, 28, 30, 34, 43, 45, 53, 55, 56, 57, 73, 77, 81, 95, 101, 116, 146, 154, 181, 193, 221n, 225n, 226n, 245n, 247n, 257n
Juana, Queen of Castile, 193

labrador(es), 19, 31, 44, 45
LB1: *Cancionero de Rennert*, 67
LB2: *Cancionero de Herberay*, 146, 184, 265n
LB3: *Cancionero de Egerton*, 129, 185, 205, 206, 207, 210, 211, 212
legajo, 94, 131, 133, 250n, 252n
Leitura Nova, 136
Lemaire, Jean, 188
Léon, Diego de, 90
Léon, Fray Luis de, 150
León, Pedro de, 155
Leonor, Queen of Portugal, 115
letra(s), 15, 36, 83, 85, 88, 90, 123, 136, 145, 152, 180, 203, 230n
letrados, 13, 20, 23, 50, 56, 57, 222n, 224n
libelli, 120
libelos, 74
Liber feudorum maior, 103, 135, 169
libramiento(s), 101, 125, 129, 130, 209, 210, 212, 247n, 249n

libre de memoria, 158
libre memorial, 161, 259n
librete, 118
librico, 128
librillo(s), 128, 246n, 257n
libritos, 153, 256n
libro cosido, 127
Libro de Alexandre, 227n
Libro de buen amor, 27, 188, 260n
libro de caja, 128
libro(s) de dibujar, 153, 257n
Libro de las cartas, 138
Libro de las Estampas or *Testamentos de los Reyes de León*, 104, 136, 251n
Libro de los privilegios, 128, 138, 139, 142
Libro de los testamentos, 104, 136, 251n
libro(s) de memoria, 148, 152, 153, 159
Libro de memorias y aniversarios del monasterio de San Pedro de Cardeña, 148
libro de traslados, 125, 142, 253n
libro (e) memoria, 148, 150, 154, 254n
libro e razo, 161
libro encuadernado, 138–9. *See also* bound book
libro horadado, 132. *See also* threaded book
libro mayor, 103, 127, 128, 132
linaje, 10, 20, 25, 41, 43, 55, 151, 152, 194, 255n, 256n. *See also* lineage
lineage, 10, 19, 20, 24, 25, 26, 28, 45, 50, 55, 152, 158, 189, 191, 192, 194, 195, 196, 228n, 229n, 230n, 231n, 255n, 256n
Llibre de franqueses i privilegis del Regne de Mallorca, 135
Llibre Verd de Barcelona, 111

López de Ayala, Pedro, 101, 136
López de Córdoba, Leonor, 156, 188
López de Haro, Diego, 67
Lucas de Iranzo, Miguel, 19, 25, 189, 222n
Lucas of Tuy, 237n
Lucena, Juan de, 16, 17, 90, 153, 239n
Lucero, Diego Rodríguez, 118, 200
Luna, don Álvaro de, 10, 16, 25, 38, 45, 46, 47, 93, 101, 188, 189, 229n
Luna, Pedro de, Archbishop of Toledo, 129

Machaut, Guillaume de, 120
manicula(e), 80, 111, 121, 237n
mano(s), 15, 16, 36, 38, 38, 45, 60, 61, 62, 63, 67, 69, 73, 77, 84, 86, 87, 88, 89, 90, 93, 100, 101, 114, 115, 119, 150, 151, 153, 155, 182, 191, 248n, 249n
Manrique, Ana, 191
Manrique, Gómez, 10, 20, 21, 26, 31, 39, 46, 47, 48, 50–3, 63, 66, 89, 130, 140, 142, 146, 181, 184, 189–96, 197, 202, 205, 207, 213, 221n, 230n, 264n, 266n
Manrique, Jorge, 89, 160, 203, 238n
Manrique, Rodrigo, 48, 95, 191
manual, 127, 230n, 248n
Manuel, don Juan, 89, 167, 261n, 263n
Manuel, doña Marina, 67
Manuel I, King of Portugal, 73, 79, 136
Manuel de Lando, Fernán, 56, 62
March, Ausias, 122
March, mossèn Pere, 71
María of Castile, Queen of Aragon, 105
Martí, Prince, 71
Martín Gaite, Carmen, 6, 7, 217

Martin Luther, 235n
Martin of Leon (San Martino), 80, 81, 97, 242n
Martínez de Burgos, Fernán, 38, 39, 66, 226n
Martínez de Burgos, Juan, 38, 66, 226n
Martínez de Toledo, Alfonso, 60, 74
master book, 127, 132
Mazuela, Juan de, 48, 193
ME1: *Cancionero de Módena*, 62, 146
mediano(s), 8, 10, 12, 14, 18, 21, 23, 25, 26, 27, 31, 33, 41, 48, 54, 55, 56, 205, 212, 213, 231n
Mejía de Jaén, Hernán, 93, 203
memorandum(-a), 85, 88, 94, 95, 118, 123, 143, 151, 156, 160, 188
memoria, 76, 123, 135, 148, 149, 150, 151, 152, 153, 154, 158, 159, 161, 163, 235n, 254n, 255n, 256n, 259n
memorial(es), 9, 84, 85, 95, 123, 139, 142, 143, 151, 156–158, 159, 160–1, 162, 163, 196, 226n, 248n, 255n, 257n, 258n, 259n
memorial book(s), 9, 162, 259n
memory, 8, 9, 10, 59, 76, 120, 123, 124, 135, 143, 145, 148–63, 172, 189, 195, 196, 197, 200, 205, 216, 218, 228n, 236n, 247n, 250n, 251n, 254n, 256n, 257n, 258n, 259n, 260n
memory book, 9, 148, 152, 153–4, 158, 159, 161–3, 218, 254n, 259n
Mena, Juan de, 16, 17, 31, 44, 53, 54, 56–7, 140, 153–4, 167, 168, 207, 209, 210, 221n, 232n
Mendoza, don Juan de, 61, 165, 233n, 260n
Mendoza, Fray Íñigo López de, 42, 48, 65, 141, 160, 234n
Mendoza, Juana de, 52, 191, 193

MH1: *Cancionero de San Román*, 41, 65, 100, 101, 113, 121, 141
MH2: Obras de Juan Álvarez Gato, 64, 66, 93, 178, 196–7, 199, 261n, 267n
Milán, Luis, 65, 72, 155
minuta, 134
Miralles, Melcior, 158, 259n
MM2: *Cancionero del Duque de Gor*, 226n
MN6: *Cancionero de Hixar*, five mss in one, 61, 63, 90, 146, 165, 253n
MN7: "Pues forzado me es partir," 141, 242n
MN8: Obras de Santillana, 113, 146
MN15: *Pequeño cancionero del Marqués de la Romana*, 121, 146
MN16: Mena, "Como por Dios la alta justicia," 38, 226n
MN17: *Cancionero de Gallardo*, 62
MN19: Copy of works by Pero Guillén and others, 62, 63, 65, 128, 129, 130, 162, 185, 206, 207, 208, 209, 210, 211, 249n
MN23: Notes by Floranes on MN33, 38, 226n
MN24: Obras de Gómez Manrique, 184, 189, 190, 265n, 266n
MN32: *Coplas de la Panadera* and Mena-Mariscal Iñigo Ortiz, 57
MN33: Partial copy of the *Cancionero de Martínez de Burgos*, 177, 211, 226n
MN34: Notes by Floranes on MN33, 226n
MN35: Notes by Floranes on MN33, 226n
MN49: Notes by Floranes on Santillana and MN33, 226n
MN54: *Cancionero de Stúñiga*, 62, 67, 90
MN55: Part of the *Cancionero de Barrantes* ca 1490, 227n
MN56: Coplas by Alvaro de Luna, Carta by Badajoz, 162
Mondragón, 49, 203
Montesino, Fray Ambrosio, 194
Montoro, Antón de, 10, 27, 31, 48, 49, 52, 53, 54–5, 62, 63, 128, 129, 130, 147, 185, 189, 194, 195, 202, 203, 205–13, 225n, 230n, 231n, 232n, 249n, 265n, 268n
MP2: *Cancionero de poesías varias*, 57, 185, 206, 208, 209, 210, 211
MP3: Obras de Gómez Manrique, 189, 190, 191–5, 266n
MR2: Part of the *Cancionero de Barrantes*, 162
MR3: Mena, *Coronación*, 162
MT1: Obras de Santillana, and others, 113, 227n

Nebrija, Antonio de, 119
NN2: Cancionero italiano, 121
nobility, 8, 10, 11, 12, 13, 14, 15, 16, 18, 19, 20, 21, 22, 23, 24, 26, 27, 29, 31, 45, 46, 47, 48, 49, 83, 134, 152, 177, 194, 196, 203, 217, 221n, 222n, 230n, 256n
nómina, 129, 132
nota, 102, 126, 137
notarial, 68, 140–2
notario, 30, 34, 38, 41, 101, 224n, 227n
notaris, 85
notary(-ies)(-al), 24, 30, 33, 34, 35, 38, 39, 40, 41, 43, 68, 101, 102, 115, 116, 117, 121, 126, 133, 140, 141, 142, 145, 156, 159, 160, 161, 162, 170, 187, 188, 201, 205, 207, 225n, 227n, 228n, 235n, 241n, 242n, 245n, 246n, 252n, 253n, 259n

Index   329

notebook, 6, 7, 87, 117, 118, 119, 163, 220n, 257n
Noya, Francisco de, 193
Núñez, Hernán, 56

open books, 131–4
Ordenações Manuelinas, 73, 76, 139
ordenador(es), 179, 180, 183, 192, 264n
ordenar, 51, 82, 90, 101, 137, 149, 155, 181, 182, 190
ordinatio, 109, 135, 168, 170, 179, 181, 201, 205, 261n
Ortiz de Estúñiga, Íñigo, 57
Ortúñez de Calahorra, Diego, 118

Pacheco, Juan, 11, 46
Palencia, Alfonso de, 24, 51, 90, 226n, 239n, 259n
Palencia, Vázquez de, 65, 234n
Palmireno, Juan Lorenzo, 119, 246n
palmo, 102, 241n
papel(es), 6, 15, 61, 62, 63, 68, 86, 88, 92, 93, 94, 95, 99, 100, 101, 114, 115, 120, 138, 150, 153, 245n, 248n, 257n
paper(s), 6, 7, 8, 9, 14, 22, 33, 35, 36, 38, 39, 40, 58, 59–77, 81, 82, 86, 87, 88, 89, 91, 92, 93, 94, 95, 96, 97, 100, 101, 102, 103, 106, 107, 108, 111, 113, 115, 117, 118, 119, 120, 121, 122, 123, 124, 125, 126, 127, 128, 129, 130, 131, 132, 133, 136, 137, 138, 139, 141, 144, 145, 146, 147, 149, 152, 153, 156, 159, 161, 162, 163, 164, 171, 175, 176, 184, 186, 187, 190, 194, 196, 213, 214, 220n, 226n, 233n, 238n, 239n, 244n, 245n, 246n, 248n, 249n, 250n, 252n, 255n, 256n, 257n, 258n

Paredes, Sancho de, 99, 118, 256n
pecia, 116, 245n
Peire Cardenal, 120, 172
Pere, Prince, 71
Pere III, King of Aragon, 240n
Pere IV, King of Aragon, 71, 83, 89, 126, 131, 235n, 250n
Pérez, Martín, 40, 170
Pérez de Guzmán, Fernán, 20, 21, 25, 45, 53, 63, 120, 151–2, 158, 188, 207, 226n, 247n, 255n, 258n
Pérez de Salinas, Ochoa, 103, 128
Pérez de Vivero, Alonso, 25
Petrarch, 107, 121, 145, 188, 262n
Petrus Riga, 172
Pimentel, Rodrigo Alonso de, Conde de Benavente, 48, 189
Pinar, Gerónimo de, 64, 234n
Pizan, Christine de, 188
pizarras, 244n
Planella, Ramón, 71
pliegos horadados, 132
pliegos sueltos, 122, 247n
PM1: Cancionero de San Martino delle Scale, 141
PN1: *Cancionero de Baena* (see also Cancionero de Baena), 15, 56, 56, 62, 63, 66, 68, 101, 121, 129, 146, 176, 177, 180
PN2: Miscelánea histórica, 162
PN10: Cancionero castellano de París "G," 206, 208, 210
Poeta, or de Valladolid, Juan, 46, 48, 52, 55, 185, 193, 194, 210, 211, 229n, 265n
Poridat de Poridades, 23, 227n
Portugal, Francisco de, 74
pregón(es), 37, 75, 236n
pregonero(s), 26, 37, 46, 185, 225n
print, 3, 4, 6, 7, 10, 30, 56, 59, 65, 67,

79, 89, 91, 92, 116, 119, 122, 146, 166, 171, 174, 178, 179, 183, 184, 185, 186, 206, 214, 215, 216, 245n, 254n, 265n
procesado, 100, 101, 102, 159, 259n
proceso(s), 95, 96, 97, 100, 101, 102, 240n, 241n, 258n
Puertocarrero, 208, 210
Pulgar, Fernando de, 15, 16, 20, 21, 30, 35, 40, 42, 51, 52, 59, 73, 136, 157, 188, 201, 220n, 230n, 266n

quadernets, 117
quaternus, 142
Queralt, Guerau de, 71
qüern(s), 114

Ramírez de Ribera, Vasco (Bishop of Coria), 140
Ramírez de Villaescusa, Alonso, 230n
RC1: *Cancionero de Roma*, 130, 184, 206, 208, 210, 265n
recordança, 248n
register(s), 9, 35, 40, 71, 85, 86, 87, 88, 93, 94, 102, 117, 123, 125, 126, 128, 131, 132, 133, 136, 137, 139, 140, 141, 142–7, 152, 156, 158, 159, 161, 163, 169, 170, 189, 192, 194, 196, 197, 199, 213, 226n, 238n, 241n, 245n, 248n, 250n, 251n, 252n, 253n, 254n, 256n, 258n, 259n, 261n
registro(s), 40, 123, 126, 127, 132, 133, 136, 137, 139, 140, 144, 145, 152, 155, 158, 159, 248n, 250n, 258n, 259n
Resende, Garcia de, 65, 68, 79, 83, 85, 149, 150, 229n, 234n
Ribera, Suero de, 41, 62
Robles, Fernando Alfonso de, 25, 222n

Rodríguez, Teresa, 54
Rodríguez de Sevilla, Manuel, 227n
Rodríguez del Padrón, Juan, 92–3
Rojas, Diego de, Marqués de Denia, 194
roldes, 99, 103
roll(s), 9, 79–81, 96–100, 101–14, 122, 164, 237n, 239n, 240n, 241n, 242n, 243n
rolos, 109, 174
Román, Comendador, 55, 62–3, 185, 208, 210, 212, 231n
Romero, 225n
Romero, Diego, 26, 225–6n
Romero, Sancho, 37, 225–6n
Romero, Valerio Francisco, 56
rótulo(s), 96–7, 103, 109, 241n
rotulus(-i), 96, 99, 102, 109, 172
rubric(s), 9, 49, 61, 64, 65, 66, 67, 68, 69, 70, 72, 82, 93, 94, 101, 109, 110, 114, 116, 120, 128, 129, 130, 131, 138, 140, 156, 164–86, 187, 188, 189, 192, 194, 196, 197, 198, 199, 201, 202, 203, 204, 205–9, 211–13, 248n, 249n, 260n, 261n, 262n, 263n, 265n, 266n
rúbrica(s), 166, 179, 184, 248n, 261n
rubricado, 95
rubrication, 9, 142, 147, 166–179, 180, 182, 183–6, 187, 188, 215, 260n, 262n, 264n, 265n
rubricator, 170, 176, 180

SA1: Obras de Santillana, 121, 266n
SA7: *Cancionero de Palacio*, 170, 227n, 265n
SA8: Obras de Santillana, 80, 113, 146
SA10: composite manuscript, two mss in one, 206, 208, 209, 210, 227n
Salinas, Pedro, 3, 219n
Salutati, Coluccio, 142, 145, 166, 253n

Sánchez Calavera, Fernán, 130
Sánchez de Arauz, Garci, 158, 228n
Sánchez de Santa María, Juan, 232n
Santa Cruz, Alonso de, 115, 137, 138
Santa Fe, Pedro de, 120, 183, 265n
Santillana, Íñigo López de Mendoza, Marqués de , 16, 36, 38, 39, 47, 48, 53, 54, 78, 80, 86, 88, 90, 95, 113–14, 141, 145–6, 168, 189, 190, 191, 193, 194, 195, 197, 201, 205, 207, 209, 210, 213, 225n, 227n, 229n, 230n, 237n, 239n, 244n, 254n, 256n, 266n, 268n
Sayol, Ferrer, 71
scribe(s), 8, 16, 35, 38, 39, 41, 71, 83, 84, 89, 90, 102, 103, 108, 111, 117, 120, 127, 139, 154, 180, 187, 228n, 242n, 245n, 247n, 251n, 253n, 264n
scroll, 80, 81, 97, 98, 103, 106, 108, 109, 135, 239n, 240n
secretario, 26, 30, 31, 34, 38, 43, 101, 224n
secretary(-ies), 13, 15, 20, 22, 23, 24, 26, 29, 34, 35, 36, 38, 39, 40, 42, 43, 44, 56, 58, 84, 86, 87, 88, 93, 96, 101, 103, 104, 125, 127, 140, 142, 150, 157, 189, 196, 224n, 225n, 226n, 228n, 247n
signum, 38, 84, 125, 226n, 237n, 245n
slander, 45, 49, 73, 74, 91, 93, 110, 185, 217, 218, 235n, 238n
SM2: *Coplas de la Panadera*, 57
So, Bernat de, 71
sobrescrito, 66, 67, 125, 185
Sosa, doña Juana de, 177, 263n
Sousa, Isabel de, 84
Suárez de Figueroa, Catalina, 36, 78, 113
SV2: Cancionero de la Colombina, 129, 185, 205, 206, 207, 208, 210, 211, 212

tablet(s) (tablas), 75, 81, 114, 153, 154, 236n, 239n, 244n, 257n
table of contents (tabla), 116, 135, 138, 140, 162, 166, 170–171, 175, 181, 182, 207, 261n, 262n
Talavera, Hernando de, Archbishop of Granada, 29, 84, 93, 127, 197, 200, 204, 205, 267n
Tapia, 208
Tapia, Juan de, 130, 250n
Tendilla, Íñigo López de Mendoza, Conde de, 73, 86–8, 89, 90, 93, 94, 119, 142–4, 194, 197, 238n, 258n
Thomas Aquinas, 168
threaded book, 9, 131–3, 250n
tira(s), 100–3, 106, 128, 135, 162, 241n
title(s) (paratextual), 7, 9, 61, 82, 114, 135, 164, 166, 167, 168, 169, 170, 171, 174, 179, 180, 182, 183, 184, 202, 208, 210, 237n, 253n, 261n, 264n, 265n
tituli, 169, 170
título(s), 82, 166, 167, 168, 170, 180, 182, 183
Toledo, Pedro de, 116
Tor, Miquel de la, 120
Torquemada, Juan de, 56
Torquemada, Tomás de, 54, 133, 227n, 228n, 254n
Torre, Alfonso de la, 107, 176
Torre, Fernando de la, 63, 64, 70, 90, 93, 95, 123, 234n
Torrellas, Pedro, 62
TP1: Obras de Santillana, 141, 261n
TP2: Partial copy of MP2, 66, 67
traslado(s), 43, 95, 102, 125, 142, 154, 155, 156, 253n
Troilus and Criseyde, 188
tumbo(s), 25, 134, 148, 169, 250n, 251n, 252n, 258n

Uc de Saint Circ, 173
Urriés, Hugo de, 184
Usatges de Barcelona, 135, 251n

Valente, Afonso de, 68, 234n
Valera, Diego de, 16, 20–1, 26, 28, 29, 157, 160, 188, 189, 203, 221n
Vallejo, Juan de, 85
Vega, Lope de, 162
Velasco, Alfonso de, 55, 210
Vicente, Gil, 146
vidas and razos, 166, 168, 172–3, 174, 188, 261n, 262n

Villegas, Sancho de, 62, 67, 90
Villena, Enrique de, Marqués de, 48, 118, 145, 166, 176, 179, 180, 253n
Vincent of Beauvais, 168, 170
Vives, Luis, 119, 246n

Zafra, Hernando de, 29, 36, 157
ZZ3: *Cancionero de Barrantes/Guadalupe*, now MM1, MN55, MR2, and MR3, 162
ZZ10: Madrid, Biblioteca Nacional, K–97, 177

TORONTO IBERIC

CO-EDITORS: Robert Davidson (Toronto) and Frederick A. de Armas (Chicago)

EDITORIAL BOARD: Josiah Blackmore (Toronto); Marina Brownlee (Princeton); Anthony J. Cascardi (Berkeley); Emily Francomano (Georgetown); Justin Crumbaugh (Mt Holyoke); Jordana Mendelson (NYU); Joan Ramon Resina (Stanford); Kathleen Vernon (SUNY Stony Brook)

1 Anthony J. Cascardi, *Cervantes, Literature, and the Discourse of Politics*
2 Jessica A. Boon, *The Mystical Science of the Soul: Medieval Cognition in Bernardino de Laredo's Recollection Method*
3 Susan Byrne, *Law and History in Cervantes' Don Quixote*
4 Mary E. Barnard and Frederick A. de Armas (eds), *Objects of Culture in the Literature of Imperial Spain*
5 Nil Santiáñez, *Topographies of Fascism: Habitus, Space, and Writing in Twentieth-Century Spain*
6 Nelson R. Orringer, *Lorca in Tune with Falla: Literary and Musical Interludes*
7 Ana M. Gómez-Bravo, *Textual Agency: Writing Culture and Social Networks in Fifteenth-Century Spain*

www.ingramcontent.com/pod-product-compliance
Lightning Source LLC
Chambersburg PA
CBHW030303080526
44584CB00012B/420